Communication Technology Update

7th Edition

Communication Technology Update

7th Edition

August E. Grant & Jennifer Harman Meadows, Editors

In association with
Technology Futures, Inc.

Focal Press
Boston Oxford Auckland Johannesburg Melbourne New Delhi

Editors
Jennifer Harman Meadows
August E. Grant
Technology Futures, Inc.
Production Editor Debra R. Robinson
Art Director Helen Mary V. Marek

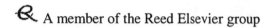

A member of the Reed Elsevier group

Recognizing the importance of preserving what has been written, Butterworth–Heinemann prints its books on acid-free paper whenever possible.

Butterworth–Heinemann supports the efforts of American Forests and the Global ReLeaf program in its campaign for the betterment of trees, forests, and our environment.

ISBN 0-240-80407-4

The publisher offers special discounts on bulk orders of this book.
For information, please contact:
Manager of Special Sales
Butterworth–Heinemann
225 Wildwood Avenue
Woburn, MA 01801–2041
Tel: 781-904-2500
Fax: 781-904-2620

For information on all Butterworth–Heinemann publications available, contact our World Wide Web home page at: http://www.focalpress.com

10 9 8 7 6 5 4 3 2 1

Printed in the United States of America

Table of Contents

Updates can be found on the
Communication Technology Update Home Page
on the Internet at
http://www.tfi.com/ctu/

Preface

In putting together this seventh edition of the *Communication Technology Update*, our focus was naturally on the individual chapters. But as we prepare to send this book to the printer, we had the opportunity to take a larger view of how communication technologies have changed in the last decade.

The biggest change in this edition from the previous one is the increasing importance of the Internet to almost all communication technologies. The fact that this was also the biggest change between the two previous editions is perhaps even more significant—the Internet has truly brought about a revolution in all areas of communication that should continue for years to come.

The mechanical process of assembling this book is perhaps the best illustration of the power and immediacy of these technologies. During production of the first edition of the book, the editorial staff was in a single city and chapters were submitted on paper, sometimes with floppy discs attached. For the current edition, the process is similar, but communication technologies are more evident. For example, in the weeks during which the individual chapters were edited, formatted, and brought together to present a coherent picture of communication technology, the editorial staff has been meeting daily and sharing redlined copies of chapters, graphics, etc. The interesting thing is that the key participants in these conversations were located in three different cities and didn't see each other face-to-face once production of the book began. Chapters were all submitted in electronic format, and there were almost no problems with compatibility of text or graphic formats among the authors and editors.

In short, the very technologies discussed in this book revolutionized the production of the book itself and allowed us to work together more efficiently and effectively than before, even though hundreds of miles separated us. One lesson we hope you take from this anecdote is that you can and should relate the discussions of the technologies presented to changes in your own lifestyle.

As always, almost every chapter was completely rewritten to present the most recent developments in each technology. The editors will also continue to maintain the *Communication Technology Update* home page (www.tfi.com/ctu) to supplement the text with updated information and links to a wide variety of information available over the Internet.

This work is the most collaborative editorial effort in the history of this book. Jennifer Meadows' role as editor is more prominent in this book than in any edition to date; indeed, she has become the driving force behind production of the book. This change is related to Augie Grant's move from academia to industry. We hope our combined industry/academic perspective results in a more comprehensive view of the evolution and roles of the technologies discussed.

This book uses a novel application of desktop publishing that allows us to have the book printed and available within one month after the last chapters are written. Because we know that changes in the technologies discussed will happen before the book is printed (no matter how fast we are), the *Communication Technology Update* home page has been created to provide up-to-date information no matter how long it has been since the chapter was written. We encourage you to contribute to the home page by sending Internet links and ideas for content.

This compilation is the product of dozens of people who have worked right up to the deadline to provide the latest developments in all areas of communication technology. We are especially grateful to the staff at Technology Futures, Inc., including Production Director Deb Robison and Art Director Helen Mary Marek. Most of all, we are grateful to our authors for their continued involvement and enthusiasm for this project.

As always, we encourage you to suggest new topics, glossary additions, and possible authors for the next edition of this book by communicating directly with us via e-mail, fax, snail mail, or voice.

Augie Grant
2Wire, Inc.
1704 Automation Parkway
San Jose, CA 95131
Phone: 408.895.1208
Fax: 408.895.1308
E-mail: agrant@2Wire.com

Jennifer Meadows
Department of Communication Design
California State University, Chico
Chico, CA 95929-0504
Phone: 530.898.4775
Fax: 530.898.5877
E-mail: jmeadows@csuchico.edu

May 2000

Communication Technology Update

7th Edition

1

The Umbrella Perspective on Communication Technology

August E. Grant, Ph.D.*

Communication technologies are the nervous system of contemporary society, transmitting and distributing sensory and control information, and interconnecting a myriad of interdependent units. Because these technologies are vital to commerce, control, and even interpersonal relationships, any change in communication technologies has the potential for profound impacts on virtually every area of society.

One of the hallmarks of the industrial revolution was the introduction of new communication technologies as mechanisms of control that played an important role in almost every area of production and distribution of manufactured goods (Beniger, 1986). These communication technologies have evolved throughout the past two centuries at an increasingly rapid rate. This evolution shows no signs of slowing, so an understanding of it is vital for any individual wishing to attain or retain a position in business, government, or education.

This text provides you with a snapshot of this evolutionary process. The individual chapter authors have compiled facts and figures from hundreds of sources to provide the latest information on more than two dozen sets of communication technologies. Each discussion explains the roots and evolution, the recent developments, and the current status of the technology as of mid-2000. In discussing each technology, we will deal not only with the hardware, but also with the software, the organizational structure, the political and economic influences, and the individual users.

Although the focus throughout the book is on individual technologies, these individual snapshots comprise a larger mosaic representing the communication networks that bind individuals together and enable us to function as a society. No single technology can be understood without understanding the competing and complimentary technologies and the larger social environment within which these technologies exist. As discussed in the following section, all of these factors

* Director of Market Research & Entertainment Programming, 2Wire, Inc. (San Jose, California).

(and others) have been considered in preparing each chapter through application of the "umbrella perspective." Following this discussion, an overview of the remainder of the book is presented.

Defining Communication Technology

The most obvious aspect of communication technology is the hardware—the physical equipment related to the technology. The hardware is the most tangible part of a technology system, and new technologies typically spring from developments in hardware. However, understanding communication technology requires more than just studying the hardware. It is just as important to understand the messages communicated through the technology system. These messages will be referred to in this text as the "software." It must be noted that this definition of "software" is much broader than the definition used in computer programming. For example, our definition of computer software would include information manipulated by the computer (such as this text, a spreadsheet, or any other stream of data manipulated or stored by the computer), as well as the instructions used by the computer to manipulate the data.

The hardware and software must also be studied within a larger context. Rogers' (1986) definition of "communication technology" includes some of these contextual factors, defining it as "the hardware equipment, organizational structures, and social values by which individuals collect, process, and exchange information with other individuals" (p. 2). An even broader range of factors is suggested by Ball-Rokeach (1985) in her "Media System Dependency Theory," which suggests that communication media can be understood by analyzing dependency relations within and across levels of analysis, including the individual, organizational, and system levels. Within the system level, Ball-Rokeach (1985) identifies at least three systems for analysis: the media system, the political system, and the economic system.

These two approaches have been synthesized into the "Umbrella Perspective on Communication Technology" illustrated in Figure 1.1. The bottom level of the umbrella consists of the hardware and software of the technology (as previously defined). The next level is the organizational infrastructure: the group of organizations involved in the production and distribution of the technology. The top level is the system level, which includes the political, economic, and media systems, as well as other groups of individuals or organizations serving a common set of functions in society. Finally, the "handle" for the umbrella is the individual user, implying that the relationship between the user and a technology must be examined in order to get a "handle" on the technology. The basic premise of the umbrella perspective is that all five areas of the umbrella must be examined in order to understand a technology.

(The use of an "umbrella" to illustrate these five factors is the result of the manner in which they were drawn on a chalkboard during a lecture in 1988. The arrangement of the five attributes resembled an umbrella, and the name stuck. Although other diagrams have since been used to illustrate these five factors, the umbrella remains the most memorable of the lot.)

Factors within each level of the umbrella may be identified as "enabling," "limiting," "motivating," and "inhibiting." Enabling factors are those that make an application possible. For example, the fact that coaxial cable can carry dozens of channels is an enabling factor at the hardware level, and the decision of policy makers to allocate a portion of the spectrum for cellular telephony is an enabling factor at the system level (political system).

Figure 1.1
The Umbrella Perspective on Communication Technology

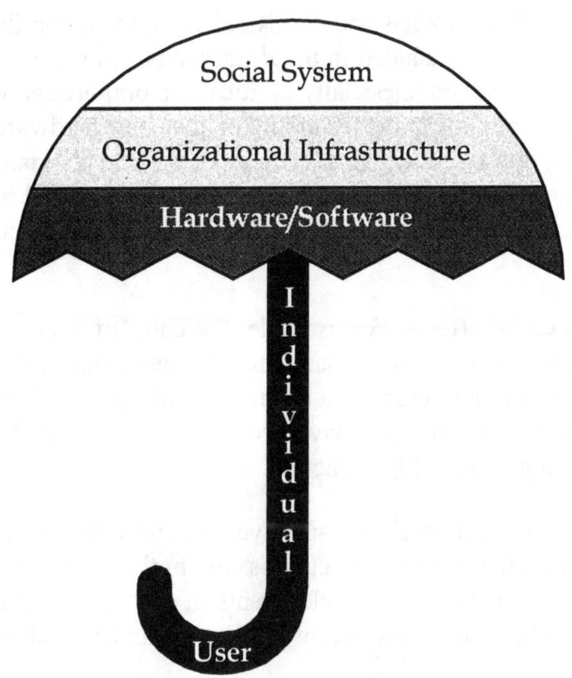

Source: August E. Grant

Limiting factors are the opposite of enabling factors. Although coaxial cable increased the number of television programs that could be delivered to a home, most coaxial networks cannot transmit more than 100 analog channels of programming. To the viewer, 100 channels might seem to be more than is needed, but to the programmer of a new cable television channel who is unable to get space on a filled-up cable system, this hardware factor represents a definite limitation. Similarly, the fact that the policy makers discussed above permitted only two companies to offer cellular telephone service in each market is a system-level limitation on that technology.

Motivating factors are those that provide a reason for the adoption of a technology. Technologies are not adopted just because they exist. Rather, individuals, organizations, and social systems must have a reason to take advantage of a technology. The desire of local telephone companies for increased profits, combined with the fact that growth in providing local telephone service is limited, is an organizational factor motivating the telcos to enter the markets for new communication technologies. Individual users who desire information more quickly can be motivated to adopt electronic information technologies.

Inhibiting factors are the opposite of motivating ones, and provide a disincentive for adoption or use of a communication technology. An example of an inhibiting factor at the software level might be a new electronic information technology that has the capability to update information more quickly than existing technologies, but does not use that capability to provide continuously updated messages. One of the most important inhibiting factors for most new technologies is the cost to individual users. Each potential user must decide whether the cost is worth the service, considering his or her budget and the number of competing technologies.

All four types of factors—enabling, limiting, motivating, and inhibiting—can be identified at the system, organizational, software, and individual user levels. However, hardware can only be enabling or limiting; by itself, hardware does not provide any motivating factors. The motivating factors must always come from the messages transmitted (software) or one of the other levels of the umbrella.

The final dimension of the umbrella perspective relates to the environment within which communication technologies are introduced and operate. These factors can be termed "external" factors, while ones relating to the technology itself are "internal" factors. In order to understand a communication technology or to be able to predict the manner in which a technology will diffuse, both internal and external factors must be studied and compared.

Each communication technology discussed in this book has been analyzed using the umbrella perspective to ensure that all relevant factors have been included in the discussions. As you will see, in most cases, organizational and system-level factors (especially political factors) are more important in the development and adoption of communication technologies than the hardware itself. For example, political forces have, to date, prevented the establishment of a world standard for high-definition television production and transmission. As individual standards are selected in countries and regions, the standard selected is as likely to be the product of political and economic factors as of technical attributes of the system.

Organizational factors can have similar powerful effects. For example, the entry of a single company, IBM, into the personal computer business resulted in fundamental changes to the entire industry. Finally, the individuals who adopt (or choose not to adopt) a technology, along with their motivations and the manner in which they use the technology, have profound impacts on the development and success of a technology following its initial introduction.

Each chapter in this book has been written from the umbrella perspective. The individual writers have endeavored to update developments in each area to the extent possible in the brief summaries provided. Obviously, not every technology experienced developments in each of the five areas, so each discussion is limited to areas in which relatively recent developments have taken place.

Overview of Book

The technologies discussed in this book have been organized into three sections: electronic mass media, computers and consumer electronics, and satellites and telephony. These three are not necessarily exclusive; for example, direct broadcast satellites (DBS) could be classified as either an electronic mass medium or a satellite technology. The ultimate decision regarding where to put each technology was made by determining which set of current technologies most closely resembled the technology from the user's perspective. Thus, DBS was classified with electronic mass media. This process also locates the discussion of a cable television technology—cable modems—in the telephony section.

Each chapter is followed by a brief bibliography. These reference lists represent a broad overview of literally thousands of books and articles that provide details about these technologies. It is hoped that the reader will not only use these references, but will examine the list of source material to determine the best places to find newer information since the publication of this *Update*.

Most of the technologies discussed in this book are continually evolving. As this book was completed, many technological developments were announced but not released, corporate mergers were under discussion, and regulations had been proposed but not passed. Our goal is for the chapters in this book to establish a basic understanding of the structure, functions, and background for each technology, and for the supplementary Internet home page to provide brief synopses of the latest developments for each technology discussed. (The address for the home page is: http://www.tfi.com/ctu.)

The final two chapters attempt to draw larger conclusions from the preceding discussions. The first of these two chapters presents a detailed statistical abstract of many of the technologies discussed, allowing you to more easily compare technologies. The final chapter then attempts to place these discussions in a larger context, noting commonalties among the technologies and trends over time. It is impossible for any text such as this one to ever be fully comprehensive, but it is hoped that this text provides you with a broad overview of the current developments in communication technology.

Bibliography

Ball-Rokeach, S. J. (1985). The origins of media system dependency: A sociological perspective. *Communication Research, 12* (4), 485-510.

Beniger, J. (1986). *The control revolution.* Cambridge, MA: Harvard University Press.

Rogers, E. M. (1986). *Communication technology: The new media in society.* New York: Free Press.

II

ELECTRONIC MASS MEDIA

Digital technologies are revolutionizing virtually all aspects of mass media. Digital video compression, interactivity, and new business opportunities are fueling an explosion in the number of mass media and the programming they provide.

The changes are most evident in multichannel video distribution services. As the following chapter indicates, cable television continues to reinvent itself, incorporating digital technology to increase channel capacity and provide new services. Chapter 5 then explains how direct broadcast satellite (DBS) services have emerged as the most aggressive competitors to cable television.

The factor shared by all of multichannel distribution services is programming. Most of these services will depend upon revenues from the pay television services explored in Chapter 3, including premium cable channels and various types of pay-per-view television.

Chapter 7 explores how digital technology is forcing the biggest change ever in broadcast television, as broadcasters must choose whether to use their new, digital frequencies to provide one channel of high-definition television, a "multicast" of up to five channels of standard definition programming, or some combination of the two. Chapter 8 then explores broadcasting's "low-definition" challenger, streaming media.

Not all technologies have fared as well. "Wireless cable" services (MMDS, discussed in Chapter 6) have had a more difficult time competing with cable television. Similarly, Chapter 4 explores how interactive television efforts have continued to disappoint inventors and investors.

Finally, Chapter 9 explains how radio is preparing for its own digital revolution. That revolution may take longer than the television revolution, but digital technology promises the same degree of change for radio as it has offered to all areas of television broadcasting.

In reading these chapters, you should consider two basic communication technology theories. Diffusion theory helps us to understand that the introduction of innovations is a process that occurs over time among members of a social system (Rogers, 1983). Different types of people adopt a technology at different times, and for different reasons. The smallest group of adopters is the innovators, the first to adopt, but they usually adopt for reasons that are quite different from those of later adopters. Hence, it is dangerous to predict the ultimate success, failure, diffusion pattern, gratifications, etc. of a new technology by studying the first adopters.

Diffusion theory also suggests five attributes of an innovation that are important to its success: compatibility, complexity, trialability, observability, and relative advantage (Rogers, 1983). In studying or predicting diffusion of a technology, use of these factors suggests that analysis of competing technologies is as important as attributes of the new technology.

A second theory to consider is the "Principle of Relative Constancy" (McCombs, 1972; McCombs & Nolan, 1992). This theoretical perspective suggests that, over time, the aggregate disposable income devoted to mass media, as a proportion of gross national product, is constant. In simple terms, people spend a limited amount of their income on the media discussed in this section, and that amount rarely increases when new media are introduced. In applying this theory to the electronic mass media discussed in the following chapters, consider which media will win a share of audience income, and what will happen to the losers.

Bibliography

McCombs, M. (1972). Mass media in the marketplace. *Journalism Monographs*, 24.

McCombs, M., & Nolan, J. (1992). The relative constancy approach to consumer spending for media. *Journal of Media Economics*, *5* (2), 43-52.

Rogers, E. M. (1983). *Diffusion of innovations, 3rd ed.* New York: Free Press.

2

Cable Television

Donald R. Martin, Ph.D.*

Cable television is one of the most established communication technologies discussed in this book, yet it continues to be one of the most dynamic technologies in terms of technological innovation and the delivery of new services and programming to viewers. For years, cable struggled for respect, trying to grow past its rural roots and the notion that its unique programming could not compete with broadcast television. Indeed, even the venerable CNN was once referred to by many in the industry as the "Chicken Noodle Network."

It is difficult to say exactly when cable gained respectability. Perhaps it was during the Persian Gulf War, when political leaders in Washington were watching Wolf Blitzer's live CNN reports from Baghdad at the same time as the rest of the country. Perhaps it was when the word "cable" became the universal term for an industry that ran its wires past almost every household in the United States and delivered an order of magnitude more programming than all the broadcast networks combined.

There is no doubt that, within the television industry, the 1999 Emmy Awards marked a watershed moment for the cable industry. Cable television programs began the evening with 134 separate nominations. The evening ended with one cable network, Home Box Office (HBO), winning more prime-time Emmy awards than any broadcast network.

This chapter will explore how the cable industry grew to its current position and where the industry is headed today. In the process, the complexity of the industry will be explored—from the local cable systems that run wires down every street in a community to the cable television programmers that provide content on every subject imaginable. (Chapter 3 explores premium television services in more depth.)

---

* Associate Professor of Communication, School of Communication, San Diego State University (San Diego, California).

Background

Cable television is a wired terrestrial-based distribution system designed to deliver multiple channels of high-quality television to subscribers for a monthly fee. With a considerable investment of money and equipment, cable infrastructure can also provide two-way services, such as data and telephony.

The technology was first deployed in the late 1940s in communities whose topography prevented them from receiving line-of-sight television signals transmitted from major cities. These primarily rural communities erected community antennas on mountaintops and distributed the received signals to television viewers in the adjacent valley communities (Baldwin & McVoy, 1988).

As with many new technologies, more than one person had the same idea at about the same time. One pioneering effort to construct what would become known as "Community Antenna Television" (CATV) was made by John Walson from Mahanoy City (Pennsylvania). At about the same time, Ed Parsons experimented with wires and an antenna to deliver Seattle television signals to Astoria (Oregon). These operations are believed to be the very first cable television systems (Crandall & Furchtgott-Roth, 1996).

In 1949, Robert J. Tarlton of Lansford (Pennsylvania), invested in a company that built a master antenna at the summit of the Allegheny Mountains to amplify television signals from Philadelphia, and then distribute them to homes in the community by coaxial cables hung on telephone poles (Baldwin & McVoy, 1988). The Lansford system became the first subscription cable system (Crandall & Furchtgott-Roth, 1996).

While the number of cable systems increased between 1950 and 1970, the number of channels on each system remained fairly static throughout that time. Since the purpose of cable technology during this period was primarily to extend the reach of regional over-the-air television stations, the universe of available programming options was limited. Some cable companies offered their subscribers text-only local weather and other information on a few channels, but most of the channels were simply extending the reach of broadcast television stations.

Cable technology remained fairly simple for the first few decades. A typical cable system consisted of three main elements: the headend, the distribution network, and the subscriber drop (see Figure 2.1) (Baldwin & McVoy, 1988).

The *headend* is the point at which all satellite- or microwave-transmitted program signals are received, assembled, and processed for transmission by the distribution network (Baldwin & McVoy, 1988).

The *distribution network* carries the program signal through the community using a system of coaxial or fiber optic cables strung along telephone or power lines or buried underground. The distribution network, often called the distribution plant, consists of two elements: the trunk system and the feeder system. The trunk consists of a large-diameter cable that leaves the headend, travels through the community, splits at various points along the route, and stops at the end of the service area. The trunk's purpose is to deliver signals to the subscriber neighborhoods; no subscribers are directly served from the trunk.

Figure 2.1
Traditional Cable TV Network Tree and Branch Architecture

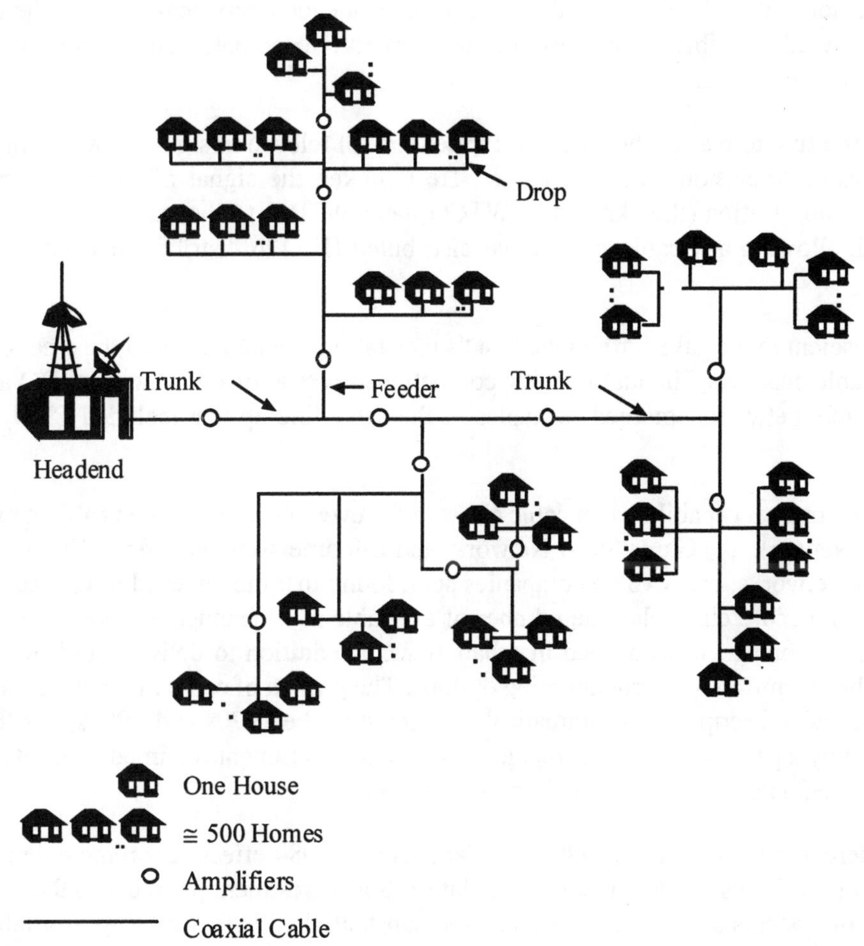

Source: Technology Futures, Inc.

Bridger amplifiers are located along the trunk at intervals between one-fourth and one-half a mile to feed the signal to the feeder system. The feeder system consists of smaller cables running along the streets in a neighborhood to which subscribers connect. The subscriber drop then takes the signal from the feeder system into the subscriber's home, where it is picked up by the television receiver (Baldwin & McVoy, 1988).

This complex system of cables, amplifiers, and receivers is necessary to make sure the television signal stays strong. As the signal travels down the cable, it loses strength. Amplifiers are set up at necessary locations along the distribution network to make sure that a person living 10 miles away from the headend gets as strong a signal as a person living next door to it.

The nature of the cable business radically changed in the early 1970s as a result of a major innovation in software (programming) rather than hardware. In 1975, Home Box Office created the first communications satellite network exclusively for cable companies. Suddenly, cable companies were no longer simple extensions of over-the-air broadcasting stations, but could offer multiple channels of unique programming to their customers. The core cable business was no longer limited to remote subscribers who could not receive an adequate broadcast signal. The unique programming offered by cable companies became attractive to potential subscribers who had adequate over-the-air service.

One of the first to realize the potential for delivery of television signals via satellite for distribution via cable television was Ted Turner. He uplinked the signal of his small, independent Atlanta television station (then known as WTCG, later WTBS) to the same satellite carrying the HBO signal, allowing any cable system that distributed HBO to distribute his station's signal as well.

This prescient move gave birth to the first "superstation," which inspired the creation of countless new "cable channels," including a few competing superstations. Following HBO and WTBS, additional cable networks emerged as staples of the cable line-up that included CNN, MTV, and ESPN.

The niche market capabilities on cable allowed for even more segmented cable networks such as the Food Network, the Game Show Network, and Lifetime, to name a few. With this proliferation of cable networks, many cable companies soon found that they needed to upgrade their plant in order to offer more channels. The number of available new channels sparked interest in cable television from consumers who lived in urban areas. In addition to delivering cleaner television pictures, cable promised new programming options. The growth of cable in urban areas caused the number of cable subscriptions to dramatically increase in the 1970s and 1980s, but the industry was hindered by a patchwork of local regulations. Local governments claimed authority over cable systems, and regulated pricing, channel line-ups, and more.

In the deregulation era of the 1980s, the Cable Act of 1984 effectively removed most of these restrictions on the industry. As a result, in the late 1980s, investment poured into the industry, and the number of systems and subscribers increased substantially. The increasing profitability of the industry led to significant consolidation, as small cable systems were purchased and operated by multiple system operators (MSOs).

Unfortunately for cable subscribers, cable prices were also deregulated, and consequently, cable rates rose rapidly and sharply. Between November 1986 and April 1991, the monthly rate for the lowest priced cable service rose by 56%. Rates for the most popular service rose by 61%—nearly three times the rate of general inflation for that time period (Albarran, 1996).

By the early 1990s, the cable industry was experiencing significant regulatory and competitive problems. Industry competitors fed consumer complaints about increasing cable subscription fees and deteriorating service to Congress. These complaints led to the Cable Television Consumer Protection and Competition Act of 1992 that re-regulated the cable industry. The Act imposed price controls and customer service requirements on the cable industry. It also delivered a major victory to broadcast television stations, allowing them to dictate to local cable systems whether they had to be included on the cable lineup ("must-carry") or whether the cable operator had to negotiate for permission to carry the station ("retransmission consent") (Albarran, 1996).

At about the same time the Act was passed, other industries arose to compete for the lucrative monthly subscriber fees that made cable very profitable. They began experimenting with alternative technologies for delivering video services, such as "wireless cable" or multichannel multipoint distribution service (MMDS), direct broadcast satellites (DBS), and video dialtone (VDT) offered by telephone companies. (See Chapter 6 for more on MMDS and Chapter 5 for more on DBS.)

The Act required cable programming networks to sell their services to these emerging competitors. To counter this potential loss of market share in the video delivery business, a number of cable companies began experimenting with using excess capacity on existing cable systems for the carriage of other telecommunications services.

The delivery of alternative services became possible when cable companies began to rebuild their systems to offer greater bandwidth and hence more video channels to their subscribers. Many of the bandwidth expanding upgrades of the 1990s reconfigured systems from traditional coaxial cable systems, with their many noise-producing downstream amplifiers, to hybrid fiber/coax (HFC) systems. The optical fiber backbones delivers greater bandwidth and better signals to feeder networks and the nodes in local neighborhoods (see Figure 2.2). Two-way interactive capacity was also engineered into many of these rebuilds, enabling the deployment of a host of new services.

In the early 1990s, some operators of these rebuilt systems conducted technical and marketing experiments with cable-delivered telephony service. The telephony experiments wrestled with technological problems such as the delivery of uninterrupted power to the cable telephony systems when there were electrical outages (necessary to preserve 911 and other emergency services). MSOs also studied marketing issues such as consumer acceptance of telephone service by a cable company. Most of the pilot studies suggested that telephony could be delivered by cable. However, despite the success of these experiments, state and federal regulators were slow to grant approval for full deployment of cable-delivered telephony service. And cable operators were even slower to make the huge investment to upgrade their systems.

A few cable operators developed data transmission capability on their rebuilt systems. HFC cable networks provide a broadband link that enables residential and business subscribers to access data at much greater speeds than are possible with traditional telephone lines. New cable modem technology offers downlink data rates of up to 10 Mb/s, compared with the much slower 28.8 Kb/s telephone modems that were the standard of the mid-1990s.

At the same time local cable systems were rebuilding their cable plants, the industry was shaken by another major change in the regulatory climate: the Telecommunications Act of 1996. Crafted to encourage competition among telecommunications providers, the Act deregulated cable rates (again). More important, it prescribed conditions in which competitors such as telephone companies could begin to enter the video distribution business and freed cable companies to deploy telephony, data, and other services. Initially, several MSOs considered making substantial investments in telephony service. However, it became apparent that other competitors, such as resellers, could quickly enter the telephone business with minimal investment. When compared with the substantial capitalization cost, the potential market share of the telephony business did not appear to be a reasonable investment.

Figure 2.2
Hybrid Fiber/Coax Cable TV System

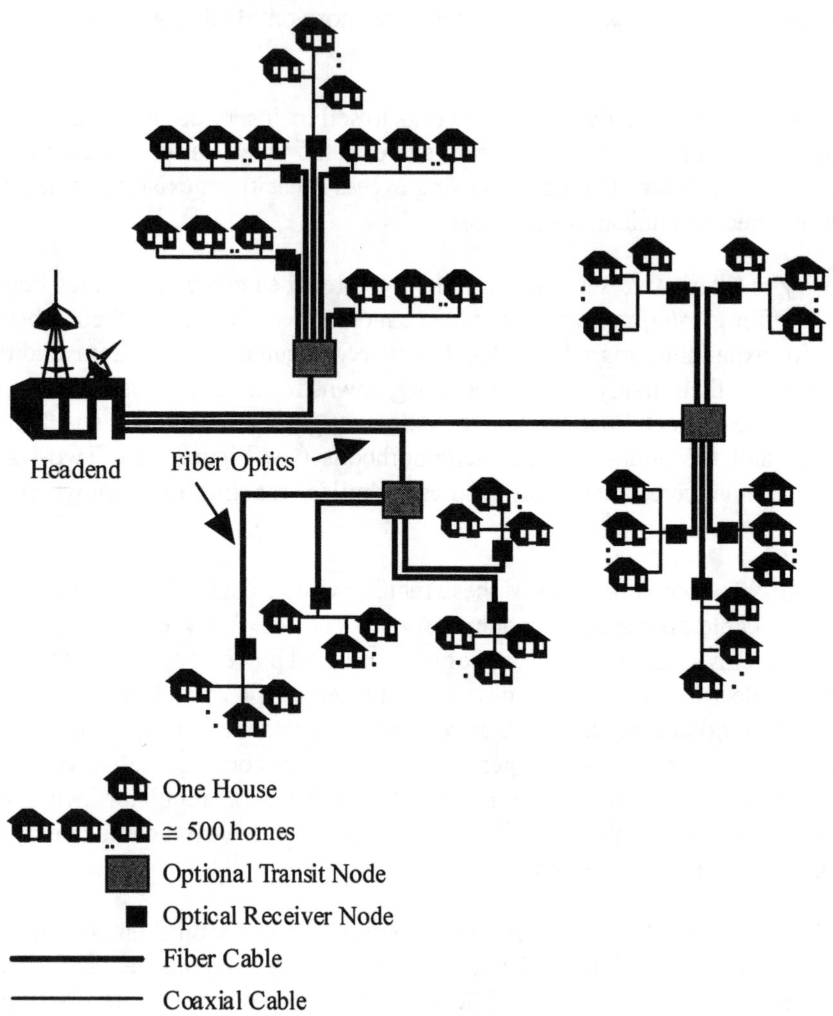

Source: Technology Futures, Inc.

However, cable had a valuable resource that other competitors lacked—a deployed broadband network into millions of residences. With the explosion of the Internet, a number of cable companies shifted their deployment of auxiliary service priorities from telephony to data. This decision was driven by the consumer need for faster Internet service, because data-intense media such as sophisticated graphics and video are very slow when exchanged through telephone modems. In particular, Time Warner and Cox successfully deployed and marketed cable modem services to residential customers.

After the initial deployment of cable modem services, some cable companies began to roll out business and residential telephone services. For example, in 1998, Cox deployed a telephone service in several of its franchise areas. Many cable companies also began to offer satellite-delivered digital audio services in the late 1990s. These add-on services provided subscribers with many high-quality music channels for distribution through home stereo systems.

Recent Developments

Digital Conversion

In 1996, the Federal Communications Commission (FCC) authorized deployment of a digital over-the-air television service in the United States with a projected date of 2006 for the retirement of the existing analog system. In 2000, about 67% of television homes in the United Sates were cable subscribers. Most of these cable systems transmit exclusively analog video signals, and there is a concern that cable, the primary television delivery system, may be a bottleneck in the rollout of a digital television system. Some cable companies have already built digital capacity into their systems. These channels are used primarily for enhanced viewing of satellite-delivered cable services and will have the capacity to distribute the signals of local digital television stations as they begin service.

Before they can do so, however, two major barriers must be overcome:

(1) The technique of transmitting digital signals over cable television is different from that used for broadcast stations. The process of converting signals from one format to the other—or of cable systems converting to a different digital modulation standard—will have to be addressed.

(2) Some agreement will be needed between broadcasters and cable companies regarding whether the cable company will transmit a broadcaster's digital signal, analog signal, or both. Broadcasters prefer cable systems to transmit both, but cable companies argue that, if they must make room for two channels per broadcaster, they will have to drop carriage of existing channels. Moreover, they claim that such a demand constitutes an "illegal taking," and there is sufficient uncertainty about it that it is unlikely that the FCC will force them to carry two separate signals for each broadcast channels.

Convergence of Industries

The Telecommunications Act of 1996 was crafted to promote competition within and among telecommunications providers and was expected to increase the number of new competitors in cable and stimulate the deployment of new services, as well as be a check on prices. However, by the late 1990s and early 2000s, it has become evident that competitors wanting access to the cable business are choosing to acquire the necessary facilities through mergers and acquisitions rather than by building new systems.

Therefore, there are new competitors in the cable industry, but the deployed facilities base has not changed significantly.

Competition for Video Delivery

The DBS industry has experienced significant growth in subscription levels in the past several years. In early 2000, about 7% of television households in the United States were DBS subscribers. Recently, Congress enacted legislation detailing the conditions under which DBS operators would be able to distribute the signals of local television stations to their subscribers. Essentially, it allows satellite operators to deliver local signals within local markets, fittingly termed "local-in-local" service.

This change portends an accelerated erosion of the traditional customer base for cable companies, as local television stations are the most highly-viewed cable system channels. On the other hand, terrestrial-based wireless cable or MMDS systems appear to be less of a competitive force to cable as the influx of capital from the regional Bell operating companies and other significant competitors has dwindled in the past several years.

Current Status

As of the end of 1999, there were 10,466 cable systems in the United States that together employed more than 130,000 people. Annual revenues for the cable television industry were $37 billion. During the 1998-1999 television season, all cable networks combined earned a 45% share of the national television audience (National Cable Television Association, 2000).

In 2000, cable reached into approximately 70% of American households. The average system distributed 57 channels of predominately analog video. There were approximately 600,000 residential subscribers to cable-based telephony services and 1.5 million households receiving data services via cable modems (National Cable Television Association, 2000).

Despite the success of cable modems, one of the problems with widespread deployment of this technology has been the complexity and cost of installing cable modems in new subscribers' homes. Consumers are often charged $100 to $150 for cable company installers to physically place hardware into their home computers and make software adjustments. However, several cable companies have been developing partnerships with retailers such as Circuit City to market "plug-and-play" cable modems that can be purchased and installed by new subscribers (Dawson, 2000).

With the growth in cable Internet access, industries traditionally not associated with cable are, in fact, merging and acquiring cable systems. Most cable systems are owned by large MSOs, many of which are vertically integrated. Time Warner, for example, has 12.6 million cable subscribers, and the company also has significant interests in cable programming and cable networks through TBS Entertainment, CNN Newsgroup, HBO, and other properties (Time Warner, 2000).

Other traditional cable companies are disappearing as some of the telecom giants are purchasing them to gain access to their broadband networks. For example, within the last several years, AT&T acquired two of the larger cable MSOs, Tele-Communications, Inc. and MediaOne, and, as of this writing, AOL (America Online) is in the process of acquiring Time Warner and all its cable properties. While these acquisitions generally produce higher stock prices for the companies, they have raised anticompetitive concerns in Congress and at some federal regulatory agencies (Albiniak, 2000).

Factors to Watch

Video Delivery

Cable will continue to dominate the residential video delivery business for the foreseeable future. Its installed base of broadband plant will make it the dominant medium in the video delivery business. However, cable's growth in this market may be diminished in the next few years as DBS continues to attract an ever-increasing subscriber base. This problem could be accelerated if cable systems are unable to add digital channel capacity by the time there is a significant consumer demand to view new digital television stations. The Yankee Group estimates that there will be 6.25 million households subscribing to digital cable in 2000 (Umstead, 2000). However, even though there is an increasing demand for digital cable, penetration rates for digital are still relatively modest when compared with all cable subscribers. DBS is already a digital technology, and its market share could increase significantly if cable companies are unable to deliver digital video in a relatively short period of time for a reasonable cost.

Telephony Services

Consumers may have many choices for both wired and wireless telephony services in the future. Cable-delivered telephony will grow to the extent that it can deliver reliable service at a cost lower than the traditional wired and wireless carriers in an area. Wired cable telephony will continue to be viable to the extent that there is a sufficient subscriber base for other services, such as video delivery or data, on the same broadband network.

Cable companies are also continuing to exploit the capacity of their fiber optic network backbones. For example, some cable operators are currently using their networks for backhaul services for PCS wireless systems and other telecommunications services, allowing a significant additional revenue stream.

Data Services

The deployed broadband network of cable companies has allowed them to take the lead in delivering high-speed online data services to residential customers. In the next few years, some cable companies plan to offer their cable modem customers an open access environment in which there will be a choice of competing Internet service providers (ISPs) on their systems. This competition may improve the services offered and keep rates in check (Hearn, 2000).

However, DSL (digital subscriber line) services now being offered by local exchange carriers and other providers will offer significant competition to cable modem technology. While cable modem systems are theoretically faster than DSL, cable modem systems often pass data at slower rates when many subscribers in a local area are using the service simultaneously. Subscribers can also use their DSL line for simultaneous telephony services. In this competitive environment, residential cable data services will continue to prosper only as long as they can offer a quality services at an attractive price.

Regardless of what happens in the residential data marketplace, cable companies should continue to realize added value from the excess capacity on their broadband networks by offering point-to-point data network services to businesses.

In summary, cable is currently the major residential broadband delivery system in the United States. In the next few years, as all delivery systems and user devices become digital, the cable industry will continue to enjoy this premier position only if it makes a rapid conversion to digital, is able to offer its services at prices below competitors, and is perceived by its subscribers as a reliable telecommunications provider. Should it be unable meet any of these stipulations, it risks eroding its market share to other wired or wireless broadband service providers.

Bibliography

Albarran, A. (1996). *Media economics: Understanding markets, industries, and concepts*. Ames, IA: University Press.

Albiniak, P. (2000, March 6). Promises, promises. *Broadcasting & Cable*, *130*, 6.

Baldwin, T., & McVoy, S. (1988). *Cable communication*. Englewood Cliffs, NJ: Prentice-Hall.

Crandall, R., & Furchtgott-Roth, H. (1996). *Cable TV: Regulation or competition?* Chicago: R.R. Donnelley & Sons.

Dawson, F. (2000, March 20). Plug-and-play modems enhance retail's role in data rollouts. *Multichannel News*, *21*, 45.

Hearn, T. (2000, March 6). Lathen: Access plans need fleshing out. *Multichannel News*, *21*, 46.

National Cable Television Association. (2000). *The cable industry at a glance: Current estimates*. [Online]. Available: http://www.ncta.com/glance.html.

Time Warner. (2000). *About Time Warner*. [Online]. Available: http://www.timewarner.com/corp/.

Umstead, R. (2000, March 6). Digital launch plus one. *Multichannel News*, *21*. 3A.

3

Pay Television Services

Larry Collette, Ph.D.*

The consumer acceptance of cable television proved that people would willingly pay directly for in-home television entertainment, a notion that now has a half-century of history behind it. Slightly newer is the notion that people would pay additional monies to see programming unbundled from the basic advertiser supported tiers, in the form of premium or pay services. Some in the industry have even suggested that such direct payments could better ensure that programming would meet the wants and desires of television viewers (Litman, 1998).

This demand model has flourished and given rise to a number of pay services across different platforms and has given promise to full-service networks that eventually can enable even more pay television offerings. This chapter discusses these pay television services as they have developed in the United States.

Premium or pay television services are made available to subscribers for a flat rate, generally around $9 to $13 per month. The price is not sensitive to the time the subscriber spends viewing television. Most cable systems offer multi-pay discounts for subscribers who take more than one of these pay services. The churn rates (disconnects) for pay television services have historically been higher than those of basic cable. This fact is perhaps a reflection of sheer numbers, as basic cable tiers offer more programming channels relative to the dollar spent by the subscriber.

The most recognized and firmly established premium services are HBO, Cinemax, The Movie Channel, and Showtime. These familiar brand names enjoy the greatest number of subscribers in the pay portion of the market, who are attracted by their lineups of motion pictures, special events, some sports, and original, regularly-scheduled series (e.g., HBO's *The Sopranos*).

In the early 1990s, "mini-pays" arrived on the scene. These services were lower priced and differentiated from the larger premium networks by programming older "classic" motion pictures. Encore, Starz!, and Flix are examples of mini-pays. These services generally charge $1 to $6 per

* Assistant Professor, University of Denver (Denver, Colorado).

month, with some thematic multiplex packages offering more channels at a higher cost. Developments such as the acquisition of more recent hit movies, multiplex packaging, and digital cable have changed the mini-pay segment of the market.

Pay-per-view (PPV) differs from pay or premium services because subscribers pay a set price for *each* individual program they order. PPV offerings include recent motion pictures, adult fare, and one-time-only events such as professional wrestling, championship boxing, and concerts. The PPV movie window follows both theatrical and home video release of the motion picture.

The means of accessing PPV differs from system to system. On some systems, subscribers dial an 800 number, while other subscribers (on digital systems) may point, click, and choose their PPV program.

Programmers and operators define the marketing success of a PPV property in terms of "buy rates," the percentage of subscribers who actually purchase a PPV program. The charges are added to the subscriber's monthly bill. Prices for movies generally are a few dollars, which is comparable to the price of a video store rental. Championship fights generally command the highest fees, with recent prices in the $50 range.

Video on demand (VOD) is the programmer's version of entertainment nirvana. Through VOD, at-home television viewers can call up a program when they want it. Although existing in both industry rumors and test beds for some time now, this technology has yet to be deployed on a wide scale. VOD not only allows the subscriber to make programming choices, it also enables them to have VCR-like functionality in their viewing experience, complete with fast forward and rewind. True VOD requires enormous memory at the service provider's server end, reliable interactivity across various points of the system architecture, and end-user set-top boxes capable of a variety of interactive functions.

Near video on demand (NVOD) is the interim step between full VOD and PPV. Here, programs are sold on a per-use basis just as with PPV, but program start times are staggered to offer greater flexibility to the viewer. For example, a movie may start at the top of the hour and at fifteen-minute increments on different channels after that, giving the movie four runs within an hour. This strategy increases the subscriber's opportunities for viewing and is calculated to increase buy rates.

Background

The earliest pay television service came from Zenith and its Phonevision system in the 1940s (Parsons & Frieden, 1998). The system used a scrambled signal that was broadcast to the viewer's home, and billing took place over a telephone line. Although Phonevision was tested in the 1950s in the Chicago area, it vanished from the scene despite the best efforts of its parent company. Other versions of pay services followed, including one backed by Paramount that used a coin box (the truest expression of direct pay TV). Aside from television sets in downtown bus terminals with coin boxes attached to them that delivered broadcast TV to bored and weary travelers, the concept of "jukebox" television never caught on. These early attempts at pay television, coin operated and otherwise, suffered from consumer indifference and political opposition. As the 1950s rang to a close, pay television was nowhere to be seen.

Other than a few isolated attempts by system operators to offer pay services on a local basis, the next attempt at pay television came in 1963 with Subscription Television, Inc. (STV). The idea behind STV was to deliver sports and motion pictures to subscribers in the Los Angeles and San Francisco areas through a combination of wireline and broadcast technologies. The service drew the ire of theater owners and broadcasters alike who feared the new service would take revenue (box office and advertiser, respectively) from them. Championing the cause of "saving" so-called free TV, these partners supported a referendum that prohibited charging for television delivered to the home. The referendum was passed by California voters and signaled caution to anyone even pondering a pay television service. Though struck down by the courts, the damage was done, as STV went into bankruptcy the following year and exited the marketplace.

After STV, the fear of "siphoning" programming product from so-called free over-the-air television to pay services gained the attention of the Federal Communications Commission (FCC), which passed a series of programming rules that were very adverse to pay television (FCC, 1970). These "anti-siphoning" rules sharply constricted the supply of available programming products and represented a substantial barrier to entry for those poised to enter the pay television market. For example, no sporting events that had appeared the previous three years on over-the-air television were permitted on pay services, nor were any regularly scheduled program series. Motion pictures had to be "captured" *within* two years or *afte*r 10 years of their theatrical release. In the movie fan's mind, a 10-year blackout period would likely seem an eternity, and pay services would be left to peddle a leftover "classic" feature film.

Media historians and economists generally acknowledge the birth of Home Box Office (HBO) in 1972 as the watershed event in the creation of pay television. Arising from the proposal of Gerald Levin and Charles Dolan at Time, Inc., HBO would offer sports and motion pictures to paying customers (Clurman, 1992). Cable operators had an incentive to help market the service by virtue of a revenue sharing arrangement created by the company. The first programs, a hockey game from Madison Square Garden and a movie, were delivered via microwave and then through cable television to 365 subscribers in Wilkes-Barre, Pennsylvania (Cable Center, 1997). In 1973, perhaps offering quiet testimony to the difficulties of obtaining programming given the restrictive FCC rules, HBO's first original production was "The Pennsylvania Polka Festival" from the Agricultural Center in Allentown, Pennsylvania (Cable Center, 1997).

The use of a microwave system to distribute signal restricted HBO to a regional New York and Pennsylvania service. Because a series of repeater towers had to be arrayed every 25 to 30 miles, the distribution costs rose appreciably with the distance traveled. The technological breakthrough necessary for creating a truly national service came with HBO's use of RCA's Satcom I satellite in 1975 (Clurman, 1992). Satellite distribution helped solidify HBO's lead in premium television services even as programming remained a problem. Relief from programming restrictions came when the Federal Circuit Court in 1977 declared the programming rules imposed by the FCC to be null and void (*HBO v. FCC*, 1977). The siphoning argument had been put to rest by the court, and the opportunity for creating new viable pay services now existed.

Soon, Showtime, The Movie Channel, and Cinemax (HBO's sister channel) entered the market. Yet, HBO enjoyed a sizable "first mover" advantage and vacuumed up a sizeable portion of the pay market, and the others were forced to play catch-up. HBO used its considerable clout in negotiating licensing deals with major motion picture distributors. Yet, the motion picture companies were accustomed to being dealt with in a respectful manner by subservient buyers, and they had played one broadcast television network against the other when it came time to negotiate the sale of their films.

In spring 1980, four major motion picture companies (Paramount, Fox, Universal, and Columbia), along with Getty Oil, announced the establishment of the Premiere Network. In an attempt to make an end run around HBO and other pay services, this company would retain an exclusive nine-month window for all its partners' motion pictures. Despite being a latecomer to the market, Premiere would attempt to compete by garnering the strategic resources under its direct control: the films that all pay services desperately needed. After complaints by HBO and Showtime that such a "refusal to deal" by these companies was an antitrust violation, the Justice Department intervened in the summer of 1980. An injunction was eventually granted, and the Premiere network never did "premiere."

HBO and others had learned an important lesson and attempted to increase their self-sufficiency by producing their own programs (e.g., HBO Films and Silver Screens) and by locking up a steady stream of motion picture products under long-term contracts, a strategy that continues to the present day.

The premium or pay service market became a duopoly when Showtime (owned by Viacom) acquired The Movie Channel, leaving Viacom and Time, Inc. (which later became Time Warner) with its HBO and Cinemax alone in the marketplace. The Spotlight Network, a smaller movie-driven pay service, had exited the market at around the same time.

Entry by newcomers is made difficult by the shortage of hit movies and the contracts that tie up movies for extended periods. Market entry in pay services has been limited to specialty type services, mini-pays such as Encore, sexually-oriented services such as Spice and Playboy, and a few sports channels that operate as regional pay television services. Playboy eventually migrated to a PPV service. The Disney Channel, which began as a pay service, has been largely converted to the basic tier as it became clear that Disney could realize greater revenues from the positioning.

One notable recent entry in the pay television market is the Starz! Network (a service of Starz Encore Group), which was launched nationwide by Tele-Communications, Inc. (TCI) in 1994. With the support of the nation's largest operator of cable systems, Starz! has been able to negotiate windows for major motion pictures that preceded the more dominant pay television networks. Starz! has since grown to become the fifth largest premium cable television service.

A fall off in the number of pay subscriptions in the early 1990s caused many multiple system operators (MSOs) to reduce the price of their pay services. The industry also responded by introducing "multiplexing," the practice of duplicating a pay network's programming across two or more channels with staggered start times on each channel.

The rise of direct broadcast satellites (DBS) in 1994 has opened up new markets for pay television networks. Encryption has made it possible for pay TV services to ensure that viewers pay for the content they watch, whether it is delivered via cable, DBS, or C-band satellite.

Premium channels achieved rapid popularity once HBO was introduced. However, pay-per-view has not garnered such consumer acceptance. A key enabler of pay-per-view television is addressable technology, which has been around since the 1970s. Addressable technology allows the service provider to communicate directly with the set-top box in each household and thus provide programming without offering it to everyone. Two large PPV services, Viewer's Choice and Request TV, were launched in 1984, and they vanished some years later.

By 1985, PPV showed growth potential, and nearly six million cable homes were capable of receiving it (Cable Center, 1997). What changed dramatically in the late 1980s was the ease of access to PPV events, as remote controls and two-way technology made the transaction simpler and enabled impulse buys. Still, PPV accounts for a relatively small amount of cable operator revenues. One reason is the explosion in VCRs and home video, which provide a reasonable substitute for PPV.

Motion pictures are the staple of PPV. However, the vision of movies premiering in theaters and on PPV at the same time never materialized. Perhaps, Universal's ill-fated experience in releasing *Pirates of Penzance* (starring Linda Rondstadt and Barry Bostwick) in this manner in 1983 provided a serious and long-remembered caution to anyone attempting such a move.

Still, PPV would seem to hold certain advantages over home video. A visit to the video shop necessitates a return trip to bring the videotape back and, even then, a customer is not assured of getting his or her first or even second choice of films.

In 1994, motion picture studios extended their home video release windows from 60 days to 90 days before permitting a PPV release of their films. This made the task of marketing PPV movies all the more difficult because people were more likely to have already seen the product before it ever got to that PPV window. At the same time, buy rates for movies were found to increase substantially with MSO tests of NVOD in place of PPV that same year (Cable Center, 1997).

The largest attempt at a special event on PPV was the "TripleCast" of the 1992 Summer Olympic Games in Barcelona, Spain by NBC and Cablevision. Offering specialized, three-channel coverage of 15 days of the Games, the full package was priced at $125, and daily buys were $30. NBC hoped to defray through PPV $100 million of the $410 million rights fee that they paid for the Games (Carter, 1992). With a hoped-for buy rate of two million or 5%—based on its projections—NBC planned to cover event heats and give less familiar sports wider coverage (Lafayette & Maddox, 1992). It became apparent that the project was in trouble when, one week from the event, only 50,000 to 60,000 TripleCast subscriptions had been sold. A series of price slashing measures in the days before the Games only increased buys slightly. The United States was in the grips of a recession, and consumers were reluctant to spend money on a package whose "prime events" were already on broadcast television for free. Also, the number of cable systems having access to the capital needed to upgrade their systems for PPV fell far below expectations due to economic conditions. NBC and Cablevision are believed to have lost between $100 million and $150 million on the venture (McClellan, 1992).

On the brighter side, much was learned about marketing and coordinating such a large PPV event. And some large MSOs had upgraded their physical plants in time for the Games, allowing them to leverage that upgrade to their advantage to carry PPV in the future.

Video on demand allows customers to request material and receive it more or less immediately. As of mid-2000, VOD has been limited mostly to tests in specific markets aimed at measuring consumer interest and activity. At the same time, these tests gave companies the needed opportunity to test the technology itself. For example, in the early 1990s, cable subscribers in TCI's Littleton (Colorado) system were able to order videos on demand (Baldwin, McVoy & Steinfield, 1996).

However, the state of the technology was primitive—roller skaters placed tapes in videocassette recorders (VCRs) when subscribers ordered their movies. Technological advances eventually led to Time Warner's Full Service Network, which began in 1994 in Orlando (Florida). This system allowed subscribers to order a broad range of programming and services on demand. The project's technology, which cost around $4,000 to $5,000 per household, ceased operation in 1997. The costs of the necessary technology have fallen dramatically since the Orlando trial, with cable system upgrades and digital set-top boxes combined now costing in the $700 to $800 range (Higgins, 2000b). However, what was learned in terms of marketing and the technology itself led those involved in the project to regard it as successful.

Your Choice TV, which sought to extend the video on demand concept beyond motion pictures to include other programming fare such as rescheduled TV shows, launched in 1992. After six years and $25 million in losses, Your Choice ceased operations. The problems in acquiring rights fees to programs and the slowness of MSOs to deploy digital technology were blamed for its eventual failure (Higgins, 2000b).

Recent Developments

The most significant recent events affecting the pay television services marketplace fall into three categories: service competition, technology, and the mega-branded multiplexing of content.

Over the long term, competition in delivery systems is expected to translate to increased opportunities for expanded pay television services for consumers. Yet, the Telecommunications Act of 1996 has not yielded the anticipated flood of competitors. With rare exception (e.g., Ameritech New Media, and BellSouth MMDS), the regional Bell operating companies (RBOCs) that had appeared very eager to deliver video services to customers have cancelled or scaled back many of their video initiatives. Tele-TV, the three RBOC partnership aimed at acquiring and producing program content, dissolved shortly after the Act was passed. Choosing instead to concentrate on protecting their core telephony business, most telcos have still continued to upgrade their physical plant (Bowermaster, 1997). The FCC itself characterized telephone company entry into video markets as being "slow to develop" (FCC, 1998).

For now, broadband, data service, and Internet access seem to have replaced delivery of video as the priority for the telcos. Still, this trend does not mean that plans for large-scale telco entry into the provision of video have been abandoned; their full service networks will be able to accommodate all forms of digitized information.

Currently, direct broadcast satellite is a formidable competitor to cable in the delivery of multi-channel television. DirecTV and EchoStar deliver many of the same pay services as cable, but have created unique marketing tiers of their own. The number of subscriptions has been climbing, and DBS has been able to offer some very user-friendly approaches in NVOD packages. No longer confined to rural areas where cable was at a distinct disadvantage, DBS has made inroads in providing pay services in urban areas as well. (For more on DBS, see Chapter 5.)

DBS services have been much more successful with pay-per-view offerings to addressable set-top boxes than cable operators. The few cable systems that have upgraded to digital have found the expanded channel capacity a boon for pay television services. Digital deployment has allowed systems to devote more channels to NVOD and been credited with giving that market segment a substantial boost. Yet, in the years that have passed since addressable technology was first introduced, only 44% of all cable homes can even order PPV. By way of contrast, DirecTV, which devotes 50 channels to PPV, has buy rates almost four times that of the average cable operator. Indicative of the changing tide toward digital services, Time Warner announced a significant $2 billion capital campaign aimed at deploying first-generation digital converters in early 2000 (Higgins, 2000a). In 1999, approximately 430,000 of Time Warner's 11.5 million subscribers ordered new digital services. The digital upgrade is expected to enable a larger package of video and interactive content from both Time Warner and its merger partner, AOL.

Other MSOs such as AT&T have taken a similar tack and increased the deployment of digital cable service. In the case of AT&T, prices for the basic digital tier are set at $10, with pay services bundled in multi-pay packages adding to the bill. Compression technologies and increased deployment of fiber to within that "last mile" of subscribers (as cable systems are upgraded) will increase bandwidth and content offerings.

At the same time, the increased number of channels has allowed the "name brand" services to expand their offerings across multiple channels and create new umbrella multi-pay packages (Richmond, 1998). HBO and Cinemax recast their services into mega-branded multiplexes "HBO the Works" and "MultiMAX." HBO repositioned HBO2 and HBO3 as "HBO Plus and "HBO Signature." "HBO Family," "HBO Zone," and "HBO Comedy" soon followed. Each service was differentiated from the other through program offerings; for example, "HBO Signature" offers more independent and vintage movies. The traditional HBO programming channel remains the same.

Time Warner's Cinemax was repackaged into a mega-branded multiplex with the addition of separate channels, "MoreMAX," "ActionMAX," and "Thriller MAX," under the "MultiMAX" umbrella. In all, these Time Warner pay services offered 24 individual program feeds within its various time zone feeds.

Showtime also created the mega-branded multiplex, "Showtime Premium Pak," as an umbrella for Showtime, the Movie Channel, Flix, and the Sundance Channel. An action alternative channel, "Showtime Extreme," was added to help target young, male viewers. Flix continued to operate as a mini-pay channel, offering classic films.

Encore, migrating away from its mini-pay roots, was revamped into more of a premium movie service through the acquisition of new movie packages. In December 1999, Encore and Starz! announced the creation of their own combined mega-branded multiplex service, "Starz! Encore Super Pak." The branding campaign for the service touted "12 channels for $12" (McAdams, 1999d). Meanwhile, Encore also continued offering its already-created six-channel Thematic

Movieplex (e.g., Westerns) estimated to reach 23.1 million subscribers through a combination of satellite and cable distribution (McAdams, 1999d).

Current Status

The number of pay television networks in the United States stands at 43, with nine PPV services (NCTA, 2000). In the cable industry, total pay revenues, including mini-pays, were $4.930 billion (NCTA, 2000). Those revenue figures have been fairly stable over the past several years, but do remain below the $5 billion mark surpassed only in 1990. This fact indicates a leveling off of revenue when the market became more crowded as pay services proliferated their brands across multiple channels. The total number of pay units sold (number of pay subscriptions) is over 49 million in the United States (NCTA, 2000).

Figure 3.1 shows the subscriber distribution for the major pay TV networks. HBO continues its dominance of the pay services market, with 26,659,000 subscribers at the end of 1999 (Cablevision, 2000). Cinemax, Time Warner's other pay service, had 14,820,000 subscribers, with Showtime claiming 13,554,000, Encore with 13,179,000, and Starz! with 9,160,000 subscribers.

Figure 3.1

Pay TV Network Subscribership

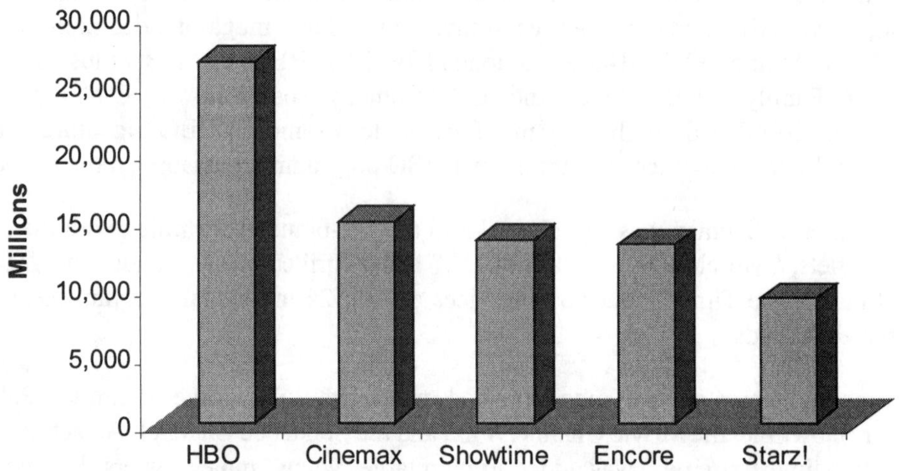

Source: Cablevision

In the PPV segment of the marketplace, Viewer's Choice, which changed its name to In Demand on January 1, 2000 (McAdams, 1999a), claims over 20 million subscribers and 1,700 affiliated systems. The next largest PPV channels are the adult-oriented Playboy, Spice, and Spice 2, each having over 10 million subscribers able to take their PPV offerings. In early 1999, Playboy Entertainment announced its $100 million acquisition of Spice Entertainment, further consolidating control in that portion of the adult PPV market (Schlosser, 1999). BET Action Pay-per-View was available to 10 million cable subscribers (Cablevision, 1999).

The average number of hours per year spent by U.S. adults viewing pay services on cable in 1998 was 64.5 hours (Papazian, 1999). Although a greater percentage of those with higher incomes were pay subscribers (42% of those with incomes over $75,000), the actual time spent viewing pay services was greatest among those with the lowest income (under $30,000) (Papazian, 1999). Ironically, this means that those less likely to afford pay services are also the group most likely to spend the time using it. The inverse also appears true, as those who have pay services are also those least likely to actually spend time using it. Also, those in the 35 to 54 age range are more likely to spend time viewing pay services than are those in younger and older demographics (Papazian, 1999). In 1999, U.S. PPV penetration was 55%, nearly four times the level it was in the mid-1980s. Still, the use of PPV remains infrequent, as only 23% of households have ever used PPV to access a program of any kind (Papazian, 1999). Even among households that use PPV, most limit their use of PPV to four times a year or less.

Movies continue to generate about half of the total revenues for PPV. Total revenues from motion picture PPV were $241 million in 1998; by comparison, home video is a $15 billion a year business (McAdams, 1999c). According to a study commissioned by Showtime, in 1999, PPV movie revenues grew dramatically to over $1 billion, owing largely to DBS and digital cable upgrades (McAdams, 1999c). The return on PPV investment is expected to increase even further once real VOD takes hold of the market.

Throughout the 1990s, boxing contributed about 50% of the $2.7 billion in revenues generated by event PPV. Wrestling accounted for around 41% of all revenues generated by event PPV in the 1990s. The WWF reported that its revenues from PPV events increased over $37 million in 1999, as compared to those of 1998 (Higgins, 1999). Over the same period, buy rates grew from 216,000 to 416,000 for WWF wrestling. In 1999, the Oscar de la Hoya versus Felix Trinidad welterweight fight became the third largest PPV event of all time, generating 1.25 million buys for a total of $64 million (McAdams, 1999b). The fight still ranked behind two Mike Tyson and Evander Holyfield fights, which remain the largest PPV events of all time.

Now there is likely to be a new competitor in the pay-per-view arena: the Internet. Long form PPV, movies, and vintage television programs came to the Internet in 2000 through Movies-Online.com and MeTV.com (Tedesco, 2000). One business plan calls for independent film titles streamed at 300 Kb/s to be priced at $2.95, with lower prices for television titles. ClickMovie.com announced plans to offer a three-tiered subscription service priced from $9.99 to $29.99 per month for films and TV programs. The viability of these new ventures will be known only over time. (For more on streaming video, see Chapter 8.)

Factors to Watch

Pay television services, much like the technologies delivering them and the marketplace in which they operate, are dynamic processes subject to a variety of dramatic changes. Listed below are a number of factors to watch in the area of pay television services:

Subscription Video on Demand—In an attempt to add incremental subscriber revenue, several premium services are planning SVOD services. For an additional monthly fee ($6 to $12 per month), subscribers can see some portion of HBO or Starz! scheduled programming at any time within the month. For example, HBO On Demand will offer 100 titles per month, 75% original programming, and 25% theatrically released motion pictures. So far, large cable operators have been noncommittal in their assessment of this concept and its potential. Starz! is planning a trial of SVOD in Atlanta on a Media One Group cable system in 2000.

Video Streaming over the Internet. Today's Internet is a pale imitation of television. In the future, however, expanded bandwidth and faster interconnect speeds will ensure the Internet's place as a multi-channel video provider. Although cable companies do not want to be reduced to mere carriers of content for others, there have been recent assurances by some major MSOs (AT&T and Time Warner) that they will allow outside ISPs "open access" to their systems regardless of content. This may be an "invitation to battle" for the pay services market. As distribution costs shrink via use of the Internet, the potential for new, emerging pay services will grow accordingly.

Increased Deployment of Digital Cable. As mentioned previously, digital cable will be instrumental to the growth of pay services, in particular video on demand. This will be especially true as cable systems are forced to meet the competitive challenges of DBS and the next generation of newcomers to the multi-channel television marketplace.

Improved Image and Sound Experience on Pay Services. HBO is currently providing HBO HDTV to some systems. The display provides a wide-screen format with a 16:9 aspect ratio that enhances the cinematic experience of the in-home viewer. At the same time, the company is also working on providing Dolby 5.1 digital sound to in-home viewers, further enhancing the movie experience.

Expanded Original Content Production. Original productions offer an economical alternative to theatrically released motion pictures and are first-run material. Regularly scheduled series like HBO's *Sex in the City* and *The Sopranos* have proven popular with audiences and critics (Grego, 1999). These programs also provide an added incentive for subscribers to keep their pay subscriptions. In addition, these self-produced programs can eventually be sold in after-markets when their cable and satellite runs are finished.

Hyperplexing. The mega-multiplexing strategy used by program services will reach new heights as the number of channels needing to be filled on delivery services continues to expand. In a continuation of their "brand proliferation" strategies, the major pay networks are likely to create an even larger array of channel offerings under larger programming umbrellas in an attempt to bolster their market share.

Increased Use of Content across Multiple Internal Windows within Large Media Companies. As an outgrowth of the mergers occurring in the communications industry, companies are likely to use their increased clout in negotiating for the rights to use content across multiple windows. For example, Viacom may negotiate a movie deal for Showtime, with the intention of also securing rights for later use over its CBS broadcast network. Producers may also benefit from the reduced transaction costs of negotiating these multi-tiered deals in one fell swoop.

Deployment of True Video on Demand. The promise of VOD seems clear. Merrill Lynch estimates that VOD will grow from a market with no revenue in 1999 to nearly $6 billion by 2009 (Higgins, 2000b). The value of cable systems will increase as a direct result of this windfall, adding to operator incentives to provide these services. Technology providers such as Diva Systems (Menlo Park, California) and Ncube owned by Oracle (Foster City, California) are among the companies vying to be equipment providers for this burgeoning market.

Increased Contentiousness in Negotiating New Windowing Rights. HBO refused a recent multi-billion dollar film package deal with Sony Pictures because Sony would not sell Internet window rights to the films. This trend seems likely to increase in the future as companies on both sides seek to hedge their bets when it comes to new windowing opportunities.

Bibliography

Baldwin, T., McVoy, S., & Steinfield, C. (1996). *Convergence: Integrating media, information & communication.* Thousand Oaks, CA: Sage.

Bowermaster, D. (1997). Phone cable firms keep their distance. *MSNBC.* [Online]. Available: http://www.msn.com/default.95p.htm.

Carter, B. (1992, July 20). NBC's Olympic hopes rise on pay-per-view. *New York Times*, C6.

Cable Center. (1997). *Milestones: A 50-year chronology of cable television.* Denver: National Cable Television Center and Museum.

Cablevision. (2000). *Cable industry data.* [Online]. Available: http://www.cablevisionmag.com/database/db_pay.asp.

Clurman, R. M. (1992). *To the end of time.* New York: Simon & Schuster.

FCC. (1970). *Second report and order*, 23 FCC 2nd 825.

FCC. (1998). *Fifth annual report on competition in video markets.* CS Dockets 98-102, released March 28. Washington, DC: Federal Communications Committee.

Grego, M. (1999, September 20). Emmy sings for HBO. *Broadcasting & Cable*, 50-51.

HBO v. FCC. (1977). 567 F. 2d 9.

Higgins, J. (1999, August 8). And now, weighing in at $250 million. *Broadcasting & Cable*, 28.

Higgins, J. (2000a, February 7). Levin has set-top fever. *Broadcasting & Cable*, 48-49.

Higgins, J. (2000b, March 8). VOD: Is it a business yet? *Broadcasting & Cable*, 24-28.

Lafayette, J., & Maddox, K. (1992, August 3). NBC finding Olympic gold despite losses. *Electronic Media*, 1, 23.

McAdams, D. (1999a, September 20). Viewer's Choice changes name. *Broadcasting & Cable*, 59.

McAdams, D. (1999b, September 27). TVKO scores record haul. *Broadcasting & Cable*, 79.

McAdams, D. (1999c, December 6). The great fight hope. *Broadcasting & Cable*, 46.

McAdams, D. (1999d, December 20). Starz!, Encore combine. *Broadcasting & Cable*, 54.

McClellan, S. (1992, August 3). Olympics draw viewers, few payers. *Broadcasting & Cable*, 3, 12.

National Cable Television Association. (1999). *Cable television developments.* Washington, DC: NCTA.

National Cable Television Association. (2000, Summer). *Cable television developments.* Washington, DC: NCTA.

Papazian, E. (1999). *TV Dimensions '99.* New York: Media Dynamics.

Parsons, P. R., & Frieden, R. M. (1998). *The cable and satellite television industries.* Boston: Allyn & Bacon.

Richmond, R. (1998). Premium nets: More screens mean more value. *Multichannel News Online.*

Schlosser, J. (1999, March 15). Adult channels mate. *Broadcasting & Cable*, 38.

Tedesco, R. (2000, January 3). Long form PPV hits the Web. *Broadcasting & Cable.*

4

Interactive Television

Paul J. Traudt, Ph.D.*

Interactive television (ITV) has long been considered the ultimate goal of matching video-based entertainment and information, with viewers actively engaging or even creating video content. The raised level of involvement is very appealing to advertisers and programmers alike. However, reaching that goal has proven to be surprisingly elusive. Indeed, the history of interactive television has so far resulted in failed efforts, despite over one billion dollars in capital investments for prototype testing. To many observers, even the words *interactive* and *television* are contradictions.

Part of the problem is that analog-based telecasting has always been *one-way* communication. As a result, traditional television audiences are habitually passive viewers for the simple reason that one-way technologies inhibit any real form of interactive communication. The TV audience has learned to watch television as a "lean back" experience, rather than a "lean-forward" one. Thus, it becomes the obligation of any evolving ITV platform to be so compelling as to overcome viewer habits.

It is instructive to look back at the various peaks of activity in the development of ITV. Simple and elegant efforts included children's programming in the 1950s. There were additional trials and failures in the 1960s, followed by full-scale but financially unsuccessful efforts in the 1970s. Lessons learned from these earlier experiments, converging communications technologies, competition, and potentially lucrative returns on capital investment contributed to an explosion of ITV projects in the early 1990s. The popularity of Internet-based entertainment, information, and shopping demonstrated the growing potential for consumer interest in some form of television on demand.

The result has been ongoing corporate mergers and alliances between entertainment, information, telephone, computer, and Internet industries in an attempt to develop the so-called "killer application." When viewed from a larger perspective, events in ITV over past years suggest that a

* Associate Professor of Media and Information Studies, Hank Greenspun School of Communication, University of Nevada (Las Vegas, Nevada).

single technological platform may never be realized. One may infer, if recent trends are any indication, that the eventual ITV landscape may be structured similarly to that of cable television—individuals subscribing to varying levels of interactivity.

Efforts in the past two or three years would suggest that the ITV industry might finally be on track in terms of interactive technologies and services. However, limited programming and revenue streams still continue to dampen claims of success.

In its current form, interactive television is the combination of traditional linear video and a range of interactive services (Feldman, 1997). The International Telecommunications Union (ITU) has defined ITV as "services in which the end-user influences in real-time or quasi-real-time the content...delivered by sending messages to the...origination point" (Baron & Krivocheev, 1996, p. 196).

The typical configuration includes linear video broadcasting accompanied by layered, transparent, or semi-transparent graphical or text elements. The ITV provider distributes HTML (hypertext markup language) information to analog or digital television receivers, set-top boxes, or personal computers (Carat Group, 2000; Swedlow, 1999). Home users connect to ITV via the Internet, a local cable television provider, direct broadcast satellite (DBS), a local telephone company, or some combination of the four. Web-based interactivity takes the form of user manipulation of graphical interfaces via remote control or cordless keyboards.

Five major service categories currently pepper the ITV landscape (Haley, 1999; Kerschbaumer, 2000; Swedlow, 1999). These five divisions are becoming increasingly fuzzy as major players converge two or more of these categories into bundled services. Some industry observers refer to all current services as *enhanced television.*

Internet-on-TV

This technology provides Windows or browser-type interactivity in the form of graphical or text enhancements over a broadcast video background. Proprietary software resides in a receiver set-top box and tuner card. HTML data are sent via the now deregulated vertical blanking interval (VBI, the part of a broadcast television signal not visible on most television receivers). Digital television will eventually eliminate the need for VBI utilization. Another version of the technology utilizes software at the cable headend. Revenues are generated by a combination of set-top box purchases, subscription, and monthly fees. Microsoft's WebTV currently dominates the Internet-on-TV domain. WorldGate, a proponent of the cable headend approach, is a distant second. AOL and Morecom are also players. WebTV advocates claim that "Surfing the Net on TV is more fun than doing it on a PC" (Swann, 1999, p. 4).

Personal TV

This time-shifting technology records video programming via integrated digital hard drives. Televiewers can capture and save up to 30 hours of programming. They can also record, pause, play back, and resume viewing network feeds. Revenues are generated via a combination of set-top box purchases and monthly service fees. TiVo and Replay are major players in personal TV services.

Interactive Program Guides or Navigators

This ITV service is provided via satellite, cable, or other set-top box (STB) system, with programming schedules and information often sent via VBI over the Internet or down telephone wires connected to the STB. Advanced digital STBs include memory components that store information about individual users and navigational preferences. Based on the data, some services recommend similar programs to viewers.

Revenues are generated in several different ways. Some services charge subscriber fees, others take in a percentage of monthly cable service fees, and some are advertising-supported. Currently, TV Guide's Web-based application and Interactive Media's Source Media are butting heads for dominance in the area of interactive programming guides, seen by some as the most competitive area of ITV development (Haley, 1999). Source Media utilizes a cable headend approach.

Enhanced TV (eTV)

The above types of ITV share the common characteristic of having content provided by programmers or advertisers, but without Web-browsing capability. ACTV's software in set-top boxes allows the viewer to have directorial control of live sporting events via remote control. For example, the user can choose from a number of ballpark cameras covering a baseball game while requesting a batter's season or game statistics. Penetration is estimated to be 15% to 20% of test-market cable subscribers by 2001 (Colman, 1999).

Another service, Wink Communications, provides proprietary software residing in an STB that interprets data sent over the VBI. Wink enhanced commercials or programs include an icon overlay, which allows users to request more information or order products and services with their remote control devices.

eTV is limited to content provided by programmers or advertisers. Revenues are generated in some combination of cable subscriptions and advertising.

Video on Demand

The "bandwidth hogs of the interactive world" (Haley, 1999), VOD requires up to four 6-MHz channels to serve the estimated 10% of cable subscribers using the service at any one time. Diva and Intertainer stream MPEG video via broadband. Video on demand services can provide programming and information almost whenever the end-user chooses, either by means of real-time interactivity or previously delivered programming in residence on hard drives at the user's end. Revenues are generated by means of subscriptions and advertising. Sea Change and TVN are other competitors.

Background

In 1953, CBS Television aired *Winky Dink,* an animated series designed to encourage interactivity between program characters and children watching the program. An inexpensive Winky Dink Kit was provided via mail or at local retailers. Plots included prompts, where children were instructed to apply a transparency to the television receiver screen and use crayons to generate visual overlays and connect dots to reveal "secret" messages. The program aired in some markets as late as the early 1970s (Swedlow, 1999).

Time Warner's short lived QUBE TV was an analog, two-way interactive cable experiment that began in 1977 and provided interactive game shows; religious, public affairs, and children's programming; and banking services (Rosenstein, 1994). Initiated in Columbus, Ohio, the service offered 36 channels and provided subscribers with a set-top box and wired remote. The service was eventually expanded to the Dallas and Pittsburgh markets. Users could indicate preferences and vote during programming generated for the system. The prototypes for this early programming became such venues as QVC, Nickelodeon, The Movie Channel, and MTV. Decline in investor support and technology limitations spelled the eventual demise of QUBE TV. However, subscriber interest was enthusiastic and pointed to the potential for ITV services (Swedlow, 1999).

Teletext and videotext services in Europe, such as the BBC's Ceefax, were also tested in the United States. GTE's Viewdata and Knight-Ridder's Viewtron provided electronic newspapers, weather, and agribusiness information via telephone lines in the late 1970s and early 1980s. Users in test markets were indifferent, in part because of buggy technology and limited options.

GTE offered interactive cable television to subscribers in Cerritos, California in the late 1980s and early 1990s. The system offered home shopping, banking, video games, and movies on demand. Less than 5% of system households subscribed to the expanded service before it was dismantled (Rosenstein, 1994).

By the mid-1990s, large corporations, consortia, and alliances were field-testing multi-service, high-speed interactive and enhanced television systems. Competition, growing Internet popularity, developments in MPEG compression technologies, and a general belief that ITV would result in lucrative returns fueled these efforts. Source Media's Interactive Channel was one of more than 40 experiments launched domestically and internationally between 1994 and 1996 (Swedlow, 1999). NetChannel, TV Answer, and ACTV's *Wheel of Fortune* were others. The plug was pulled on most of these efforts, but lessons were learned about what ITV consumers did and did not prefer.

Some of the efforts during this time were creative and incredibly simple in concept. Telemorphix, formed in 1992-1993 in the San Francisco Bay area, was a weekly live television program aired via a leased cable channel. The program, *21st Century Vaudeville*, portrayed animated characters. Home viewers would dial in via telephone and wait their turn to provide extemporaneous dialogue for the animated characters on screen (Swedlow, 1999).

Time Warner's QUBE II, better known as the Full Service Network, was launched in 1994 in Orlando (Florida) at a cost of $250 million. By May 1997, Time Warner had announced its intent to end the costly service, citing as major reasons the high cost-per-household hardware and difficulties in shifting user interests from home videocassettes. At the time, industry observers predicted that this would be the last full-scale effort at proprietary-based ITV because of booming Internet and Web-based possibilities (Goldstein, 1997; Shiver, 1997; Marriage of convenience, 1997). Time Warner put positive spin on the failed experiment, claiming that they knew the venture would never be lucrative. The company also learned that ITV consumers wanted simple interactive options and particularly liked video on demand services.

Most recent advances in ITV reflect lessons learned from previous attempts. The cable and telephone industries continue in their efforts to marry Internet with television in addition to providing high-speed broadband data and Internet access.

Microsoft Corporation entered ITV in April 1997 with their acquisition of WebTV, effectively leading the charge to move personal computing and Web surfing out of the home office and into home video centers (Marriage of convenience, 1997). Bill Gates was convinced that the corporation's future depended on acquiring a larger share of the ITV market (Desmond, 1997). By November 1997, WebTV claimed 200,000 subscribers, up 100% since Microsoft's acquisition the previous April.

In January 1998, Microsoft agreed to provide ITV technologies to TCI for the production of Internet-ready set-top boxes, and included an order for five million of the boxes preloaded with the Windows CE operating system to provide e-mail, Internet access, and video on demand to subscribers. Oracle Corporation reacted with efforts to expand into enhanced TV and to speed up slow-loading Web pages (Clark, 1997). At the same time, Bill Gates also invested $1 billion in Comcast, endorsing cable system bandwidth as a digital distribution medium in an attempt to establish Microsoft's operating system as the standard for use in digital set-top boxes (Lesly, et al., 1997). Web TV's interface includes a home page for television and a Web home page with a grid containing television program listings. Clicking on a program title switches the user to that program. A "Plus" version of the system provides access to e-mail and other browser-based services.

Proponents and industry analysts predict the deployment of digital set-top boxes within one to three years. Meanwhile, in establishing its new open standard, the cable television industry effectively dictated the technical requirements for ITV, including specifications for writing software in both Windows CE and Java operating system languages for servers and STBs. The new standard effectively redefines the platform for ITV competitors and makes proprietary software and hardware obsolete.

Internationally, British Interactive Broadcasting was producing programming for enhanced TV in the form of interactive commercials with immediate-response icons (Rogers, 1997). Flextech, a British subsidiary of U.S.-based Tele-Communications, Inc. (TCI), was in discussions with Microsoft about forming an interactive television alliance.

Current Status

Capital investments in ITV have continued to skyrocket in recent years, with proponents hoping to strike "geysers of cash" with one dominant technology and programming for an entire industry (Haley, 1999, p. 18). In 1993, Bell Atlantic's then chairman predicted "absolutely explosive" growth in ITV given advances in storage, logic, and digitalization (Jessell, 1993, p. 18). So far, the source for this "geyser of cash" continues to be companies and investors for two reasons:

(1) A standard evolved for how signals were to flow back-and-forth between interactive service providers and home users. The Advanced Television Enhancement Forum (ATVEF), comprised of 14 of ITV's major players (including CNN, Discovery, Microsoft, PBS, Sony, and TCI), was established in 1998 to develop platform standards and specifications

(Pegg, 1999a). The Internet, with TCP/IP protocols, is now the widely accepted open standard for ITV. Every company with investments in television, the Internet, information, and entertainment is hedging its bets by buying into collaborative ITV efforts.

(2) Revenue streams are taking shape. According to Andrew Serovitz, former president of the Association for Interactive Media, "no one five years ago was sure how interactive TV services would pay for themselves. Now, the Internet has proven that interactive business models can succeed. [T]he Web's popularity has greatly reduced the risk in developing [ITV]" (Freed, 1998a, p. 18).

Research based on 2,500 interviews indicated that 13 million U.S. adults were using the Internet while viewing television in November 1998. The same study showed that eight million people were telesurfing at least once a week (Freed, 1998b). Nearly 48 million American households will be online by 2001, predicted to increase to nearly 57 million by 2002 (Tedesco, 1999). Major players anticipate being able to convert this growth market to telesurfing. Forrester Research, in assessing ITV's potential, predicts "$11 billion in annual advertising revenues, $7 billion in e-commerce sales, and $20 billion from subscriptions in 2004. Another $3.1 billion could come from distributing movies on demand (Spyglass, 2000; Haley, 1999). However, the ITV industry has a long way to go before posting returns of this magnitude.

Five ITV services currently dominate the landscape. All these services now offer expanded user options, faster system speeds, and service tiers and bundles. Major players in Internet-on-TV, for example, are making moves to include personal TV as part of their services, and personal TV companies are moving quickly to make VOD part of their offerings.

Internet-on-TV

WebTV now offers an expanded line of services and features (see Table 4.1). Microsoft licensed the G2 Internet streaming technology owned by its competitor RealNetworks in late 1999. RealNetworks dominated the streaming media market. In return, RealNetworks agreed to provide WebTV with G2 format upgrades for over 800,000 WebTV subscribers. RealNetworks also agreed to develop new RealPlayer G2 for the Windows CE platform, thus allowing Microsoft to incorporate the application in future Internet appliances so they can work with WebTV (Beacham, 1999).

EchoStar now uses WebTV's personal TV service as part of its DISHPlayer satellite TV system. The system has a 17.2 GB hard drive and features a satellite TV system, digital video recorder, game player, and Internet browser (Mossberg, 2000). DirecTV has teamed with both Wink and TiVo to combine satellite television with personal TV features. In 1999, AOL invested heavily in DirecTV's parent company to create a joint service. Philips Electronics will build a set-top box using the WebTV Network Personal TV service. Notable among its features will be digital video recording, live TV pause, and Internet capability. EchoStar has announced that the same service is available as a feature on its DISHPlayer satellite TV system (Beacham, 2000).

Table 4.1
WebTV Internet-on-TV Packages

Package Name	WebTV Classic Internet	WebTV Plus Interactive	WebTV Personal TV for Satellite	WebTV Plus for Satellite
Required Hardware	Internet Terminal	Internet Receiver	DISHPlayer Satellite Receiver	DISHPlayer Satellite Receiver
Suggested Price	$99	$199, $249 with Wireless Keyboard	$199 or $299 DISHPlayer 300 or 500	Same as Personal TV Package
Hardware Manufacturers	Philips, Manavox, Sony	Philips, Magnavox, Sony, RCA	EchoStar	EchoStar
Internet Access & Multiple Accounts	Yes	Yes	No	Yes
Chat & Discussion Groups	Yes	Yes	No	Yes
WebTV Centers & Searches	Yes	Yes	No	Yes
Surfwatch & Favorites	Yes	Yes	No	Yes
Page Builder	Yes	Yes	Yes	Yes
2-Day Searchable Listings	No	Yes	No	Yes
7-Day Searchable Listings	No	No	Yes	No
Web Picture-In-Picture	No	Yes	No	Yes
Interactive TV Programming	No	Yes	No	Yes
One-Touch VCR Recording	No	Yes	Yes	Yes
Program Remind	No	Yes	Yes	Yes
Auto Program Recording	No	Yes	Yes	No
Instant Replay	No	No	Yes	No
Record, Skip, Fast Forward, Rewind	No	No	Yes	No
Digital Picture & Sound	No	No	Yes	Yes
TV Pause (Live & Recorded)	No	No	Yes	No

Source: WebTV

To some extent, these efforts reflect an effort to compete with cable television's ability to offer high-speed Internet access—something as yet unavailable via DBS. Cable television's bandwidth and high-speed return path also overshadow DBS' reliance on telephone lines as a return path (Pegg, 2000).

Program Guides

Navigator software providing the gateway to digital television is seen by some as the hottest area in all of ITV (Haley, 1999). Program guides, viewed as portals to TV content, are projected to penetrate "55 million homes and create $3.2 billion in advertising revenues in the next five years" (Spyglass, 2000, p. 1). Whoever controls navigation also controls audience exposure to selected programming and information services. Program guide developers are approaching the issue in a number of different ways.

Guide Plus+ is an interactive navigator that allows for up to two days of program listings organized by genre. It utilizes Gemstar software that resides at the home (television receiver, VCR, or set-top), and updated programming information is sent via a television signal's vertical blanking interval four times a day. The service is free to customers, and revenues are generated via advertising support. Interactive Channel is a two-way service utilizing set-top boxes. This service includes a number of features, including an Internet TV Web browser, CableMail, and VOD navigator. End users use a remote to call up information stored at the cable headend. The service is part of a cable operator's digital service, with local cable operators paying $0.50 per month per subscription.

TV Guide Interactive is a navigator that allows subscribers to create their own program listings, to create program schedules and reminders, and to exercise parental control over programming. The software is located in set-top boxes and is updated continuously via server. The cost is included in monthly digital-tier subscriber fees. System operators pay TV Guide a few pennies per subscription per month (Haley, 1999).

Video on Demand

Cable operators providing near-VOD, movies, and other material offered every 15 to 30 minutes have seen pay-per-view revenues increase via adoption of up to 50 digital movie channels. Many operators worry that VOD is still unproven, citing cumbersome servers and modulators and complicated integration with cable system engineering (Higgins, 2000). Most VOD systems offer up to 400 titles, with recent movie hits driving most of the business.

Enhanced TV

Wink Communications has the leading one-way, enhanced TV system available on cable systems in the south, east, and west (see Figure 4.1). Participating cable companies download Wink software to set-top boxes via the VBI. A remote control device is used to access information when a Wink icon appears in the corner of the television screen. Services include enhanced weathercasts, sports statistics during baseball games, and listings of local retailers during commercials. Wink has an agreement with DirecTV to provide reception of enhanced programs, advertising, and virtual channels via DBS. DirecTV units enhanced with Wink capability were to be available in over 26,000 retail stores by summer 1999, and activated during the third quarter. Considered "service at the low end" of ITV, the service is free to customers (Pegg, 1999b, p. 14). According to

Allan Thygesen, Wink's programming vice president, the technology "is not Internet browsing-over-TV. We enhance the TV that viewers already love to watch" (Reveaux, 1999, p. 20).

Figure 4.1
One-Way Enhanced TV System

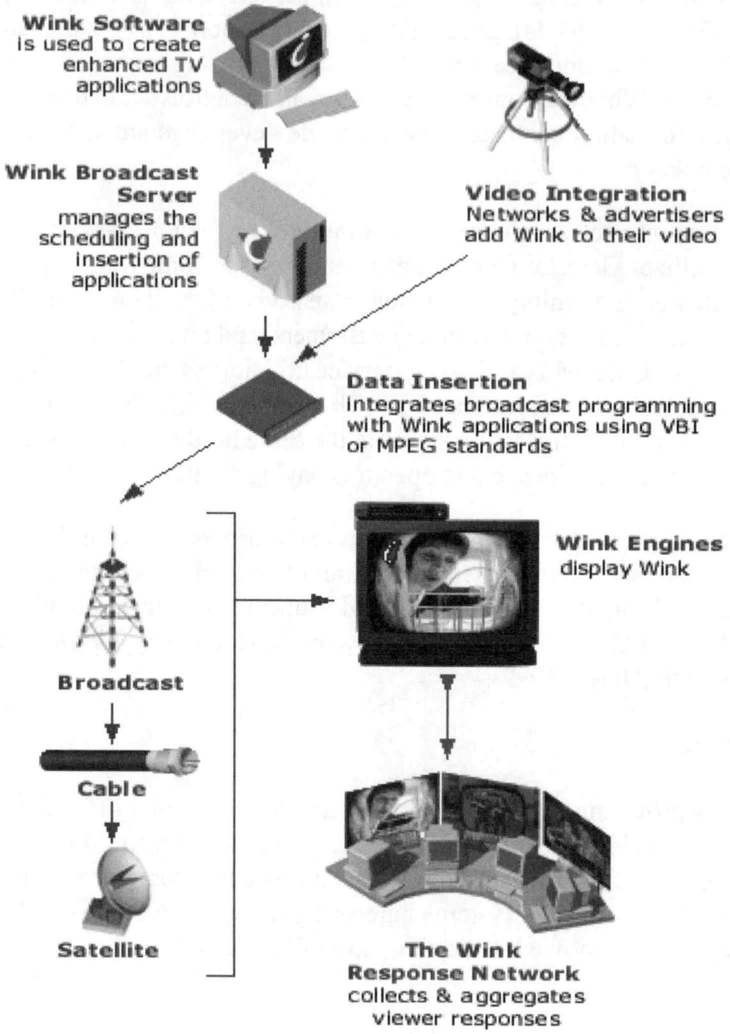

Wink Software is used to create enhanced TV applications

Wink Broadcast Server manages the scheduling and insertion of applications

Video Integration Networks & advertisers add Wink to their video

Data Insertion integrates broadcast programming with Wink applications using VBI or MPEG standards

Broadcast

Cable

Satellite

Wink Engines display Wink

The Wink Response Network collects & aggregates viewer responses

Source: Wink Communications

Factors to Watch

Despite 25 years of development and testing, ITV penetration remains very low. WebTV penetration is approaching 1% of all U.S. households. PowerTV's digital set-top operating system is currently deployed in 1.5 million households (PowerTV, 2000).

Forrester Research optimistically predicts that ITV services will generate $11 billion in advertising, $7 billion in commerce, and $2 billion in subscription revenues by 2004. Program and navigational guides are expected to reach into 55 million homes and create $3.2 billion in advertising revenues during the same timeframe. Jupiter Communications has conducted research suggesting that enhanced television might be well suited for news and movies (Spyglass, 2000).

In an attempt to join the ITV gold rush, dozens of additional companies have joined ATVEF to further standardize technological platforms and licensing relationships. Key to ITV market domination might be the one application that combines the most popular of current ITV applications with seamless integration via powerful browser-type navigators. "People don't want to put a lot of effort into interactivity in front of the TV," says industry researcher Josh Bernoff. "What Wink and WebTV offer is that, while you are watching a program, you can pop something up, interact with it, and put it away" (Pegg, 1999b, p. 14).

Perhaps industry developers are beginning to realize that 50 years of learned, passive televiewing behavior must be enhanced with moments of interactivity as linear video continues to dominate the scene. The question is whether or not such cherry picking behaviors on the part of most users will generate, over the long term, enough revenues via subscriptions, monthly fees, and T-commerce (television commerce) to offset the tremendous costs of developing ITV infrastructure and programming. Current projections for fiscal glory and substantial household penetration within five years may be unrealistic. Meanwhile, cable system operators continue to oppose Federal Communications Commission (FCC) "open access" mandates to their franchise networks on the part of Internet service providers. Ongoing legal struggles between these industries may slow the expansion of ITV services.

In 1999, the industry expanded its marketing efforts to promote ITV via major print and electronic media to foster greater awareness and recognition of major brands and services beyond the core of early adopters. Such efforts will probably contribute to ITV's market penetration, but certainly not at levels approaching 30% of U.S. households within the next four years. Hundreds of millions of dollars annually will be spent in an effort to carve out market shares and establish dominant protocols. Whatever the eventual platforms for ITV, most industry players agree that users will not interact with remotes or keyboards for long unless there is compelling content, and, "for content to flow freely, there have to be standards" (Pegg, 1999a, p. 16).

Bibliography

Baron, S., & Krivocheev, M. (1996). *Digital image and audio communications.* New York: Van Nostrand Reinhold.

Beacham, F. (1999, August 11). WebTV opens gates for streaming. *TV Technology, 17,* 1, 12.

Beacham, F. (2000, February 9). TV getting personal. *TV Technology, 18,* 24.

Carat Group/Edinburgh University Interactive Television. (2000). *What is interactive TV.* [Online]. Available: http://www.itvnews.com/whatis/.

Clark, D. (1997, August 13). Oracle plans to integrate TV programs with data from the World Wide Web. *Wall Street Journal,* B7.

Colman, P. (1999, October 25). Power to the headend. *Broadcasting & Cable, 129,* 67, 70.

Desmond, E. (1997, November 10). Set-top boxing. *Fortune, 136,* 91-93.

Feldman, T. (1997). *An introduction to digital media.* London: Routledge.

Freed, K. (1998a, July 13). Interactive TV hits its stride. *TV Technology, 16,* 18, 22.

Freed, K. (1998b, November 30). What is a tele-webber? *TV Technology, 16*, 16.
Goldstein, S. (1997, May 17). Time Warner proves its RIP for VOD; Direct mail, DVD tentative bedfellows. *Billboard, 109*, 58.
Haley, K. (1999, September 6). New direction: Forget the superhighway; Many roads lead to interactive TV. *Broadcasting & Cable, 129*, 18-19, 22, 24, 26, 28, 30, 32, 34, 36.
Higgins, J. (2000, March 6). VOD: Is it a business yet? *Broadcasting & Cable, 130,* 23, 26, 28.
Jessell, H. (1993, November 8). Face-to-face with Ray Smith. *Broadcasting & Cable, 123,* 18, 20, 22, 24.
Kerschbaumer, K. (2000, April 12). Ballmer: Place your bets. *Broadcasting & Cable*, *130*, 12-14.
Lesly, E., Cortese, A., Reinhardt, A., & Hamm, S. (1997, November 24). Let the set-top wars begin. *Business Week, 3554,* 74-75.
Marriage of convenience. (1997, November). *Time Digital*, 60-64.
Mossberg, W. (2000, March 26). DishPlayer offers lower-cost way to enjoy digital television. *Las Vegas Review Journal*, 3L.
Pegg, J. (1999a, April 21). Interactivity needs standards. *TV Technology, 17,* 16.
Pegg, J. (1999b, April 21). Major interactive players emerge. *TV Technology, 17,* 14, 16.
Pegg, J. (2000, March 8). What's next for DBS. *TV Technology, 18*, 1, 14.
PowerTV, Inc. (2000, April 12). *PowerTV and Prasara technologies to merge, creating comprehensive software solution for interactive television.* [Online]. Available: http://www.powertv.com/press/prasara.html.
Reveaux, T. (1999, February 24). A wink at interactive television. *TV Technology, 17,* 20, 52.
Rogers, D. (1997, May 29). Buying through the box. *Marketing, 27,* 1.
Rosenstein, A. (1994). Interactive television. In A. Grant (Ed.). *Communication technology update, 3rd ed.* Newton, MA: Focal Press.
Shiver, J. (1997, June 2). Time Warner's interactive TV project blinks. *Los Angeles Times*, D1.
Spyglass, Inc. (2000, March 23). Set-top box/interactive TV. *Inside interactive TV.* [Online]. Available: http://www.spyglass.com/tvexpert/n_statistics.html.
Swann, P. (1999, April). Caught in the Web TV. *TV Online, 4*, 4.
Swedlow, T. (1999, July). *Enhanced television: A historical and critical perspective*. [Online]. Available: http://www.itvt.com/etvwhitepaper.html.
Tedesco, R. (1999, March 8). Who'll control the video streams? *Broadcasting & Cable, 129*, 20-22, 24.

Websites

Enhanced TV
- www.actv.com
- www.wink.com

Hybrid
- www.ictv.com

Internet-On-TV
- www.aol.com
- www.morecom.com
- www.webtv.net
- www.wgate.com

Personal TV
- www.replaytv.com
- www.tivo.com

Program Guides
- www.gemstar.com
- www.sourcemedia.com
- www.tvguide.com

Video On Demand
- www.DIVAtv.com
- www.intertainer.com
- www.schange.com
- www.tvn.com

5

Direct Broadcast Satellites

Ted Carlin, Ph.D.*

What is the best way to get multichannel video programming today? The answer is becoming more complicated. Cable television is upgrading to digital service that offers more channels, better picture quality, and extras such as Internet service. But recent cable price hikes, primarily due to increased programming costs, have given consumers more reason to consider another option for their television service. Already frustrated by the typical cable problems of signal quality, channel capacity, and customer service, higher cable prices in 1999 and early 2000 have pushed many consumers to search for an alternative.

Waiting for these consumers are the communication technologies of direct broadcast satellite (DBS) systems, wireless cable systems (MMDS and LMDS), home satellite dishes (HSDs), telephony, and the Internet. With the increasing development of digital production, transmission, and reception equipment, the multichannel video program distribution field is getting crowded as companies race to develop digital systems that will entice consumers away from firmly-established cable television providers. The digital revolution in television technology is starting to give consumers what the Federal Communications Commission (FCC) has always promoted: a level playing field of multichannel video program distribution services from which consumers can pick and choose.

This chapter will focus on the fastest-growing multichannel video program distribution service: DBS systems. Providers of DBS systems are actively competing with cable television systems for subscribers, and currently offer the most comprehensive programming alternatives for consumers.

* Assistant Professor of Radio/Television, Department of Communication and Journalism, Shippensburg University (Shippensburg, Pennsylvania).

Background

As originally conceived in 1962, satellite programming was never intended to be transmitted directly to individual households. After the FCC implemented an "Open Skies Policy" to encourage private industry to enter the satellite industry in 1972, satellite operators were content to distribute programming between television networks and stations, cable programmers and operators, and business and educational facilities (Frederick, 1993). The FCC assigned two portions of the Fixed Satellite Service (FSS) frequency band to be used for these satellite relay services: the low-power C-Band (3.7 GHz to 4.2 GHz) and the medium-power Ku-band (11.7 GHz to 12.2 GHz).

In late 1975, Stanford University engineering professor Taylor Howard was able to intercept a low-power C-band transmission of the Home Box Office (HBO) cable network on a makeshift satellite system he designed (Parone, 1994). In 1978, Howard published a "low-cost satellite-TV receiving system" how-to manual. Word spread rapidly among video enthusiasts and ham radio operators, and, by 1979, there were about 5,000 of these television receive-only (TVRO) HSD systems in use. Demand for cable television programming in areas not yet wired for cable television fueled growth of the TVRO industry that distributed "backyard" dishes and decoder boxes throughout the United States.

Today, these 6- to 12-foot TVRO satellite dishes are still commonplace throughout the United States, especially in rural areas not served by cable television services. As of mid-2000, there were just under 1.8 million in use (FCC, 2000). The large receiving dish is required to allow proper reception of the low-power C-band transmission signal. A number of factors—the high cost of the HSD system (around $2,000), the large size of the dish, city and county zoning laws, and the scrambling of C-band transmissions by program providers—have prevented HSD systems from becoming a realistic, national alternative to cable television for multichannel video programming.

In the 1980s, a few entrepreneurs turned to the medium-power Ku-band to distribute satellite transmissions directly to consumers. By utilizing unused transmission space on existing Ku-band relay satellites, these companies would be the first to create a direct-to-home (DTH) satellite transmission service that would use a much smaller receiving dish than existing HSDs (Whitehouse, 1986). The initial advantages of the DTH systems over HSD systems included the higher frequencies and the higher power of the Ku-band, which resulted in less interference from other frequency transmissions and stronger signals to be received on the smaller 3-foot dishes.

Many factors proved to be the primary contributors to the failure of these medium-power Ku-band DTH services in the 1980s, including:

- High consumer entry costs ($1,000 to $1,500).
- Potential signal interference from heavy rain and snow.
- Limited channel capacity (compared to existing cable systems).
- Restricted access to available programming (Johnson & Castleman, 1991).

DTH ventures by Comsat, United Satellite Communications, Inc., Skyband, Inc., and Crimson Satellite Associates failed to get off the ground during this period.

Also between 1979 and 1989, the World Administrative Radio Conference (WARC) of the International Telecommunications Union (ITU) authorized and promoted the use of a different section of the FSS frequency band. The ITU, as the world's ultimate authority over the allocation and allotment of all radio transmission frequencies (including radio, TV, microwave, and satellite frequencies), allocated the high-power Ku-band (12.2 GHz to 12.7 GHz) for "multichannel, nationwide satellite-to-home video programming services in the Western Hemisphere" (Setzer, Franca & Cornell, 1980, p. 1). These high-power Ku-band services were to be called direct broadcast satellite services. Specific DBS frequency assignments for each country, as well as satellite orbital positions to transmit the frequencies, were allocated at an ITU regional conference in 1983 (RARC, 1983).

A basic description of a DBS service, based on the ITU's specifications, was established by the FCC's Office of Plans and Policy in 1980:

> A direct broadcast satellite would be located in the geostationary orbit, 22,300 miles above the equator. It would receive signals from earth and retransmit them for reception by small, inexpensive receiving antennas installed at individual residences. The receiver package for a DBS system will probably consist of a parabolic dish antenna, a downconverter, and any auxiliary equipment necessary for encoding, channel selection, and the like (Setzer, Franca & Cornell, 1980, p. 7).

The FCC then established eight satellite orbital positions between 61.5°W and 175°W for DBS satellites. Only eight orbital positions are available for DBS because a minimum of nine degrees of spacing between each satellite is necessary to prevent the interference of signal transmissions. The FCC also assigned a total of 256 analog TV channels for DBS to use in this high-power Ku-band, with a maximum of 32 DBS channels per orbital position. Only three of these eight orbital positions (101°W, 110°W, and 119°W) can provide DBS service to the entire continental United States. Four orbital positions (148°W, 157°W, 166°W, and 175°W) can provide DBS service only to the western half of the country, while the orbital position at 61.5°W can provide service only to the eastern half.

The FCC received 15 applications for these DBS orbital positions and channels in 1983, accepted eight applications, and issued conditional construction permits to the eight applicants. The FCC granted the construction permits "conditioned upon the permitee's due diligence in the construction of its system" (see 27 C.F.R. Sect. 100.19b). DBS applicants had to do two things to satisfy this FCC "due diligence" requirement:

(1) Begin construction or complete contracting for the construction of a satellite within one year of receiving the permit.

(2) Begin operation of the satellite within six years of the construction contract.

The original eight DBS applicants were: CBS, Direct Broadcast Satellite Corporation (DBSC), Graphic Scanning Corporation, RCA, Satellite Television Corporation, United States Satellite

Broadcasting Company (USSB), Video Satellite Systems, and Western Union. During the 1980s, some of these applicants failed to meet the FCC due diligence requirements and forfeited their construction permits. Other applicants pulled out, citing the failures of the medium-power Ku-band DTH systems, as well as the economic recession of the late 1980s (Johnson & Castleman, 1991).

In August 1989, citing the failures of the eight DBS applicants to launch successful services, the FCC revisited the DBS situation to establish a new group of DBS applicants (FCC, 1989). This new group of applicants included two of the original applicants, DBSC and USSB. They were joined by Advanced Communications, Continental Satellite Corporation, Direcsat Corporation, Dominion Satellite Video, EchoStar Communications Corporation, Hughes Communications, Inc., and Tempo Satellite Services.

From 1989 to 1992, not one of these DBS services was able to launch successfully. In addition to raising capital, most were awaiting the availability of programming and the development of a reliable digital video compression standard. Investors were unwilling to invest money in these new DBS services unless these two obstacles were overcome (Wold, 1996).

Cable operators, fearing the loss of their own subscribers and revenue, were placing enormous pressure on cable program networks to keep their programming off the new DBS services. Cable operators threatened to drop these program networks if they chose to license their programming to any DBS service (Hogan, 1995). These program networks were seen as the essential programming needed by DBS to launch their services because the DBS companies had little money or expertise for program production of their own (Manasco, 1992).

In late 1992, DBS companies had the programming problem solved for them through the passage of the Cable Television Consumer Protection and Competition Act. The Act guaranteed that DBS companies would have access to cable program networks, and it "[forbade] cable television programmers from discriminating against DBS by refusing to sell services at terms comparable to those received by cable operators" (Lambert, 1992, p. 55). This provision, which has since been upheld in the Telecommunications Act of 1996, finally provided DBS companies with the program sources they needed to attract investors and future subscribers.

The other obstacle—the establishment of a digital video compression standard—was solved by the engineering community in 1993 when MPEG-1 was chosen as the international standard. By using MPEG-1, DBS companies could digitally compress eight program channels into the space of one analog transmission channel, thus greatly increasing the total number of program channels available service to consumers. (For example, the FCC had assigned DirecTV 27 analog channels. Using MPEG-1, DirecTV could actually provide their subscribers with 216 channels of programming.) In 1995, DBS companies upgraded their systems to the improved broadcast-quality version, MPEG-2.

With these obstacles behind them, two of the DBS applicants, Hughes Communications and USSB, were the first to launch their DBS services in June 1994. Under the leadership and direction of Eddie Hartenstein (DirecTV) and Stanley Hubbard (USSB), Hughes established a subsidiary, DirecTV, to operate its DBS system, and then agreed to work with USSB to finance, build, deploy, and market their DBS systems together (Hogan, 1995). Hughes launched three satellites to the 101°W orbital position from 1993 to 1995. Both companies then signed contractual agreements with Thomson Consumer Electronics to use Thomson's proprietary Digital Satellite System (DSS) to transmit and receive DirecTV and USSB programming (Howes, 1995).

DSS uses an 18-inch receiving dish to receive the high-power Ku-band digital transmissions, a VCR-sized integrated receiver-decoder (IRD), and a multi-function remote control. Consumers purchased the DSS receiving equipment from satellite retailers, consumer electronic stores, or department stores. They then had the choice of purchasing programming on a monthly or yearly basis from DirecTV, USSB, or both.

To ensure its availability in rural areas, DirecTV signed an exclusive agreement with the National Rural Telecommunications Cooperative (NRTC) that allowed NRTC affiliates the right to market and distribute DirecTV in rural markets. A number of affiliates implemented the agreement in 1996, offering sales, installation, billing, collection, and customer service. Consolidation among affiliates became commonplace, and two affiliates emerged as the major players: Pegasus Satellite Television and Golden Sky Systems, Inc. By January 2000, Pegasus had acquired just over 1.13 million households, Golden Sky had signed up over 350,000 households, and they had announced their own merger. The combined operations of Pegasus and Golden Sky will serve subscribers in 41 states and reach approximately seven million rural households. The 1.2 million subscribers will make Pegasus the third largest provider of DBS and the eighth largest multichannel video programming distributor (MVPD) in the United States. Pegasus is the only MVPD focused exclusively on rural and underserved areas of the country (Golden Sky, 2000).

After the successful launch of DirecTV and USSB in 1994, the FCC tried to force the other DBS applicants to bring their DBS services to the marketplace. In late 1995 and early 1996, the FCC once again reevaluated the DBS applicants for adherence to its due diligence requirements. After several hearings, the FCC revoked the application of Advanced Communication Corporation, and stripped Dominion Satellite Video of some of its assigned channels for failing to meet the requirements.

The FCC denied appeals by both companies, and auctioned the channels in January 1996. MCI and News Corporation, working together in a joint venture, obtained the Advanced DBS channels, while EchoStar obtained the Dominion channels to add to its previously assigned channels. EchoStar also acquired the channels from two other applicants, DirecSat and DBSC, through FCC-approved mergers in 1995 and 1996 (FCC, 1996).

In another merger, R/L DBS, a subsidiary of Loral Aerospace Holdings, acquired the DBS channels of Continental Satellite Corporation. Continental was forced to turn over the channels to R/L DBS after failing to meet previous contractual obligations with Loral Aerospace for the launching of Continental's proposed DBS satellite (FCC, 1995). R/L DBS has yet to announce or launch its DBS service.

On March 4, 1996, EchoStar launched its high-power DBS service, the Digital Sky Highway (DISH) Network, using the EchoStar-1 satellite at 119°W, and became the third DBS applicant to successfully begin operations. Similar to DirecTV, EchoStar launched a second satellite in September 1996 to the 119°W orbital position, which increased its number of available program channels to 170 on the DISH Network.

The DISH Network does not use the same DSS transmission format used by DirecTV. Instead, it uses the international satellite video transmission standard, Digital Video Broadcasting (DVB), which was created after the DSS standard. Like DSS, the DISH Network's DVB equipment utilizes MPEG-2 for digital video compression. What this means is that DISH Network subscribers can receive only DISH Network transmissions, and DirecTV subscribers can receive only DirecTV transmissions. The DVB system employs an 18-inch receiving dish to receive its high-power Ku-band digital transmissions, a VCR-sized integrated receiver-decoder, and a multi-function remote control like DirecTV.

Similar to DirecTV, the DISH Network requires subscribers to purchase the DVB system, and then pay a separate amount for monthly or yearly programming packages. Also, like DirecTV, the DISH Network offers professional installation of the DVB equipment or a do-it-yourself installation kit. Currently, both companies offer professional installation for $99, and sometimes for free as part of various promotional strategies. Because of competition, this price has dropped over $100 since 1996 for both companies. The DISH Network sells its equipment directly from the factory via the Internet and an 800 phone number. (DirecTV began selling its merchandise over the Internet in 1999.) Equipment can also be purchased from authorized satellite retailers and mass merchandisers.

The cable television industry did not ignore the implementation and growth of these DBS companies. In 1994, Continental Cablevision, intent on establishing a cable "headend in the sky" for consumers living in non-cabled areas of the United States, was able to enlist the support of five other cable operators (Comcast, Cox, Newhouse, Tele-Communications, Inc., and Time Warner), and one satellite manufacturer (GE Americom) to launch a successful medium-power Ku-band DTH service (Wold, 1996). The service, named Primestar, transmitted 12 basic cable channels from GE Americom's medium-power K-1 satellite to larger 3-foot receive dishes. Primestar offered far fewer channels than any of the cable operators' own local cable systems, so they believed that their cable subscribers would not be interested in Primestar as a replacement for cable service. (Primestar is not a true DBS service because it does not use FCC-assigned high-power DBS channels, although most consumers are unaware of this discrepancy.)

As DirecTV and USSB began to prove that DBS was a viable service in late 1994, Primestar decided to change its focus and expand and enhance its offerings to compete directly with DBS. Primestar converted its 12-channel analog system to a proprietary DigiCipher-1 digitally compressed service capable of delivering about 70 channels. In 1997, Primestar moved its service to GE Americom's medium-power GE-2 satellite at 85°W, and increased its channel capacity to 160. To differentiate itself from the DBS companies, Primestar decided to market its service just like a local cable TV service does by leasing the equipment and the programming packages together in one monthly fee. Subscribers were not required to purchase the Primestar dish, IRD, and remote, although equipment purchase was an option.

Recent Developments

As of 1998, there were four companies operating DTH services in the continental United States. There were three DBS services (DirecTV, USSB, and the DISH Network) and one DTH service (Primestar).

Also at this time, Primestar was attempting to acquire the 28 high-power DBS channels licensed to MCI/News Corp., along with two high-power DBS satellites under construction, in an effort to expand and improve its service. Primestar also proposed to reorganize its ownership structure and sought to transfer control of the DBS channels licensed to Tempo from its subsidiary TSAT to Primestar. Acquisition of these DBS assets would have given Primestar the capacity to become a high-power DBS operator, allowing its customers to use smaller antennas and increasing its channel capacity. In May 1998, the U.S. Department of Justice sued to block this transaction because of concern that the cable ownership of Primestar, combined with the assignment of MCI/News Corp.'s DBS channels, would substantially lessen competition and tend to create a monopoly in markets for the delivery of multichannel programming services. Subsequently, Primestar announced that it would withdraw its petition to acquire MCI/News Corp.'s DBS channels. Instead, it would focus on its medium-power business and seek financing to support that business.

Less than six months later, in January 1999, Primestar was once again the subject of takeover rumors. However, this time, Primestar was the company to be acquired. The buyer: its chief rival of more than five years, DirecTV. Citing financial considerations and regulatory roadblocks for expansion, cable TV's venture into DTH service ended on April 28, 1999. In the deal, DirecTV's parent company, Hughes Electronics Corporation, obtained the 2.3 million-subscriber Primestar business and two Tempo high-power satellite assets in two separate transactions valued at about $1.82 billion. Primestar received approximately $1.32 billion for its medium-power DTH business, comprised of 4,871,000 shares of General Motors Class H common stock and $1.1 billion cash. Hughes paid $500 million cash for the Tempo high-power satellite assets (DirecTV, 1999a). Together, the transactions provided DirecTV with:

(1) Eleven high-power DBS channels at 119°W, from which the in-orbit Tempo satellite can deliver programming on a national basis at any time.

(2) A second Tempo satellite, already built, which can be launched at a future date or used as a backup satellite to secure uninterrupted service for DirecTV subscribers.

(3) Increased revenues immediately from over 2.3 million existing Primestar subscribers, and ongoing revenues from these subscribers as they are transitioned to DirecTV.

DirecTV planned to operate the medium-power Primestar business from Primestar's headquarters in Denver for a period of two years, during which time it hoped to transition Primestar subscribers to the high-power DirecTV service.

During this same time period, Hughes was in active negotiations with the Hubbard family to further strengthen the market position of DirecTV by acquiring USSB. On April 1, 1999, the FCC approved the USSB takeover, and the lineup of new DirecTV programming packages was introduced to consumers on May 24. Existing USSB customers were converted to the new, comparable DirecTV programming packages by July 1999. The consideration Hughes paid for USSB's business and assets was comprised of cash and shares of GM stock valued at approximately $1.6 billion (DirecTV, 1998b).

The merger allowed Hughes to combine its DirecTV business with USSB's assets and business at the 101°W orbital slot. The acquisition also included three channels at 110°W, from which DirecTV planned to launch a lineup of Spanish-language services. The acquisition enabled

DirecTV to achieve cost savings for the combined businesses through the consolidation of duplicative operations (billing, customer service, remittance processing, and broadcasting centers), and it increased its average revenue per subscriber. The transaction also allowed DirecTV to provide a simplified consumer offering and expand its 185-channel programming lineup to more than 210 channels via the addition of premium multichannel movie services such as HBO and Showtime.

The combination of DirecTV and Primestar, along with USSB, advanced DirecTV's position as the premier digital MVPD in the United States. The completion of these transactions created a DirecTV that featured:

- More than seven million U.S. subscribers.
- More than 370 entertainment channels delivered through five high-power DBS spacecraft: DBS-1, -2, and -3, a high-power Tempo satellite, and DirecTV 1-R (launched in mid-1999).
- The broadest distribution network in the DBS industry, combining more than 26,000 points of retail sale with Primestar's rural and small urban-based distribution network.
- High-power DBS frequencies at each of the three orbital slots that provide full coverage of the continental United States: 101°W, 110°W, and 119°W.
- The opportunity for DirecTV to begin "local-into-local" broadcast signal carriage.
- For the first time, DBS service to Hawaii, a state not previously served by any DBS or DTH provider.

Using aggressive pricing strategies for programming and DVB equipment, the DISH Network reached one million subscribers faster than any other DBS/DTH service in December 1997 (Hogan, 1998a). Marketing itself as the best value in satellite television, the DISH Network offered its equipment for only $199, plus installation. DirecTV responded by lowering its DSS prices, resulting in a price war that continues today.

In January 1997, EchoStar's Chairman, Charlie Ergen, and News Corp.'s Rupert Murdoch, shocked the DBS industry by announcing a partnership to deliver a new DBS service, complete with local broadcast stations, to the continental United States. Called ASkyB, this new DBS service would have combined the DISH Network channels with MCI/News Corp.'s unused DBS channels to deliver over 500 channels to subscribers. Citing strategic management differences with Ergen, and EchoStar's unstable financial picture, Murdoch unexpectedly pulled out of the project only a few months later in May 1997 to pursue other ventures, including the purchase of The Family Channel and the proposed DBS venture with Primestar described above.

Then, in mid-1998, the DISH Network once again announced plans to transmit local broadcast stations back into their local markets. By early-1999, the DISH Network finalized a new agreement with MCI to acquire its 28 DBS channels at 110°W. This provided the DISH Network with enough channel capacity to provide local-into-local service in major U.S. television markets. On May 19, 1999, the FCC granted the application of MCI and EchoStar for transfer of MCI's license to construct, launch, and operate a DBS system at the 110° W location. Then, on June 16, 1999,

Echo-Star was also granted authority to temporarily relocate one of its satellites to a new orbital slot in order to improve DBS service to Alaska and to initiate service to Hawaii.

On May 17, 1999, the FCC granted Dominion Video Satellite, Inc. authority to commence operation of a DBS service using the EchoStar III satellite currently in orbit at 61.5°W. This authorization waived Dominion's due diligence requirement to build and launch a satellite of its own because the lease arrangement with EchoStar was viewed as an efficient method of commencing service, as long as Dominion maintains control over the programming (FCC, 1999). Dominion then launched its SkyAngel religious programming service in the fall of 1999 with 16 channels of video and 10 channels of audio for $9 per month. Subscribers must use DISH Network equipment to access SkyAngel's programming. Table 5.1 summarizes the status of the DBS licensees.

Table 5.1
U.S. DBS Licensees

Orbital Position	61.5°W	101°W	110°W	119°W	148°W	157°W	166°W	175°W	Total
Satellites Deploy-ment	ESIII	DBS1R, DBS2, DBS3, BBS4S	DBS1 ESV ESV1 ESVIII	ESI, ESII, ESIV, ESVII, Tempo1, Tempo2					
DirecTV Channels		32	3	11		27			46
DISH Channels	11		29	21	24			32	107
SkyAngel	8								8
R/L DBS	11								22
Unas-signed	2				8	32	32	10	73
Total	32	32	32	32	32	32	32	32	256

(1) ES = EchoStar; DBS & Tempo = DirecTV
(2) R/L DBS surrendered its 11 channels at 166°W to the FCC on September 15, 1998.
(3) SkyAngel has applied to the FCC for eight channels at 166°W.
(4) Delivery of EchoStar VII and VIII is expected in December 2001, and delivery of EchoStar IX expected in 2002.

Source: T. Carlin

Current Status

United States DBS/DTH

The United States is the world's number one user of DBS/DTH services. As of March 2000, there were 11.74 million DBS/DTH subscribers in the United States. Table 5.2 summarizes the subscription figures for the industry since July 1994.

Table 5.2
U.S. DBS/DTH Subscribers

Date	Total DTH	DirecTV	DISH	Primestar
7/1/94	70,000			70,000
12/1/94	390,000	200,000		190,000
7/1/95	1.15 mil.	650,000		500,000
12/1/95	1.98 mil.	1.1 mil.		880,000
7/1/96	2.95 mil.	1.6 mil.	75,000	1.27 mil.
12/1/96	4.04 mil.	2.13 mil.	285,000	1.60 mil.
7/1/97	5.04 mil.	2.64 mil.	590,000	1.76 mil.
12/1/97	5.95 mil.	3.12 mil.	965,000	1.90 mil.
7/1/98	6.60 mil.	3.45 mil.	1.14 mil.	2.01 mil.
12/1/98	8.48 mil.	4.36 mil.	1.85 mil.	2.27 mil.
7/1/99	9.93 mil.	5.42 mil.	2.57 mil.	1.94 mil.
12/1/99	10.97 mil.	7.84 mil.	3.13 mil.	
3/1/00	11.74 mil.	8.24 mil.	3.50 mil.	

Source: SkyReports

Most of these 11.74 million subscribers are located in rural areas that are not served by a local cable system. According to the FCC, total MVPD household penetration in the continental United States, including 66.7 million cable television subscribers and 1.78 million TVRO users, is just about 81% as of June 1999. This means that "the U.S. is running out of unserved homes to pitch, particularly in the boonies" (DBS knockin', 1998, p. 2). Cable television is still the dominant technology for the delivery of video programming to consumers in the MVPD marketplace, although its market share continues to decline. As of June 1999, 82% of all MVPD subscribers received their video programming from a local franchised cable operator, compared to 85% a year earlier (FCC, 2000a).

It also means that DirecTV and the DISH Network are aggressively pursuing current cable customers for their services. Various new marketing campaigns by DirecTV and the DISH Network are constantly implemented to attack the cable industry's most observable weaknesses: rate hikes (due largely to increased programming costs), lack of channel variety (due to limited analog systems), and customer service problems (due to past monopolistic practices).

While attacking these cable industry problems, the DBS companies are also trying to solve two main issues impacting the DBS industry at present: (1) multiple television setups in the home and (2) subscriber access to local broadcast television stations and broadcast networks. Both issues have been considered major impediments to the development and growth of DBS systems as true competitors for cable television customers (Hogan, 1998c).

Multiple TV Setups

When DBS companies began operations in the mid-1990s, the goal was to get a basic one-TV DBS system into as many rural subscriber homes as possible (Boyer, 1996). Due to declining costs, increased technology, and a new effort to attract cable customers, the focus has shifted to providing more user-friendly DBS service. According to Bill Casamo, executive vice president for DirecTV, the cost of a second receiver has always been a barrier to entry for some first-time subscribers. "As we go more into cabled markets, that becomes more of a factor," because cable customers are accustomed to seeing cable in multiple rooms of the home (Hogan, 1998e, p. 18).

As a result, DirecTV and the DISH Network have started marketing multiple TV setups for new subscribers. There are now three types of dishes available to consumers. They include:

(1) A single-LNB (low noise block) dish, which receives programming from one satellite orbital location.

(2) A dual-LNB dish, which also receives programming from one satellite orbital location.

(3) A multi-location dish, which receives programming from multiple satellite orbital locations.

The single-LNB model is the most basic and allows only one DBS receiver to be connected. It receives signals only from the DBS provider's primary orbital location. That means it can receive most of the mainstream programming but not some of the less common programming such as foreign language programming or high-definition television (HDTV) programming. Also, depending on a subscriber's location, a single-LNB model may not allow the reception of local network affiliates, since not all local affiliates are broadcast from the primary orbital location. Use of a single-LNB dish requires a single coaxial cable run from the dish into the home to connect the single receiver, so it is the easiest to install.

The dual-LNB model receives programming from a single orbital location, but viewers can connect one, two, or more receivers to it. This may be a better choice for those who want only the most common mainstream programming but who want more than one receiver connected either now or sometime in the future. Like the single-LNB models, dual-LNB also may not allow the reception of local network affiliates. If subscribers want more than one receiver in the household, two coaxial cables must be run from the dish into the home.

Multi-location dishes are required for those who want to receive signals from multiple satellite locations. These dishes have only recently become available, but they are becoming more popular, especially in those areas where local affiliates are not carried on the primary orbital location. Like the dual-LNB model, they actually have two antennas, but they are focused at different positions in the sky. Multi-location dishes require up to four coaxial cables to be run from the dish into the home.

Local-into-Local

The second issue, access to local broadcast television stations, has been much more difficult to overcome for DBS in urban and most suburban communities. A provision in the Satellite Broadcasting Act of 1988 prohibited DBS subscribers who live *within* the Grade B coverage area of local broadcast television stations from receiving any broadcast television stations or broadcast networks via their DBS system (SHVIA, 1988). Subscribers in these areas were forced to connect an over-the-air television antenna to their DBS system, or subscribe to a local cable TV system, to receive any broadcast television stations. (DBS providers are allowed to provide broadcast stations that are available via satellite to those subscribers living *outside* of local station Grade B coverage areas [i.e., rural, non-cabled areas], and each offers various á la carte packages of independent stations and network affiliates.)

This issue was finally resolved when The Satellite Home Viewer Improvement Act of 1999 (SHVIA) was signed into law on November 19, 1999. SHVIA significantly modified the Satellite Home Viewer Act of 1988, the Communications Act, and the U.S. Copyright Act (SHVIA, 1999). SHVIA was designed to promote competition among MVPDs such as DBS companies and cable television operators, while, at the same time, increasing the programming choices available to consumers.

Most significantly, for the first time, SHVIA permitted DBS companies to provide local broadcast TV signals to *all* subscribers who reside in a local TV station's market (also referred to as a designated market area [DMA]), as defined by Nielsen Media Research. This ability to provide local broadcast channels is commonly referred to as local-into-local service.

The DBS company has the option of providing local-into-local service, but is not required to do so. DirecTV and the DISH Network are already providing this service in selected markets. By the end of 2000, they have indicated that they will provide local television broadcast signals in markets serving more than 50% of all American households (SHVIA, 1999).

In addition, beginning January 1, 2002, a DBS company that has chosen to provide local-into-local service will be required to provide subscribers with *all* of the local broadcast TV signals that are assigned to that DMA and that ask to be carried on the satellite system. However, a DBS company will not be required to carry a local broadcast TV station that substantially duplicates the programming of a local broadcast TV station that is already being carried. A DBS company also will not be required to carry more than one local broadcast TV station that is affiliated with a particular TV network unless the TV stations are licensed to communities in different states.

SHVIA also permits satellite companies to provide distant network broadcast stations to eligible satellite subscribers in unserved areas. (A distant signal is one that originates outside of a satellite subscriber's local television market, the DMA.) The FCC created a computer model for DBS companies and television stations to use to predict whether a given household is served or unserved. Congress incorporated this model into SHVIA, but also required the FCC to improve the accuracy of the model by modifying it to include vegetation and buildings among the factors to be considered. The DBS company, distributor, or retailer from which subscribers obtain their satellite system and programming are to be able to tell subscribers whether the model predicts that they are served or unserved. (The FCC does not provide these predictions). If unserved, the subscriber would be eligible to receive no more than two distant network affiliated signals per day for each

TV network. For example, the household could receive no more than two *ABC* stations, no more than two *NBC* stations, etc.

SHVIA also permits DBS companies to distribute a national PBS (Public Broadcasting System) signal to all subscribers —served and unserved—until January 1, 2002. After that date, DBS companies may choose to provide the local PBS affiliate or another noncommercial station within a local market or may provide the national PBS signal to subscribers that are eligible to receive distant signals.

There are five other issues involving DBS that have attracted much attention in the last several months: retransmission consent agreements, antenna restrictions, public interest obligations, transmission of HDTV signals, and Internet access.

Retransmission Consent Agreements

In order to deliver local-into-local service, DBS companies were mandated by the SHVIA to seek retransmission consent agreements with television broadcast stations. SHVIA required the FCC to revise the existing cable television rules surrounding retransmission consent agreements to encompass all MVPDs. SHVIA prohibits a TV station that provides retransmission consent from engaging in exclusive contracts for carriage or failing to negotiate in "good faith" until January 1, 2006, allowing DBS companies time to fully implement their local-into-local services.

In March 2000, the FCC, by First Report and Order (FCC, 2000b), established a two-part test for good faith negotiations. The first part consists of a brief, objective list of procedural standards applicable to television broadcast stations negotiating an agreement. The second part allows an MVPD to present facts to the FCC that constitute a TV station's failure to negotiate in good faith. The order directs the FCC staff to expedite resolution of good faith and exclusivity complaints, and notes that the burden of proof is on the MVPD complainant.

Antenna Restrictions

As directed by Congress in Section 207 of the Telecommunications Act of 1996, the FCC adopted the Over-the-Air Reception Devices Rule concerning governmental and non-governmental restrictions on subscribers' ability to receive video programming signals from DBS systems, wireless cable providers, and television broadcast stations. The rule (47 C.F.R. Section 1.4000) has been in effect since October 14, 1996. It prohibits restrictions that impair the installation, maintenance, or use of antennas to receive video programming. The rule applies to video antennas including DBS dishes that are less than one meter (39.37") in diameter (or of any size in Alaska), TV antennas, and wireless cable antennas. The rule prohibits most restrictions that: (1) unreasonably delay or prevent installation, maintenance, or use; (2) unreasonably increase the cost of installation, maintenance, or use; or (3) preclude reception of an acceptable quality signal.

The rule applies to viewers who place video antennas on property they own and that is within their exclusive use or control. This includes condominium owners and cooperative owners who have areas of exclusive use, such as a balcony or patio, in which to install the antenna. The rule applies to townhomes and manufactured homes, as well as to single-family homes.

On November 20, 1998, the FCC amended the rule so that it will now apply to rental property where the renter has exclusive use, such as a balcony or patio (OTARD, 1999). Exclusive use means an area of the property that only the renter may enter and use to the exclusion of other residents. For example, a condominium or apartment may include a balcony, terrace, deck, or patio that only the renter can use, and the rule applies to these areas. The rule does not apply to common areas, such as the roof, hallways, walkways, or exterior walls of a condominium or apartment building. In essence, this amendment greatly increases the potential customers available to receive DBS service by preventing local governments, condominium boards, and others from passing rules that limit or prohibit antenna construction.

Public Interest Obligations

Seeking to further level the competitive environment between DBS and cable, the FCC also adopted rules (FCC 98-307) implementing Section 25 of the Cable Television Consumer Protection and Competition Act of 1992, which imposed certain public interest obligations on DBS providers. The statute requires DBS companies to set aside 4% of their channel capacity exclusively for noncommercial programming of an educational or informational nature. DBS companies cannot edit program content, but must simply choose among qualified program suppliers for the reserved capacity.

As of January 2000, DirecTV offered nine channels of public interest programming to fulfill this obligation: Clara+Vision, C-SPAN, EWTN, Inspirational Life, NASA TV, PBS YOU, Star-Net, TBN, and WorldLink TV. The DISH Network offered 14 channels: BYUTV, C-SPAN, Educating Everyone, EWTN, FSTV, HITN, Mayerson Academy, NASA TV, PBS YOU, Research TV, TBN, UCTV, Universityhouse Channel, and WorldLink TV.

DBS companies must also comply with the political broadcasting rules of Section 312 of the Communications Act, granting candidates for federal office reasonable access to broadcasting stations. They must also comply with Section 315's rules granting equal opportunities to federal candidates at the lowest unit charge.

Transmission of HDTV Signals

DBS HDTV hardware has just recently become available. DirecTV and the DISH Network have unique approaches to their HDTV equipment plans. DirecTV HDTV reception is provided by new multi-location dishes and receivers able to receive both the standard compressed NTSC signals and HDTV signals from the satellites. Further, they can receive and decode terrestrial digital broadcasts as well. Both the NTSC and HDTV signals can be sent to either an HDTV-ready set (for best picture and proper aspect ratio) or a standard conventional set for display. This is quite important as a single satellite receiver can also function as a digital TV receiver for over-the-air reception of digital ATSC broadcasts.

The DISH Network is currently using a different approach in which an external adapter can be added to their Model 5000 receiver, which converts the satellite HDTV signals into a signal that resembles an over-the-air ATSC broadcast. This signal can then be fed into an ATSC receiver (either external receiver or an HDTV set with built-in receiver capability). This approach is a temporary solution and will be supplemented by the DISH HD receiver in 2000. The DISH HD receiver is to deliver HDTV programming onto a 16:9-ratio HDTV screen and supports both 720p and 1080i HD formats. This system is intended to provide seamless switching between HDTV and standard TV with accompanying Dolby Digital surround sound.

Internet Access

Similar to cable systems and phone companies, DirecTV and the DISH Network have been actively upgrading their systems to provide Internet services to subscribers. Because DBS signals are sent as digital information, the systems can send video, audio, and computer data in any combination to the receivers. Each DBS channel has a large amount of bandwidth, some of which DBS companies are using for data services such as Internet or interactive TV services.

DirecTV's first Internet offering, DirecPC, provides DirecTV subscribers with Internet service through a separate second dish and receiver. Non-DirecTV subscribers can also buy a DirecPC satellite system to be their Internet provider. Internet requests are sent from the user's PC through phone lines to the DirecPC Network Operations Center (NOC). However, before the request leaves the user's PC, the DirecPC software attaches a "tunneling code"—essentially an electronic addressing mask—to the request. That code instructs the Internet service provider (ISP) to forward the request to the DirecPC NOC instead of the server at the site requested. Once the NOC receives the customer's request, the tunneling code is stripped away, and the request is then forwarded by multiple T-3 lines to the appropriate site, and the desired content is retrieved. The NOC then uploads the information to the DirecPC satellite, which beams it down to the customer's DirecPC dish and into his or her PC in a matter of seconds (DirecPC, 2000).

DirecPC was followed by the DirecDuo system in 1999, which allows subscribers to receive up to 400 Kb/s Internet access and over 200 DirecTV channels on a single elliptical dish. Also in 1999, Hughes Electronics formed a strategic alliance with America Online to develop and market a combination DirecTV and AOL service to consumers beginning in 2000.

Working with Microsoft WebTV Networks to compete with the DirecPC systems, the DISH Network released DISHPlayer 500, the world's first interactive satellite TV system with a built-in 17.2 gigabyte hard drive capable of simultaneously recording and playing back full-quality digital video. The disk drive provides a number of enhanced features for customers who subscribe to WebTV Personal TV service, including up to 12 hours of digital video recording without a videotape, freezing a "live" TV program using TVPause, and instantly replaying or skipping ahead a few seconds to a favorite movie scene or the next big play of a sporting event. The TV listings allow users to view programming choices seven days in advance and quickly search and find programs to watch or automatically record with a touch of a button (Trowbridge, 2000).

In addition, WebTV Personal TV service provides subscribers with an Instant News channel that delivers the latest headlines, business, weather, sports, and stock information they can customize right on their TVs. Instant News is updated throughout the day via satellite, requiring no Internet connection. DISHPlayer 500 also includes popular games such as You Don't Know Jack and Solitaire. WebTV Plus provides subscribers with an avenue to explore the Internet from the comfort of their living rooms through an easy-to-use Web browser with access to chat rooms, newsgroups, and six e-mail accounts. A discount of $5 per month is applied when the customer subscribes to both WebTV Personal TV and WebTV Plus.

In terms of programming, DirecTV and the DISH Network have been able to acquire all of the top cable networks, sports channels and events, and pay-per-view (PPV) events as envisioned by the 1992 Cable Act. What differentiates one service from the other is how the program services are priced, packaged, and promoted. Each service has the following: on-screen program guides, parental control features, preset PPV spending limits, instant PPV ordering using the remote control and a phone line hookup, favorite channel lists, equipment warranties, and 800 phone numbers for customer service.

DirecTV. Programming on DirecTV consists of packages of basic cable channels and premium movie channels, as shown in Table 5.3. It also offers individual PPV movies, concerts, and sporting events as available through DirecTicket (i.e., movies for $2.99; boxing for $14.95). Using the remote control, subscribers can search the interactive program guide to access desired channels or to request PPV events. DirecTV also offers unique packages of college and professional sports (MLB Extra Innings, MLS Shootout, NBA League Pass, NFL Sunday Ticket, NHL Center Ice, and ESPN College Basketball and Football). It also offers 31 CD-quality, commercial-free digital audio channels as part of its Total Choice packages.

Table 5.3
DirecTV Programming Packages

1	Select Choice Basic Package *Around 45 popular channels of news, sports, and entertainment programming.*	$19.99 a month
2	Total Choice Basic Package *Around 100 popular channels of news, sports, and entertainment programming.*	$29.99 a month
3	Total Choice Sports Value Package *Includes the Basic Package plus six national and 20 regional sports channels*	$39.99 a month
4	Total Choice Movies Value Package - STARZ! *Includes Total Choice, 4 STARZ!, 8 Encore, and the Independent Film Channel.*	$39.99 a month
5	Total Choice Movies Value Package - Showtime *5 Showtime, 2 TMC, FLIX, Sundance Channel, and Total Choice Basic Package.*	$44.99 a month
6	Total Choice Movies Value Package - HBO *5 HBO, 2 HBO Family, 3 Cinemax, and Total Choice Basic Package.*	$47.99 a month
7	Total Choice Movies Value Package - HBO/STARZ! I *Total Choice, plus HBO and STARZ!*	$49.99 a month
8	Total Choice Movies Value Package - HBO/STARZ! II *Total Choice, plus HBO and STARZ!, and Cinemax.*	$57.99 a month
7	Total Choice Movies Value Package - HBO/STARZ!/SHO *Total Choice, plus HBO, STARZ!, Showtime, Cinemax, and TMC.*	$72.99 a month
6	Total Choice Platinum Value Package *All of the available channel packages in one.*	$80.99 a month

Source: DirecTV

DirecTV is also offering a Spanish-language service, DirecTV Para Todos, to subscribers. DirecTV Para Todos offers more than 22 Spanish language national and international channels including Univision, Discovery en Español, FOX Sports World Español, Galavisión, MTV S, TVN Chile, and Canal Sur, among others. DirecTV Para Todos offers two bilingual packages: (1) Opcion Especial, which provides 15 Spanish-language and 24 English-language channels for $19.99 a month; and (2) Opción Extra Especial, which provides up to 23 Spanish-language and 77 English-language channels for $31.99 a month. Both packages include seven Music Choice channels of commercial-free Spanish-language music.

To see any DirecTV programming, subscribers must purchase a DSS equipment package available from a variety of retailers and have the DSS system installed. DirecTV has authorized 17 different companies to manufacture the DSS equipment (including GE, HNS, Panasonic, RCA, Sanyo, and Sony), hoping to entice consumers with familiar, reliable brands. Prices vary according to individual retailers (including Best Buy, Circuit City, Sears, and Wal-Mart), the brand name chosen, and the complexity of the DSS system selected. Equipment prices can range from $149 to over $499, plus $99 to $199 for installation (see www.directv.com for the most current information).

The DISH Network. The DISH Network is to satellite TV what Saturn is to automobiles: The service is highly practical, and the packages are the most inexpensive yet comprehensive (see Table 5.4). With its large channel capacity and deployed satellites, The DISH Network is marketing itself as the only satellite service to deliver over 500 video and audio channels to subscribers. This capacity has also allowed the DISH Network to be the first satellite provider to supply local channels to selected cities via a second dish antenna. Most of these locals can now be obtained with a dual-LNB dish. The optional second dish can also provide subscribers with international programming in 10 languages, specialty religious or science programming, and data services.

Despite the fact that receiver specifications were developed exclusively by EchoStar, the units offer a wide range of options and are competitively priced to similar DSS units. Some of the higher-grade receivers offer such unparalleled features as RF remotes, timed remote control of VCRs, seamless integration with off-air signals, and local listings in the channel guide, on-screen Caller-ID, and even an integrated D-VHS recording deck. All receivers are feature-upgradeable via satellite. Equipment is available directly from EchoStar (via phone or the Web) or through local distributors. Like DirecTV, programming sign-up and/or changes are implemented immediately via a 24-hour 800 number. Technical support for installation or hardware issues is also available (see www.dishtv.com for the most current information).

Table 5.4
DISH Network Programming Packages

	Package	Price
1	America's Top 40 CD *The basic DISH package which includes 40 basic cable channels, plus The Disney Channel.*	$19.99/month
2	America's Top 100 CD *Includes all of America's Top 40 channels, plus additional international and national channels, one regional Fox SportsNet affiliate, and the 30-channel DISH CD digital audio service.*	$28.99/month
3	Movie Packages *Subscribers can select up to four movie channel packages from HBO, Cinemax, Showtime Networks, and Starz/Encore. One package is $10.99, two packages are $19.99, three packages are $27.99, and four packages are $34.99.*	$ per number of packages selected per month
4	Multisport Package *Available only to America's Top 100 subscribers, this sports package includes all of the Fox Sportsnet affiliates and five other regional sports networks.*	$4.99/month
5	DISHpix *A la carte offerings of selected channels.*	$ per number of packages selected per month
6	DISH-on-Demand PPV Movies	$2.99 to $3.99 per month

Source: EchoStar

International DBS/DTH

Although other countries have used satellites to transmit television signals to stations and cable systems, Japan was the first country to launch a DBS service in 1984 (Otsuka, 1995). In October 1996, Japan's largest satellite operator, JSAT, launched the country's first digital DBS system, PerfecTV. In March 1998, PerfectTV merged with a competitor, News Corporation's JSkyB. The combined digital DBS service, SkyPerfecTV, delivered about 200 channels to its subscribers.

DirecTV Japan, a digital DBS competitor launched by Hughes Electronics Corporation in December 1997, ended service in 1999 by folding its operations into SkyPerfectTV. This adds about 400,000 subscribers to SkyPerfectTV's base of 1.7 million. As part of the transaction, Hughes and other shareholders of DirecTV Japan received an equity stake in SkyPerfecTV (SkyReport, 2000b).

STAR TV was launched in 1991 in Hong Kong, and is still the driving force for television in the rest of Asia. Within six months of STAR TV's launch, eight million viewers had tuned in. As of early 2000, Star Choice covered 53 countries, spanning an area from Egypt to Japan and the

Commonwealth of Independent States to Indonesia, reaching an estimated audience of 300 million viewers (STAR TV, 2000). STAR TV offers both subscription and free-to-air television services, using AsiaSat 1 as its primary satellite platform with additional services available on the AsiaSat 2 and Palapa C2 satellites.

In Europe, satellite consortiums SES Astra and Eutelsat continue to dominate the European MVPD market, grabbing 92% of the total MVPD households, including cable TV. According to SES Astra, one out of every two households in Europe receives satellite television. SES Astra delivers 880 digital and analog radio and television channels to over 77 million households (SES-Astra, 2000). Eutelsat's Hot Bird Satellite TV service delivers 550 channels to over 75 million households (Eutelsat, 2000).

Primary competition to SES Astra and Eutelsat has been from a number of recent national/regional DTH systems including: News Corporation's England-based BSkyB, France's CanalSatellite and TPS, Germany's DF-1, Italy's Telepiu, Norway's Canal Digital, and Spain's CSD and Via Digital. Digital DBS/DTH operators have done extremely well in France, Japan, and Malaysia, but have been under-performing in Germany and Italy (Global warmings, 1999).

In Latin America and Canada, DBS systems are also expanding. In Latin America, as deregulation and privatization of the telecommunications markets continues to spread throughout the region, the result has been fierce competition in satellite services. The leader is Galaxy Latin America, which provides a Latin American DirecTV service to over 300,000 subscribers in 12 countries, including Brazil, Costa Rica, Mexico, and Panama (DirecTV, 2000). Galaxy Latin America, which commenced service in mid-1996, is a multinational company consisting of Hughes Electronics Corporation, Venezuela's Cisneros Group of Companies, Brasil's Televisao Abril, and Mexico's MVS Multivision.

Using a Telstar 5 satellite owned by Loral Skynet, EchoStar recently launched a medium-power DTH service, SkyVista, to serve the Caribbean, Virgin Islands, Puerto Rico, Hawaii, and Alaska. Customers can purchase a complete SkyVista satellite system for a suggested retail price of $399, which includes a 90-cm dish antenna and an EchoStar Model 1000 satellite TV receiver with an on-screen programming guide and V-chip technology for parental control. SkyVista offers a 20-channel "Best of Satellite" package for $19.99 per month (i.e., The Disney Channel, The Weather Channel, MTV, CNN, Discovery, etc.). SkyVista also offers HBO for an additional $9.99 per month and, for eligible subscribers, five network broadcast channels for $5.99 per month. Ethnic, sports, and adult channels are also available (SkyVista, 2000).

In Canada, Star Choice Television is one of two firms that have satellite systems in operation. Star Choice serves over 345,000 subscribers and is optimistic that it can continue to attract consumers to its combination of local TV outlets and cable-like programming from the United States and Canada (Star Choice, 2000). The other company, Bell ExpressVu, uses EchoStar's DISH Network's equipment to operate a 200-channel DBS service. Subscribers must purchase the DVB equipment and a basic tier of programming, and then can add a wide range of specialty programming tiers including the Sports Bar, Kids Size, the Network Platter, and Film Feast. Now owned by the largest telecom company in Canada (BCE, Inc.), Bell ExpressVu serves over 440,000 subscribers. Both Bell ExpressVu and Star Choice offer programming in English and French (Bell ExpressVu, 2000).

Factors to Watch

What was once an industry in search of reliable distribution technology and attractive programming is now an industry focused on brand awareness, marketing strategies, and strategic alliances. DirecTV and the DISH Network are consistently using promotions such as "freeviews" of various channels and sponsorships of major events (i.e., WWF and college basketball tournaments) and concerts (i.e., Shania Twain and Tom Petty).

DirecTV has established strategic alliances with SMATV and MMDS services, such as CS Wireless, Wireless One, and Heartland Communications, to provide DirecTV to multiple dwelling units (MDUs) such as apartments and townhomes. In addition, DirecTV has also formed a distribution alliance with GTE, SBC Communications, and Bell Atlantic to allow these telecom companies to offer DirecTV program packages through their phone lines via digital set-top converter boxes to MDUs and single-family homes.

Other recent DirecTV alliances include several with various business establishments (i.e., bars, restaurants, hotels, hospitals, private offices, malls, and fitness clubs) to provide customized packages of DirecTV services. Clients include Applebee's, Ruby Tuesday's, American Airlines, and Marriott Hotels.

For the first time in the commercial airline industry, airline passengers will soon have the opportunity to privately view live television programming on flat screens installed in each seat back. DirecTV Airborne programming will debut in 2000 on select aircraft from Alaska Airlines and two new airlines, Legend Airlines and JetBlue Airways. DirecTV plans to offer up to 24 channels on the new in-flight programming service. The DirecTV Airborne service will be viewed on in-flight entertainment equipment supplied by LiveTV. The low-profile LiveTV antenna, located in the top center of the aircraft's fuselage, maintains constant communication with the three DirecTV satellites located at the 101° W orbital slot (DirecTV, 1999b).

Getting more involved in interactive enhancements, DirecTV and Wink Communications announced an alliance to provide Wink-enhanced DirecTV receivers in July 1999. Wink's technology allows advertisers and television networks to create interactive enhancements to accompany traditional television advertisements and programs. By clicking their remote controls during an enhanced ad or program, viewers receive program-related information such as local weather and sports updates and product samples and coupons. Purchases can be made instantly and are free of charge (Wink, 2000).

In January 2000, DirecTV announced a partnership with TiVo personal television to get even more involved in interactive television. The new DirecTV/TiVo combination digital receiver will store up to 30 hours of recorded programming. Additional features include high-quality all-digital video, high-end Dolby Digital (AC-3) audio, live TV pause, instant replay, slow motion, frame forward/back, variable speed rewind, and fast forward (TiVo, 2000).

The rapidly growing DISH Network, which has attacked cable television from its inception, is actively promoting its local-into-local broadcast television packages to consumers as the best way to challenge cable television's "monopoly" status in MVPD services.

The DISH Network has also jumped into the interactive services arena quite aggressively. Working with interactive services company, OpenTV, the DISH Network will develop and produce a number of applications and interactive TV services which will include video replay, interactive TV advertising, and entertainment services such as movie information and music news. The digital VCR-type product will also have the ability to record OpenTV-enabled interactive programs and services, allowing the viewer to interact even with recorded TV programs. OpenTV set-top box software has been shipped with or installed in more than 6.1 million digital set-top boxes worldwide, including British Sky Broadcasting (BSkyB) in the United Kingdom and TPS in France (OpenTV, 2000).

In another partnership, the DISH Network, Microsoft, and Gilat-to-Home, Inc. have agreed to jointly offer consumers the first two-way, high-speed satellite Internet access, along with hundreds of channels of the DISH Network, via a single small satellite dish by the end of 2000 (Gilat, 2000). Under the agreement, EchoStar will distribute the Gilat-to-Home satellite Internet service powered by MSN along with DISH Network satellite TV service through its more than 20,000 retailers nationwide. DISH Network customers will benefit by having "always-on" access to the Internet via MSN Internet Access and an MSN-Gilat-to-Home co-branded portal.

EchoStar will streamline the consumer experience by providing complete installation of the satellite television receiver, a two-way Internet terminal, and a multi-location dish antenna at the consumer's home. EchoStar will provide installation for consumers through its nationwide DISH Network Service. The oblong multi-location dish, approximately 24" by 36", will be capable of receiving and sending Internet data, while at the same time receiving DISH Network channels.

Like DirecTV, the DISH Network has also started to provide its equipment and programming to MDUs. In early 2000, it signed an agreement with Castle Cable Services, Inc., a private cable TV service provider, to install the first of EchoStar's quadrature amplitude modulation (QAM) systems for MDUs in San Francisco. Residents of Avalon Towers will be able to select various packages of the DISH Network's digital television services—up to 500 channels—as well as Castle Cable's 60-channel analog service (Castle Cable, 2000).

Used for the last five years in Europe, QAM offers an affordable way to provide MDUs with hundreds of channels of programming, and residents can receive their personalized DISH Network programming using shared dishes instead of individual dish antennas. QAM technology can be applied using a building's existing cable, which saves thousands of dollars in installation expenses. The QPSK signal used by the direct broadcast satellite industry is converted to QAM, a digital modulation scheme used by the cable television industry. This conversion, which takes place at the headend, allows the delivery of DISH Network programming using only 300 MHz of bandwidth.

Not content with just developing these types of services to attack cable TV's leadership position, DirecTV and the DISH Network have recently been attacking each other—in court. The DISH Network started the battle in February 1999 with a suit in federal court claiming that DirecTV and its manufacturing partners were preventing the sale of DISH Network systems in top national retail chains. DirecTV filed a countersuit claiming that EchoStar is abusing its trademarks and misleading consumers on the availability of NFL football games (SkyReport, 2000a). The lawsuits are further evidence of the intense competition that dominates this dynamic industry.

Regardless, the immediate future for the DBS/DTH industry will be filled with marketing campaigns attacking cable's pricing and service, while showcasing the virtues of digital quality and program variety. DBS companies will continue to use all avenues of marketing—network TV spots, infomercials, direct mail, in-store retail promotions and demonstrations, and event sponsorships—to entice new subscribers. The big winner will be the consumer, as diversity of choice, a long-standing goal of U.S. communications policy, becomes a reality in the world of MVPD services.

Bibliography

Bell ExpressVu. (2000, March). About Bell ExpressVu. [Online]. Available: http://kusat.cyberus.ca/ev/.

Boyer, W. (1996, April). Across the Americas, 1996 is the year when DBS consumers benefit from more choices. *Satellite Communications*, 22-30.

Castle Cable, Inc. (2000). *New developments*. [Online]. Available: http://www.castlecable.com.

Colman, P. (1998, March 9). DBS gets local help. *Broadcasting & Cable, 128* (10), 61-63.

DBS knockin' on cable's doors. (1998, March). [Online]. Available: http://www.mediacentral.com/magazines/cableworld.

DirecPC. (2000). *What is DirecPC?* [Online]. Available: http://www.direcpc.com/consumer/what/what.html.

DirecTV. (1998a, February 24). *DirecTV Japan announces new line-up*. [Online]. Available: http://www.newsbytes.com.

DirecTV. (1998b, December 14). *Hughes to acquire USSB*. [Online]. Available: http://www.directv.com/press/pressdel/0,1112,36,00.html.

DirecTV. (1999a, January 22). *Hughes to acquire Primestar*. [Online]. Available: http://www.directv.com/press/pressdel/0,1112,28,00.html.

DirecTV. (1999b, September 28). *Live television to be available to airline passengers via DirecTV Airborne television entertainment service*. [Online]. Available: http://www.directv.com/press/pressdel/0,1112,231,00.html.

DirecTV. (2000, March). *What is DirecTV?* [Online]. Available: http://www.directvamericas.com/english/flash/en_w1wha.html.

DTH subscriber counts. (2000, March). [Online]. Available: http://www.skyreport.com/dthsubs.htm.

EchoStar. (1998, March). *Statement by Charlie Ergen in support of effective competition in video markets*. [Online]. Available: http://www.dishtv.com/.

Eutelsat. (2000, March). *TV channels*. [Online]. Available: http://www.eutelsat.org/tvchannels/index.html.

Federal Communications Commission. (1989). *Memorandum opinion and order*. MM Docket No. 86-847.

Federal Communications Commission. (1995). *Memorandum opinion and order*. MM Docket No. 95-1733.

Federal Communications Commission. (1996, February 14). *Status report*. Report No. SPB-37.

Federal Communications Commission. (1998a, November 19). *Implementation of Section 25 of the Cable Television Consumer Protection and Competition Act of 1992/Direct Broadcast Satellite Public Interest Obligations*. Report and Order, MM Docket 93-25.

Federal Communications Commission. (1998b, December 17). *Annual assessment of the status of competition in markets for the delivery of video programming*. Fifth Annual Report, CS Docket No. 98-102.

Federal Communications Commission. (1999, May 14). *Dominion Video Satellite, Inc. Application for minor modification of authority to construct and launch order and authorization*. Report and Order, CS Docket No. 98-102.

Federal Communications Commission. (2000a, January 14). *Annual assessment of the status of competition in markets for the delivery of video programming*. Sixth Annual Report, CS Docket No. 99-230.

Federal Communications Commission. (2000b, March 14). *Implementation of the Satellite Home Viewer Improvement Act Of 1999/Retransmission Consent Issues, Good Faith Negotiation, and Exclusivity*. First Report & Order, FCC 00-99.

Forrester, C. (1997, November). Digital satellite in Europe: An expanding powerhouse. *Via Satellite*, 44-54.

Frederick, H. (1993). *Global communications & international relations*. Belmont, CA: Wadsworth.

Gilat Networks. *What is Gilat-To-Home*? [Online]. Available: http://www.gilat2home.com/faq/index.html.

Global warmings. (1999, October). *Multichannel News Online*. [Online]. Available: http://www.multi-international.com/weekly/1004-4.htm.

Golden Sky Systems, Inc. (2000). *Golden Sky to merge with Pegasus Communications Corporation in $1 billion transaction*. [Online]. Available: http://www.gssdirectv.com/ press/60X-story.html.

Hearn, T. (1998, February 9). Primestar expects consent decree. *Multichannel News, 19* (6), 1, 62.

Hogan, M. (1995, September). U.S. DBS: The competition heats up. *Via Satellite*, 28-34.

Hogan, M. (1998a, January 19). Demand remained strong for DBS in 1997. *Multichannel News, 19* (3), 33.

Hogan, M. (1998b, January 26). Primestar faces life with delays, uncertainty. *Multichannel News 19* (4), 1, 62.

Hogan, M. (1998c, February 2). Digital cable not immediate threat, says DBS. *Multichannel News, 19* (5), 12.

Hogan, M. (1998d, February 16). Primestar sets March roll-up, April launch. *Multichannel News, 19* (7), 1, 62.

Hogan, M. (1998e, March 2). DBS discounts 2nd receivers. *Multichannel News, 19* (9), 3, 18.

Howes, K. (1995, November). U.S. satellite TV. *Via Satellite*, 28-34.

Johnson, L., & Castleman, D. (1991). *Direct broadcast satellites: A competitive alternative to cable television?* Santa Monica, CA: Rand.

Kessler, K. (1998, March). The Latin American satellite market. *Via Satellite*, 17-26.

Lambert, P. (1992, July 27). Satellites: The next generation. *Broadcasting & Cable, 124* (31), 55-56.

Manasco, B. (1992, April). The U.S. multichannel marketplace in the year 2000. *Via Satellite*, 44-49.

OpenTV. (2000). *What is OpenTV?* [Online]. Available: http://www.opentv.com/whatis/.

Otsuka, N. (1995). Japan. In L. Gross. (Ed.), *The international world of electronic media*. New York: McGraw-Hill.

Over-the-Air Reception Devices Rule (OTARD Rule). (1999, January). 47 C.F.R. Section 1.4000.

Parone, M. (1994, February). Direct-to-home: Politics in a competitive marketplace. *Satellite Communications*, 28.

Regional Administrative Radio Conference. (1983). *Final report and order*. Geneva: International Telecommunications Union.

Responses cool to EchoStar. (1998, March 2). *Multichannel News, 19* (9), 55.

Ribbing, M. (1998, March 8). Pizza-size satellite dish starts to deliver. *The Baltimore Sun*, 1D, 3D.

Satellite Home Viewer Act of 1988. (1998). 17 U.S.C. § 119.

Satellite Home Viewer Improvement Act of 1999. (1999, November 19). 17 U.S.C. § 122.

SES-Astra. (2000, March). *Coverage statistics*. [Online]. Available: http://www.ses-astra.com/satellites/coverage/ index_poll.htm.

Setzer, F., Franca, B., & Cornell, N. (1980, October 2). *Policies for regulation of direct broadcast satellites*. Washington, DC: FCC Office of Plans and Policy.

Setzer, F., & Levy, J. (1991, June). *Broadcast television in a multichannel marketplace*. Washington, DC: FCC Office of Plans and Policy, Working Paper No. 26.

SkyReport. (1998a, March). *JSkyB, PerfecTV to combine operations on May 1*. [Online]. Available: http://www.skyreport.com/jskyb.htm.

SkyReport. (1998b, March). *Primestar presses ahead with medium power*. [Online]. Available: http://www.skyreport.com/213star.htm.

SkyReport. (1998c, March). *Tee-Com executives face shareholder lawsuit*. [Online]. Available: http://www.skyreport.com/tee-com.htm.

SkyReport. (1998d, March). *TSAT/Primestar restructuring inches along*. [Online]. Available: http://www.skyreport.com/213star.htm.

SkyReport. (2000a, March). *DirecTV files counterclaim against DISH*. [Online]. Available: http://www.skyreport.com/ skyreport/mar2000/031500.htm#dish.

SkyReport. (2000b, March). *SkyPerfectTV gets DirecTV Japan*. [Online]. Available: http://www.skyreport.com/skyreport/mar2000/030300.htm#dtv.

SkyVista. (2000, March). *EchoStar, Loral Skynet team up to launch new satellite television service*. [Online]. Available: http://www.skyvista.com/talktous/press153.htm.

Star Choice. (2000, March). *About Star Choice*. [Online]. Available: http://www.kusat.com/sc/index.htm.

STAR TV. (1998, March). *Star TV background*. [Online]. Available: http://www.startv.com/startv/sales/back.html.

STAR TV. (2000, March). *What's new?* [Online]. Available: http://www.startv.com/eng/index.html.

TiVo, Inc. (2000). *What is TiVo?* [Online]. Available: http://www.tivo.com/what/faq.html.

Trowbridge, D. (2000). *DISH Network by EchoStar*. [Online]. Available: http://www.qtm.net/~trowbridge/DBS-intro.htm.

Whitehouse, G. (1986). *Understanding the new technologies of the mass media*. Englewood Cliffs, NJ: Prentice-Hall.

Wink Communications. (2000). *What is Wink?* [Online]. Available: http://www.wink.com/contents/whatiswink.shtml.
Wold, R. N. (1996, September). U.S. DBS history: A long road to success. *Via Satellite*, 32-44.

Websites

DBS/DTH industry news and WWW links
http://www.dbsdish.com/
http://www.skyreport.com/
DirecTV's official WWW site
http://www. DirecTV.com/
The DISH Network's official WWW site
http://www.dishtv.com/

6

Wireless Cable Systems: MMDS & LMDS

Ted Carlin, Ph.D.*

Much has happened in the telecommunications industry over the past five years, from the explosion of the World Wide Web to the refinement of the direct broadcast satellite (DBS) industry, the beginnings of local phone competition, and the evolving world of wireless communications. Many of these advances were made possible by passage and implementation of the Telecommunications Act of 1996. Its competitive effects continue to ripple throughout the industry, benefiting consumers and businesses more than ever.

Access to the World Wide Web is changing the look and feel of wireless cable, which was first envisioned to be exclusively devoted to distributing multichannel television programming. However, recent purchases of several wireless cable systems by MCI WorldCom are sure signs that the business model for wireless cable has shifted toward telecommunications and, more specifically, to the Internet. In fact, MCI WorldCom's own Website details elaborate plans for consumer and business access to the World Wide Web through these wireless cable networks. Once a stagnant video delivery service relegated to rural and inner-city niche markets, wireless cable has been revived to be a new competitor in the broadband services arena.

Background

Multipoint multichannel distribution service (MMDS), also known as wireless cable, is a broadband service that delivers addressable multichannel television programming, Internet access, data transfer services, and other interactive services over a terrestrial microwave platform. The systems are growing rapidly, and now serve five million customers in 90 nations, with over one million customers in some 250 U.S. systems (FCC, 2000).

* Assistant Professor, Radio/Television, Department of Communication and Journalism, Shippensburg University (Shippensburg, Pennsylvania).

Operators broadcast multiple channels of television or related services at microwave frequencies from an antenna located on a tower, tall building, or mountain. The signals are received by a small microwave dish, typically about 16" × 20" in size (perhaps larger in outlying areas). A block downconverter integrated into or mounted on an antenna mast translates the signals into the band utilized by standard cable TV. A set-top converter identical in function to a standard cable TV box is located near the TV receiver (Whitehouse, 1986).

Wireless cable has superior signal quality when compared with wired cable because the signals are not required to go through miles of cable and multiple stages of signal amplification. In addition, service interruption is significantly reduced, as there is nothing that can fail between the transmit facility and the receive site. Modern MMDS transmission equipment has proven to be very reliable. Also, frequency agile equipment is available that will automatically replace a failed transmitter or receiver within seconds (Harter, 1999).

With the advent of digital transmission and data transfer capabilities over MMDS systems, the expansion of channel capacity, and the inclusion of the local channels, wireless cable offers a competitive alternative to all other broadband delivery systems.

System Architectures

The basic wireless cable system consists of a transmitter site that provides a broadband signal to multiple customer receive sites distributed in the line-of-sight service area. Depending on the transmitter's output power, it can provide service to an area within a radius of 10 to 35 miles. A basic system is designed to consider the local terrain, structures, and foliage, and still provide a high-quality signal to as many customer receive sites in its service area as possible from a single transmitter (Harter, 1999).

A more complex system configuration may be required in an area with several hills. Such topography would result in "shadowed" areas (e.g., areas where line-of-sight reception is blocked) if served by one transmitter. By strategically positioning several transmitters (signal booster stations) in the service area, the shadowed areas can be reduced. The receive antennas in each of the smaller areas are pointed to the polarized transmit antenna serving that area. The receive antennas are polarized in the same direction (horizontal or vertical) as the transmit antenna serving them (Web ProForums, 2000).

System Elements

The major elements of a wireless cable system include: (1) the transmit site, (2) the signal path, and (3) the receive site. The components of the transmit and receive sites are discussed here. The major components of the transmit site include the tower and transmit antenna, the signal processing equipment, and the billing system (Cornelius, 1998).

Tower and Transmit Antenna. The tower supporting the transmit antenna is strategically located to provide the greatest height in the service area. This may be on a tall building or hill. The transmitted signal strength may be designed to any one of several standard transmitter output powers: 10, 15, 20, 50, or 100 watts. As the signal travels from the transmitter to the antenna, it loses power as it passes through the combiners, filters, waveguides, associated jumpers, and connectors. The transmit antenna typically broadcasts signal energy omnidirectionally, with a slight downward tilt for optimum delivery to the customers in the immediate surrounding service area (Cornelius, 1998).

Figure 6.1
Wireless Multiple-Channel Distribution System

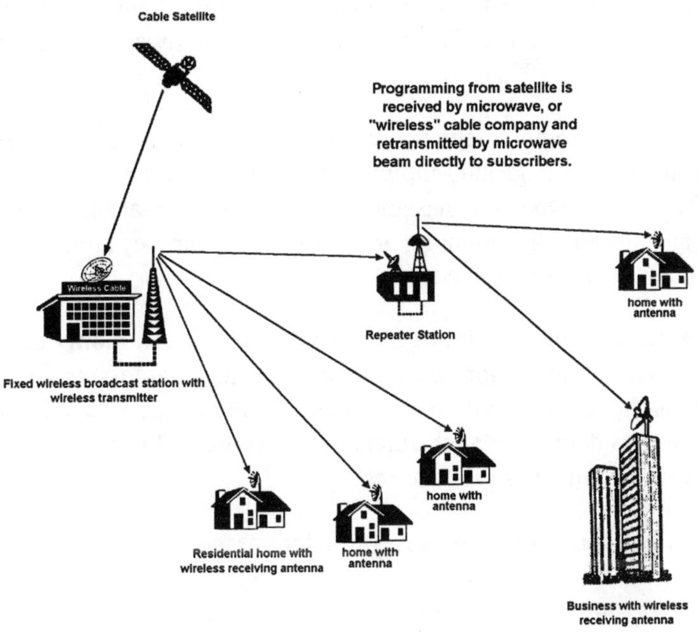

Source: R. B. Woodward

In special cases, an antenna may be designed with a transmit pattern to conform to an irregular-shaped topography in a service area. For example, a transmit antenna may be located near a body of water. Here, the antenna is designed with a cardioid transmit pattern to direct, in this case, most of its energy away from the water and toward the customers on land. This transmit pattern provides customers with higher signal levels than would have been delivered with an omnidirectional transmit antenna because the energy is concentrated and radiated over a smaller area (Cornelius, 1998).

Signal Processing Equipment. Signals from all sources are scrambled, as in a typical cable TV headend, and each signal is delivered to a transmitter for the individual channel. Then, each transmitter is fed into a channel combiner/filter. The combiner/filter feeds the main transmission line serving the transmit antenna at the top of the tower (Web ProForums, 1998).

The video and data information carried on the transmitted signal is originally imported from a number of program sources via satellites, microwaves, over-the-air TV broadcasts, or locally originated studios or tape decks. TVRO (television receive only) equipment may be located either at the transmitter site, or at some distance where the signals are transported by way of microwave, cable, or fiber. Incoming signals from satellites at the transmit site may have been previously encoded. These signals must be decoded at the transmit site and then scrambled before transmission on the MMDS system. The scrambled channels are ultimately received by the customers and descrambled by the set-top converter in each home.

Billing System. The computerized billing system is connected to the outgoing scramblers at the transmit site to easily facilitate the various programming options and configurations available to the customer, each with a different monthly billing rate.

Receive sites may be installed in either homes or multiple dwelling units (MDUs) such as apartments or commercial buildings. Components include hardware, antennas, downconverters, and set-top converters (Harter, 1999).

Installation Hardware. Some antennas are roof-mounted using a mast, while others may be attached to a small tower that is ground-mounted near the structure. To accomplish this, specific hardware is used for installing the antennas and the downconverter at the receive site. The primary items are the antenna mast and the mounting apparatus (e.g., chimney straps, tripod, or base or side mounts) used to attach the mast to the receive site.

Receive Antennas. There are two types of receive antennas commonly used by wireless cable systems: (1) microwave antennas for receiving the wireless cable signals, and (2) VHF/UHF antennas for receiving the local over-the-air channels (Whitehouse, 1986). However, some analog wireless cable systems do deliver local channels via microwave. In those systems or if the system is digital, only microwave antennas are required.

Microwave receive antennas perform several functions:

- They capture the airborne signal.
- They increase the signal received.
- They provide directivity.
- They provide polarized selectivity (Whitehouse, 1986).

Receive antennas capture the airborne microwave signal from the transmit antenna and deliver the signal to the frequency block downconverter. Signal gain is accomplished by manufacturing the antenna so that its size is several multiples larger than a signal's wavelength. The greater the number of wavelength multiples encompassed within the antenna, the greater the gain. The greater the gain of an antenna, the larger its size and expense.

Directivity is achieved by constructing the antenna to focus incoming signals from one direction only. This function allows an antenna to amplify the signal originating from the direction the antenna is pointed, while attenuating signals originate from all other directions.

To accomplish polarized selectivity, the antenna feed is positioned to receive the signal either in the vertical or horizontal plane. This allows the antenna to selectively receive the signal from the appropriate polarized transmit antenna, while rejecting a signal from an adjacent, cross-polarized undesirable transmitter.

Downconverters. The downconverter provides frequency conversion and more gain to the received microwave signal without significantly degrading it with electrical noise (snow). The downconverter receives the input signal at microwave frequencies (2.154 GHz to 2.681 GHz) and blockconverts them to a lower frequency range usable by a set-top converter (116 MHz to 128 MHz for MDS channels and 222 MHz to 408 MHz for MMDS/ITFS channels). While some downconverters are designed to convert 33 channels (which includes MDS and MMDS/ITFS channels), others are designed to convert only the 31 MMDS/ITFS channels (Cornelius, 1998).

All downconverters are mounted close to the antenna to eliminate any cable loss from the cable between the antenna and the downconverter. To accomplish this, most manufacturers integrate the downconverter within the antenna feed. Because aeronautical navigational radar, personal communication services (PCS), cellular telephones, and microwave ovens produce signals with frequencies near those of the wireless cable signal, special filters may be required to eliminate interference from these sources. These filters may be incorporated within the downconverter or mounted externally to it (Harter, 1999).

Set-Top Converters. In addition to commonly providing downconverter powering, wireless cable set-top converters incorporate a tuner with two input ports, one for the over-the-air signals and the other for the scrambled wireless cable channels. Within the set-top converter is circuitry that descrambles the channels when authorized to do so by the billing computer at the transmit site. Some manufacturers provide "whole-house" downconverters with an integrated descrambler that simultaneously decodes all received channels, eliminating the need for individual set-top converters at multiple outlets within the customer's home. These may be integrated into the antenna or may be installed as stand-alone units. They are powered by a stand-alone power supply (Harter, 1999).

Industry Regulation

Regulatory Jurisdiction

Wireless cable, unlike traditional cable TV, requires no easements to operate and thus requires no franchise. Cable TV gets its authority in the form of a franchise from the local city government to use its community easements (rights-of-way) for the construction of the cable system. Wireless cable, in contrast, is regulated solely by the Federal Communication Commission.

Channel Allocations

The predecessor to MMDS was multipoint distribution service (MDS). MDS began in the mid-1970s with the earlier allocation by the FCC of two 6 MHz channels between 2.150 GHz and 2.162 GHz. In some markets, the 6 MHz MDS second channel is replaced by a 4 MHz MDS 2-A channel. Although these channels were intended for business data, they were more commonly used to deliver TV programming on a single-channel subscription basis (FCC, 1998). It was, in fact, on these early MDS channels that Home Box Office (HBO) was shown. Later, HBO moved to satellite transmission, and thus began the move to TVRO reception of premium programming.

In order to be competitive in the multichannel environment, MDS operators needed more bandwidth and wanted to use an additional 31 channels between 2.500 GHz and 2.686 GHz. These channels were originally designated to educational institutions for the Instructional Television Fixed Service (ITFS). From these, eight channels in the E and F groups were reallocated in the early 1980s for use by wireless cable and officially dubbed MMDS channels by the FCC (FCC, 1998).

ITFS Channels

The remaining 23 ITFS channels, in groups A, B, C, D, G, and H, were made available to wireless cable operators from educational license-holders on a lease basis. In exchange for access to ITFS channels, the FCC required that wireless cable operators broadcast up to 40 hours of educational programming per week. With the addition of these channels, wireless cable now had the potential to deliver 33 analog channels of television (FCC, 1998).

Protected Service Areas

Before wireless cable frequencies were auctioned by the FCC in 1996, each licensee was granted authority to operate certain MMDS channels and awarded a protected service area (PSA). A PSA typically had a radius of 15 miles, and was protected from signal interference from the transmissions from nearby stations. Once the basic trading area (BTA) rules were adopted, the incumbent PSAs were expanded to a 35-mile radius (FCC, 1997).

Basic Trading Areas

In 1996, the nation was divided by the FCC into newly-defined basic trading areas, and these 493 markets were auctioned to the highest bidders. BTAs cover the entire United States; however, within any BTA, incumbent licensees continue to operate as before. Some incumbent licensees purchased a surrounding BTA in order to expand their service area and/or protect the systems from possible interference by other new license holders (FCC, 1997).

Recent Developments

Digital Transmission

By 1996, technological advances had made possible digital compression of video channels in the broadband communication industry. As one measure to foster competition, in July 1996, the FCC issued a declaratory ruling saying that wireless cable operators could digitize their licensed MDS and ITFS (MMDS) channels as long as adjacent wireless cable systems experienced no interference from the process of the analog-to-digital conversion. With this FCC declaratory ruling and the advances in digital technology, wireless cable could now deliver between 100 and 200 virtual channels of video from their 33 analog channels (FCC, 2000).

Yet, because the 33-channel analog capacity of MMDS systems is generally not competitive with that of most cable systems, MMDS subscribership declined before these digital upgrades could take place. As of 1999, the number of homes with a serviceable line-of-sight to an MMDS operator's transmission facilities was 62,500,000, and the number of homes actually capable of receiving an MMDS operator's signal was 35,750,000. The total number of MMDS video subscribers fell from 1 million to 821,000 between June 1998 and June 1999, a decrease of 17.9%. Of the 821,000 subscribers in 1999, 721,000 were analog MMDS subscribers and the other 100,000 were subscribers to digital MMDS services (FCC, 2000).

High-Speed Internet Access and Data Transmission

In October 1996, the FCC cleared the way for wireless cable operators to use their spectrum for one-way, high-speed, digital data applications including Internet access. The Wireless Cable Association International, along with over 100 companies, petitioned the FCC in March 1997 to grant the industry the right to use MMDS spectrum for two-way access. Two-way authorization would effectively enable voice, video, and data over wireless cable. Following almost two years of deliberation, the FCC granted the request for use of two-way transmissions on MMDS and ITFS frequencies, and the new rules took effect on January 24, 1999 (Corazzini & Goodwyn, 2000). Some of the cities that have systems offering high-speed access to the Internet over wireless cable as of mid-2000 are Washington (DC), Las Vegas, Lakeland (Florida), Colorado Springs, Santa Rosa (California), and Nashua (New Hampshire).

Two MMDS operators, Nucentrix Spectrum Resources, Inc. (formerly Heartland Wireless Communications, Inc.) and Wireless One, Inc., have announced joint ventures with DirecTV. According to these agreements, the MMDS operator will combine its MMDS frequencies with DirecTV's satellite video programming so that consumers can receive local broadcast and other channels with MMDS frequencies, in addition to DirecTV's full video service, through a direct broadcast satellite DBS dish. The local MMDS operator handles installation of and subscription to both services. This service is offered to both single-family homes and MDUs. Many MMDS operators view MDUs as underserved by cable operators and as a possible source for rapid revenue growth. Nucentrix reports that it has begun to offer its joint MMDS/DirecTV service in 41 markets (FCC, 2000).

From 1988 through 2000, MCI WorldCom and Sprint have purchased a significant number of MMDS operators (about 60 % of the total MMDS licenses). Sprint has acquired WBS America, People's Choice TV Corporation, American Telecasting, Inc., Videotron Hollard B.V., Wireless Cable of Florida, and Transworld Telecommunications, Inc. These properties give Sprint the potential of offering two-way communication services to almost 30 million households nationwide. Sprint put all these companies—and its 200 MHz worth of licenses—into a new Broadband Wireless Group, and then declared its intention to launch high-speed Internet and Sprint's Integrated On-Demand Network (ION) over MMDS (MCI WorldCom, 2000c).

MCI WorldCom has purchased CAI Wireless, which is also majority owner of CS Wireless, Wireless One, and Southern Wireless Video, Inc. With these acquisitions, MCI WorldCom has the ability to offer communication services to over 50 million homes. Sprint and MCI WorldCom intend to use this spectrum as a "last mile" connection to homes for the provision of high-speed Internet access (MCI WorldCom, 2000c).

Current Status

The rejuvenation of MMDS started when the FCC decided to allow two-way digital signals on this band. To say that this was a smart choice is an understatement: One-way TV delivery over MMDS had never caught on, and the spectrum was essentially lying fallow. The second important factor was separate decisions by both Sprint and MCI WorldCom to seriously move into MMDS, spending over $1 billion each for existing MMDS licenses. Currently, each company owns about 30% of MMDS licenses in the United States, or 60% combined (MCI WorldCom, 2000c).

The third factor was the October 1999 announcement that MCI WorldCom and Sprint would merge to form a new global communications company—WorldCom—that would be a complete competitor and offer a full range of "all-distance" products and services to compete against the "huge monopoly local phone companies, AT&T and its cable networks, and against massive foreign competitors" (MCI WorldCom, 2000a). The proposed merger, at this writing still seeking approval from the FCC, the Justice Department, 16 U.S. states, and the European Union, would have a combined spectrum coverage of two-thirds of the United States and pass 54 million households.

Why would Sprint or MCI WorldCom purchase so many MMDS licenses? The answer lies in the hidden possibilities of MMDS. Just because this spectrum can cover up to 35 miles with a single tower does not mean MMDS operators have to do so. Actually, they can build low-power MMDS transmitters within a few miles of each other, "cellularizing" the coverage area in order to boost its carriage capacity. In this configuration, for example, MMDS is a perfect fit for Sprint, which has multiple PCS towers covering small cells. This means these same towers can be re-used for MMDS, creating strong synergy between Sprint PCS and MMDS operations.

MCI has already begun field-testing their MMDS services in Jackson (Mississippi), Baton Rouge, and Memphis. Each trial is designed to test the functionality of the service, as well as determine possible price points and service bundles for business and residence customers. "Warp One" service for businesses features digital subscriber line (DSL) connections speeds at $500 to $600 per month, while "Warp 310" for residential customers provides unlimited usage at 310 Kb/s speeds for about $39.95 per month. Additional MCI WorldCom trials in Boston and Dallas will also begin during 2000, while Sprint plans to test MMDS systems in 20 markets by the end of 2000. Together, they hope to begin offering commercial MMDS service in 100 markets by the end of 2001 (MCI WorldCom, 2000b).

Another promising development in the wireless cable industry has been the FCC's creation of local multipoint distribution services (LMDS). Using two spectrum auctions in 1998 and 1999, the FCC allocated 1.3 GHz in two spectrum blocks of the 27.5 to 31.3 GHz band to LMDS. At about 17 times the size of the MMDS frequency bands, LMDS offers plenty of broadband service potential, yet, there are some propagation restrictions that limit the usability of this high frequency spectrum: (1) attenuation of the signals by rain and snow, (2) attenuation of the signals due to blockage from obstacles such as buildings or foliage, (3) the need for very directional receive antennas, and (4) a transmission radius of 3 km to 5 km (Lawrence, 1999). Therefore, LMDS systems are initially being configured in a manner similar to cellular telephone systems using multiple cells within the transmission area.

There appear to be four primary uses for the LMDS licenses as providers continue to solve the propagation difficulties:

(1) High-speed Internet access, by far, the most significant application mentioned in the trade press.

(2) Intranet services in metropolitan areas for commercial data transfer.

(3) Local loop telephony.

(4) Point-to-point backhaul for cellular, PCS, and local area network (LAN) services (Lawrence, 1999).

Each of these areas possess specific opportunities to create cost-effective solutions for groups of niche users.

For example, one company, NEXTLINK, is providing high-quality, broadband communications services to businesses over fiber optic and LMDS facilities across the United States. Currently, NEXTLINK is providing service in 49 markets. The company is the largest holder of LMDS spectrum in North America, with licenses covering 95% of the population in the top 30 markets in the United States. NEXTLINK plans to use wireless capabilities to complement and extend the reach of its local fiber optic networks in the markets in which NEXTLINK has spectrum. Additionally, NEXTLINK is acquiring exclusive rights to use certain fibers and a conduit throughout a 16,000-mile, high-speed, fiber optic backbone network that will connect over 50 cities in the United States and Canada. The network should be complete in 2001, with NEXTLINK turning on segments of the network during 2000 (Pappalardo, 2000).

Factors to Watch

It appears that the ever-increasing consumer and business demand for high-speed, two-way Internet access may be just the application that allows MMDS and LMDS to proliferate in the United States. With the recent acquisitions by companies such as AT&T, Sprint, MCI WorldCom, McCaw, Bell Atlantic, and SBC Communications, the notion of MMDS and LMDS as strictly multichannel video programming distribution has quickly disappeared in favor of digital information transfer.

However, there are some concerns facing the industry. The first concern is rolling out these services: Building fixed-wireless broadband systems takes time, even for the most cash-rich of companies. The second is the resistance of incumbent local telephone companies to opening up their networks. One reason established telcos are likely to resist aiding fixed-wireless broadband providers is because they themselves realize the potential of MMDS and LMDS. In fact, MMDS and LMDS may well be the last-mile solution of the 21st century, notwithstanding the limits of wired networks, DSL, and cable modems.

The most complicated, yet the most important question may be, "How are we going to make a profitable business of this opportunity?" Every market possesses specific residential and business needs, which may include single products such as Internet service, or a bundled array of products such as data services, telephony, and videoconferencing. With the right combination of products, market coverage, and creative management, MMDS and LMDS may indeed be technologies whose time has finally come.

Bibliography

Corazzini, R., & Goodwyn, S. (2000). *FCC approves use of two-way transmissions for MDS and ITFS.* [Online]. Available: http://www.commlaw.com/pepper/Memos/MMDS/2wayrdmds.html.

Cornelius, J. (1998). *A brief tutorial of FDMA, TDMA, and CDMA modulation techniques in a two-way digital wireless cable environment.* [Online]. Available: http://www.haa.com/technical.html.

Federal Communications Commission. (1997, July 25). *FCC announces grant of MDS authorizations.* [Online]. Available: http://www.fcc.gov/Bureaus/Mass_Media/Public_Notices/MDS_Notices/pnmm7146.pdf.

Federal Communications Commission. (1998, January). *Notice of Proposed Rulemaking: MDS and ITFS licensees to engage in fixed two-way transmissions.* FCC 97-360.

Federal Communications Commission. (2000, January 14). *Annual assessment of the status of competition in markets for the delivery of video programming.* Sixth Annual Report. CS Docket No. 99-230.

Harter, G. (1999, October 25). *Why wireless?* [Online]. Available: http://www.haa.com/technical.html.

Lawrence, R. (1999, March). *Is LMDS viable in the commercial marketplace?* [Online]. Available: http://www.haa.com/mag4.html.

MCI WorldCom. (2000a). *Merging to face the future.* [Online]. Available: http://www.WorldCom-merger.com/merger_facts/facts_face_future.htm.

MCI WorldCom. (2000b). *MCI WorldCom details MMDS plans.* [Online]. Available: http://www.WorldCom-merger.com/mmds/tele3.htm.

MCI WorldCom. (2000c). *WorldCom merger information.* [Online]. Available: http://www.WorldCom-merger.com/mmds/mmds_main.htm.

Pappalardo, D. (2000). *NEXTLINK puts money down on LMDS.* [Online]. Available: http://www.nextlink.com/ra/news/archive/news/networld/index.html.

Web ProForums. (2000, April). *Wireless broadband modems tutorial.* [Online]. Available: http://www.webproforum.com/wire_broad/topic10.html.

Whitehouse, G. (1986). *Understanding the new technologies of the mass media.* Englewood Cliffs, NJ: Prentice-Hall.

Woodward, R. B. (1995). Wireless cable (MMDS). In A. E. Grant (Ed.). *Communication technology update*, 4th edition. Boston: Focal press.

Websites

MCI-Sprint merger, http://www.WorldCom-merger.com/.
MMDS at the FCC, http://www.fcc.gov/mmb/vsd/.
Wireless Cable Association, http://www.wirelesscabl.com.
Wireless Communication Association, http://www.wcai.com/.

7

Advanced Television

Peter B. Seel, Ph.D. & Michel Dupagne, Ph.D.*

The technologically advanced nations of Japan, the United States, Canada, and the European Union are in the midst of a very expensive conversion from analog to digital television (DTV) broadcasting. It is the most significant change in global broadcast standards since color images were added in the 1950s and 1960s. However, the dream of one world television standard to replace the incompatible analog NTSC, PAL, and SECAM regional standards has failed to materialize. In fact, new competition is emerging among the proponents of various digital standards to enlist non-aligned nations to adopt one system over others.

It is only a matter of time until digital TV broadcasting is adopted around the world. For better or worse, television is a ubiquitous global medium. The glow of the tube can be found from rustic cabins near the Arctic Circle to open-air shelters on remote Pacific islands. Television is a primary source of news and entertainment for viewers from New York to Hong Kong. The diffusion of digital television will provide a movie-friendly display with a wider image, improved sound quality, and the higher resolution needed for projecting a sharp image on a screen. However, the most significant change is that it will be a computer-friendly technology—the display will be equally capable of showing the movie of the week on NBC, or presenting up-to-the-minute sports scores from ESPN.com. The remote control of the near future may have two primary buttons at the top—"TV" and "Internet."

One key attribute of digital technology is "scalability"—the ability to produce audio/visual quality as good (or as bad) as the viewer desires (or will tolerate). This term does not refer to program content quality—that factor will still depend on the creative ability of the writers and producers. Within the constraints of available transmission bandwidth, digital TV facilitates the dynamic assignment of sound and image fidelity in a given transmission channel. Within the quality universe of DTV, there are a wide variety of possible display options. The three most common are:

---

* Dr. Seel is Associate Professor, Department of Journalism and Technical Communication, Colorado State University (Fort Collins, Colorado). Dr. Dupagne is Associate Professor, School of Communication, University of Miami (Coral Gables, Florida).

- HDTV (high-definition television).
- SDTV (standard-definition television).
- LDTV (low-definition television).

High-definition television represents the highest pictorial and aural quality that can be transmitted over the air. It is defined by the Federal Communications Commission (FCC) in the United States as a system that provides image quality approaching that of 35mm motion picture film, that has an image resolution of approximately twice that of conventional National Television System Committee (NTSC) television, and has a picture aspect ratio of 16:9 (FCC, 1990). At this aspect ratio of 1.78:1 (16 divided by 9), the television screen is wider in relation to height than the 1.33:1 (4 divided by 3) of NTSC. This format is closer in aspect ratio to the wide-screen images seen in movie theaters that are 1.85:1 or even wider. Figure 7.1 compares a 16:9 HDTV set with a 4:3 NTSC display. Note that the higher resolution of the HDTV set permits the viewer to sit closer to the set, which results in a wider angle of view. As explained below, digital HDTV signals must be transmitted on a separate channel from a television station's analog signal.

Figure 7.1
Wider Viewing Angle with HDTV

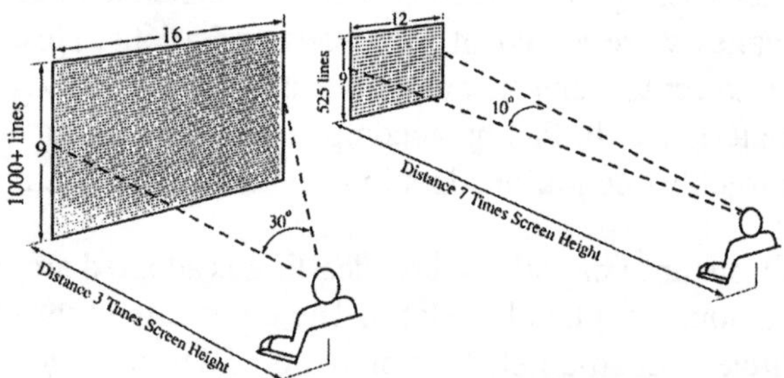

Source: M. Dupagne & P. Seel

SDTV, or standard-definition television, is another type of DTV that can be broadcast *instead of* HDTV. Digital SDTV transmissions will offer lower resolution than HDTV, but they will be available in both narrow- (1.33:1) and wide-screen (1.78:1) formats. Using digital video compression technology, it will be feasible for U.S. broadcasters to transmit four to six SDTV signals instead of one HDTV signal in the allocated 6 MHz digital channel. Thus, a television station would be able to retransmit a daytime soap opera while simultaneously broadcasting a newscast, a sporting event, a stock market update, and a children's program in SDTV. Some stations may reserve true HDTV single-channel programming for evening prime-time hours. The development of multi-channel SDTV broadcasting, called "multicasting," is an unintended consequence that was unforeseen by early HDTV researchers.

LDTV is low-definition television. With scalable digital technology, image and sound quality can be compressed to the point that obvious defects can be seen and heard (as with streamed Internet media with a slow dial-up modem). Some cable operators are using digital compression to expand their channel offerings without extensively rebuilding their facilities. They sometimes use compression ratios as high as 12:1, thereby fitting 12 digital channels in the space formerly required by a single analog channel. While such image quality is acceptable for a talk show, it may reveal obvious visual defects for fast-paced sports programming. The digital world can provide programming variety at the expense of presentation quality.

Video streamed over the Internet is another example of LDTV. Video streaming refers to the process of delivering video clips in real time to online users (Pavlik, 2000). The video is heavily compressed, and the frame rate is often 15 fps (frames per second) or less. A typical result is video playing in a small window with visible quality degradation. This shortcoming is primarily due to the limited bandwidth of dial-up connections and the Internet itself. As higher-speed Internet access becomes more widely available in the home, streamed audio and video quality will improve. (For more on video streaming, see Chapter 8.)

Background

In the 1970s and 1980s, Japanese researchers at NHK developed two related analog HDTV systems: a "Hi-Vision" analog *production* standard with 1,125 scanning lines and 60 fields (30 frames) per second, and a "MUSE" *transmission* system with an original bandwidth of 9 MHz designed for direct broadcast satellite (DBS) distribution to the Japanese home islands. Japanese HDTV transmissions began in 1989 and steadily increased to a full schedule of 17 hours a day by October 1997 (http://www.nhk.or.jp).

In 1986, Japan and the United States sought to have the Hi-Vision system adopted as a world HDTV production standard by the CCIR, a subgroup of the International Telecommunication Union (ITU), at a Plenary Assembly in Dubrovnik (Yugoslavia). However, European delegates lobbied for a postponement of this initiative that effectively resulted in a *de facto* rejection of the Japanese technology. European governments and their high-technology industries were still recovering from Japanese dominance of the videocassette recorder (VCR) market, and resolved to create their own distinctive HDTV standard that would be intentionally incompatible with Hi-Vision/MUSE (Dupagne & Seel, 1998).

Subsequently, a European R&D consortium, EUREKA EU-95, developed a system known as HD-MAC that featured 1,250 wide-screen scanning lines and 50 fields (25 frames) displayed per second. This analog 1,250/50 system was used to transmit many European cultural and sporting events, such as the 1992 summer and winter Olympics in Barcelona (Spain) and Albertville (France). However, since the 1990 development of digital HDTV technology in the United States, HD-MAC has been supplanted by a series of alternative digital video broadcasting (DVB) standards for cable, satellite, and terrestrial broadcasting.

After it became apparent that there would be no single world HDTV standard in the wake of the rejection of the Japanese system at Dubrovnik, the FCC began a series of policy initiatives in 1987 that led to the creation of the Advisory Committee on Advanced Television Service (ACATS). This committee was comprised of 25 electronic media executives charged with investigating the policies, standards, and regulations that would facilitate the introduction of advanced television (ATV) services in the United States (FCC, 1987). ACATS played a central role in the standardization of this technology in the United States. Over 1,000 volunteers from television, cable, and other telecommunications organizations worked for eight years on three ACATS subcommittees to create a testing program to identify an ideal ATV broadcast transmission scheme.

Testing analog ATV systems was about to begin in 1990 when the General Instrument Corporation announced that it had perfected a method of digitally transmitting an HDTV signal. This announcement had a bombshell impact, since many broadcast engineers were convinced that digital television transmission would be a technical impossibility until well into the 21st century (Brinkley, 1997). The other participants in the ACATS competition quickly developed digital systems that were submitted for testing. At the end of the first round of tests, the Advisory Committee decided that none of the digital proponent systems could be identified as superior to the others and called for a second round of tests. Before that testing could take place, the three competitors in the testing process with digital systems—AT&T/Zenith, General Instrument/MIT, and Philips/Thomson/Sarnoff—decided to merge into a common consortium known as the Grand Alliance. With the active encouragement of the Advisory Committee, they combined elements of each of their ATV proponent systems in 1993 into a single digital Grand Alliance system for ACATS evaluation.

The FCC made a number of key decisions during the ATV testing process that defined a national transition process from NTSC to an advanced broadcast television system. One such ruling was the Commission's *First Report and Order* of August 24, 1990, which outlined a *simulcast* strategy for the transition to an ATV standard (FCC, 1990). This strategy required that U.S. broadcasters transmit both the new ATV signal and the existing NTSC signal concurrently for a period of time, at the end of which the NTSC transmitter would be turned off. Rather than try and deal with the inherent flaws of NTSC, the Commission decided to create an entirely new television system that would be incompatible with the existing one. This was a decision with multibillion-dollar implications for broadcasters and consumers, since it meant that all existing production, transmission, and reception hardware would have to be replaced with new equipment capable of processing the ATV signal.

The Commission also proposed a transition window of 15 years from the adoption of a national ATV standard to the shutdown of NTSC broadcasting (FCC, 1992). This second ATV *Report and Order* caused consternation on the part of the broadcast television industry at what they perceived as too short a transition period. The nation was in the midst of a recession, and broadcasters were concerned about the financial cost of replacing their NTSC facilities. The transition cost for U.S. broadcasters has been estimated to range from $2 million for simply retransmitting the network signal to $10 million for complete digital production facilities (Ashworth, 1998).

The Grand Alliance system was successfully tested during the summer of 1995, and a digital television standard based on that technology was recommended to the FCC by the Advisory Committee on November 28 (ACATS, 1995). In May 1996, the FCC proposed to adopt the *ATSC Digital Television (DTV) Standard* based upon the work accomplished by the Advanced Television Systems Committee (ATSC) in documenting the technology developed by the Grand Alliance consortium (FCC, 1996a). The proposed ATSC DTV standard specified 18 digital transmission variations, as noted in Table 7.1. Stations would be able to choose whether to transmit one channel of high-resolution, wide-screen HDTV programming, or four to six channels of standard-definition programs during various day parts. The type of television owned by the viewer would dictate whether he or she is watching the program on a digital set in high or low resolution and in 16:9 wide-screen or 4:3 narrow-screen, or watching it on an older 4:3 analog set (see Table 7.2).

Table 7.1
U.S. Advanced Television Systems Committee (ATSC) DTV Formats

Format	*Active Lines*	*Horizontal Pixels*	*Aspect Ratio*	*Picture Rate**	*U.S. Adopter*
HDTV	1,080 lines	1,920 pixels/line	16:9	60I, 30P, 24P	CBS & NBC (60I)
HDTV	720 lines	1,280 pixels/line	16:9	60P, 30P, 24P	ABC (30P)
SDTV	480 lines	704 pixels/line	16:9 or 4:3	60I, 60P, 30P, 24P	Fox (30P)
SDTV	480 lines	640 pixels/line	4:3	60I, 60P, 30P, 24P	None

* In this column, "I" indicates interlace scan in *fields*/second and "P" means progressive scan in *frames*/second.

Source: ATSC

Table 7.2
International Terrestrial DTV Standards

System	**ISDB-T**	**DVB-T**	**ATSC DTV**
Region	Japan	Europe	North America
Modulation	OFDM	COFDM	8-VSB
Aspect Ratio	1.33:1, 1.78:1	1.33:1, 1.78:1, 2.21:1	1.33:1, 1.78:1
Active Lines	480, 720, 1080	480, 576, 720, 1080, 1152	480, 720, 1080*
Pixels/Line	720, 1280, 1920	varies	640, 704, 1280, 1920*
Scanning	1:1 progressive, 2:1 interlace	1:1 progressive, 2:1 interlace	1:1 progressive, 2:1 interlace*
Bandwidth	6-8 MHz	6-8 MHz	6 MHz
Frame Rate	30, 60 fps	24, 25, 30 fps	24, 30, 60 fps*
Field Rate	60 Hz	30, 50 Hz	60 Hz
Audio Encoding	MPEG-2 AAC	MUSICAM/Dolby AC-3	Dolby AC-3

* As adopted by the FCC on December 24, 1996, the ATSC DTV image parameters, scanning options, and aspect ratios were not mandated, but were left to the discretion of display manufacturers and television broadcasters (FCC, 1996b).

Source: P. Seel & M. Dupagne

Note that the standard allows for both interlace and progressive scanning. Companies representing elements of the U.S. computer industry (e.g., Apple, Intel, and Microsoft) sought to have all interlace scanning variations dropped from the proposed standard, contending that they would impede the convergence of television and computing technology (Dupagne & Seel, 1998).

Interlace scanning is a form of signal compression that first scans the odd lines of a television image onto the screen, and then fills in the even lines to create a full video frame in every 1/30 of a second. The present NTSC standard uses interlace scanning to reduce the bandwidth needed for broadcast transmission. While interlace scanning is spectrum-efficient, it creates unwanted visual artifacts that can degrade image quality. Progressive scanning—where each complete frame is scanned on the screen in only one pass—is utilized in computer displays because it produces fewer image artifacts than interlace scanning. The computer industry group was willing to contest this fundamental standardization issue because the advent of digital transmission meant that every DTV set could function as a computer display and vice versa. DTV receivers would be capable of displaying small-font text (e.g., from the World Wide Web) that would be illegible on a conventional NTSC television set.

In December 1996, the FCC finally approved a DTV standard that resolved the conflict in a remarkable way—it deleted *any* requirement to transmit any of the 18 transmission video formats listed in Table 7.1 (FCC, 1996b). It also did not mandate any specific aspect ratios or any requirement that broadcasters transmit true HDTV on their digital channels. While the computer group did not succeed in forcing interlace transmission out of the standard, neither did the Commission stipulate it. Instead, the decision was left to the discretion of the licensees. The FCC resolved the image aspect ratio controversy by leaving this basic decision up to broadcasters as well. They will be free to transmit digital programming in narrow- or wide-screen ratios as they wish.

Other aspects of the ATSC standard remained unchanged, such as the multi-channel Dolby AC-3 audio system. The AC-3 specifications call for a surround-sound, six-channel system that will approximate a motion picture theatrical configuration that has speakers screen-left, screen-center, screen-right, rear-left, and rear-right, with a subwoofer for bass effects. This audio system will enhance the diffusion of "home theater" television systems that have multiple speakers and either a video projection screen, direct-view CRT monitor, or flat-panel display mounted on the wall like a painting.

Recent Developments

Since 1997, the U.S. DTV focus has shifted from standardization matters to DTV implementation considerations, such as programming mix, station conversion, and consumer adoption. In April 1997, the FCC (1997a) issued a *Report and Order* that defined how the United States would make the transition to DTV broadcasting. The Commission set December 31, 2006 as a target date for the phase-out of NTSC broadcasting. However, the U.S. Congress passed a bill in 1997 that would allow television stations to continue to operate their analog transmitters as long as more than 15% of the households in a market cannot receive digital broadcasts through cable or DBS, or if 15% or more local households lack a DTV set or a converter box to display digital images on their older television sets (Balanced Budget Act, 1997). This law gave broadcasters some breathing room if consumers do not adopt DTV technology as fast as set manufacturers would like.

To demonstrate their good faith in an expeditious DTV conversion, the four largest commercial television networks in the United States (ABC, CBS, NBC, and Fox) made a voluntary commitment to the FCC to have 24 of their affiliates in the top 10 markets on the air with a DTV signal by November 1, 1998 (Fedele, 1997). All of their affiliates in the top 30 markets were expected to broadcast digital signals by November 1, 1999. Table 7.3 outlines the rest of the proposed rollout.

Table 7.3
U.S. Digital Television Broadcasting Phase-in Schedule

Phase	# of Stations	Market Size	Type of Station	DTV Transmission Deadline	Percentage of U.S. HHs*
1	24	Top 10	Voluntary	November 1, 1998	- -
2	40	Top 10	Network Affiliates	May 1, 1999	30%
3	80	Top 30	Network Affiliates	November 1, 1999	53%
4	~ 1,037	All	All Commercial	May 1, 2002	~100%
5	~ 365	All	Non-commercial	May 1, 2003	- -
6	~ 1,500	All	All	December 31, 2006	Planned NTSC reversion date

* Television households capable of receiving at least one local DTV broadcast signal.

Sources: FCC and *TV Technology*

Also in April 1997, consistent with the Telecommunications Act of 1996, the Commission assigned each broadcaster a new channel for DTV operations (FCC, 1997b). In the first *Table of Allotments*, over 93% of broadcasters received a frequency that would cover at least 95% of their current NTSC service area. In February 1998, the FCC modified these DTV allotments and established the final DTV core spectrum between channels 2 and 51 (FCC, 1998a).

In 1998, the first HDTV receivers went on sale in the United States at prices ranging from $5,000 to $10,000 or more (Brinkley, 1998). Many of these early models were only "HDTV-capable" and required a separate $1,000 or more decoder box to receive analog signals. Since January 2000, however, more sets have been available as fully-integrated models (decoder included). According to the Consumer Electronics Association, 134,402 DTV sets were sold in the United States from August 1998 to December 1999 (http://www.dtvweb.org).

In July 1998, the FCC (1998b) launched a *Notice of Proposed Rule Making* to address the contentious issue of carriage of local broadcast digital signals during the DTV conversion. It proposed seven must-carry options, running the gamut from the Immediate Carriage Proposal to the No Must Carry Proposal. At issue was not so much whether the (analog) must-carry rules, upheld constitutionally in *Tuner Broadcasting System v. F.C.C.* (1997), were applicable to the digital world as whether the Commission could mandate cable operators to carry *both* analog and digital signals until 2006 or later. A decision was expected in 1999, but, as of March 2000, the Commission had yet to act on this important issue.

Current Status

United States

In late February 2000, 81% (1,376 stations) of all eligible stations in all markets had filed DTV construction permit applications with the FCC (FCC, 2000b). As of early March 2000, 121 broadcast stations in the United States were transmitting a DTV signal, covering 62% of all television households (http://www.nab.org), and more are going on the air with each passing month. As noted in the Table 7.3, all commercial stations must air a DTV signal by May 1, 2002. Non-commercial stations (e.g., primarily PBS affiliates) will have an additional year. A little-known fact about the digital conversion is that about 25% of the 2,200 low-power television (LPTV) stations will be displaced to other channels or will simply have to shut down permanently (Kerschbaumer & McConnell, 2000).

Because over 65% of U.S. television households receive their signals from a cable provider, digital conversion of cable facilities is essential to pass along DTV programming from broadcasters. However, the cable industry has been ambivalent about providing HDTV/SDTV service for two fundamental reasons:

(1) They would have to provide an extra DTV channel for every existing analog channel under FCC simulcast rules, constraining system capacity without increasing revenue.

(2) SDTV technology would permit terrestrial broadcasters to transmit four to six of these channels simultaneously over the air, in effect transforming them into multi-channel providers in direct competition with cable news and sports channels.

For these reasons, the cable industry has been decidedly unenthusiastic about the advent of DTV in the United States. DTV proponents, such as broadcasters and consumer electronics manufacturers, insist that cable systems must carry DTV simulcasts of broadcast programming for the new format to succeed. Given that over two-thirds of American households watch television and receive local television signals via cable, at some point, cable providers will have to provide DTV programming. The key phrase is "at some point." Cable systems would prefer to wait until there is a larger DTV audience to justify upgrading their plants and allocating channel capacity for digital simulcasts. Broadcasters argue that there will be no incentive for consumers to buy DTV sets if they cannot see digital programs—a classic chicken-and-egg conundrum. The FCC cable/DTV must-carry decision should resolve this issue.

In spring 2000, under pressure from the FCC, television set manufacturers and the cable industry resolved a major technological stumbling block inhibiting cable carriage of DTV signals—the connection between the cable box and the DTV set (McConnell, 2000a). These groups issued a set of joint specifications that outlined how DTV signals would pass through the digital set-top box and be displayed on the television set. Still unresolved are significant cabling and copy-protection standards—the latter is an important issue for Hollywood studios since an unencrypted digital signal could allow the piracy of perfect copies of televised and pay-per-view films.

Another technological issue affecting the DTV transition concerns problems with indoor reception of DTV signals. In urban areas such as New York City, DTV signals reflect off tall buildings and create multi-path reception problems for small indoor "rabbit-ear" antennas. The Sinclair Broadcast Group filed a petition with the FCC in October 1999 asking that the Commission reconsider its selection of the 8-VSB (vestigial sideband) modulation technology for DTV transmission (McConnell, 1999). Sinclair's tests indicated that the European COFDM (coded orthogonal frequency division multiplexing) modulation scheme was superior to 8-VSB DTV for indoor reception in urban areas. However, the FCC rejected the Sinclair petition in February 2000 after its Office of Engineering and Technology reviewed the evidence and decided that the merits of COFDM did not justify modifying the DTV transmission standard (FCC, 2000a). As of mid-2000, this issue is not dead. Further testing is required on recently-developed 8-VSB decoder chipsets that promise to resolve indoor reception problems. If they do not, COFDM may re-emerge as an alternative transmission technology.

The major television networks are dealing with the chicken/egg DTV diffusion dilemma by providing a gradually increasing amount of DTV programming. Japanese consumer electronics companies are underwriting production and analog-to-digital conversion costs on ABC (Panasonic) and CBS (Mitsubishi) in an effort to jump-start DTV set sales. ABC has transmitted feature films, *Monday Night Football*, and the 2000 Super Bowl in HDTV (Dickson, 1999). Both CBS and NBC are simulcasting some prime-time programming and major sporting events in HDTV, including the NCAA Final Four basketball championship series aired on CBS. Just as televised sporting events such as boxing were instrumental to the diffusion of analog television, broadcasters are counting on sports mega-events such as the Super Bowl and the Final Four tournament to drive DTV set sales.

Viewers outside the top 30 markets who would like to see DTV prior to 2002 have another access option—direct broadcast satellite. Both DirecTV and EchoStar are now transmitting HDTV programming from Home Box Office (HBO) and the Discovery Channel to their 11 million subscribers (Bowser, 2000). HBO has been simulcasting HDTV versions of its films since 1998. DirecTV and EchoStar were the first providers of SDTV-quality digital programming to the nation, several years before the FCC terrestrial DTV standard was issued.

In summary, the U.S. conversion to DTV is on schedule, but only on the part of terrestrial and satellite broadcasters. The cable industry has been slow to respond for the reasons outlined above. DTV manufacturers have tried to jump-start DTV set sales by subsidizing program conversion in high-definition formats on ABC and CBS. Prices for wide-screen DTV sets that include a digital tuner dropped below $4,000 in early 2000. The COFDM/8-VSB indoor reception controversy has been resolved for the time being, but it could be a major issue if 8-VSB signal modulation does not work as promised for over-the-air reception. Japanese and European broadcasters are using variants of COFDM for terrestrial DTV broadcasting

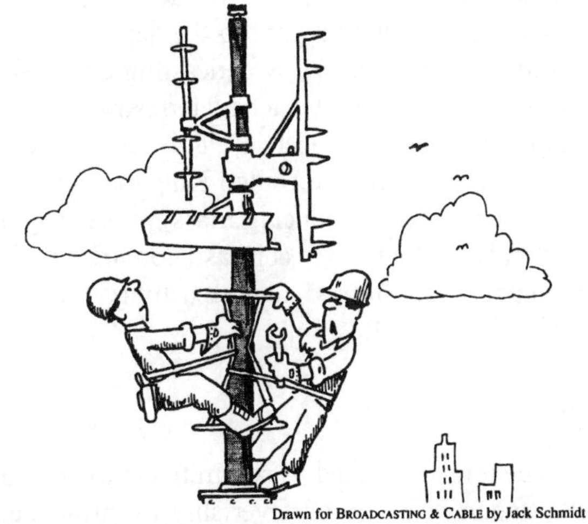

Drawn for BROADCASTING & CABLE by Jack Schmidt

"They're not sure of what to do with their new spectrum so they ordered an HDTV-SDTV-multiplexing combination antenna..."

Reprinted with permission.

with success, and their innovations could influence digital transmission technology in the United States.

Europe

Established in 1993 to spearhead digital specification efforts in Europe, the Digital Video Broadcasting (DVB) project has become an international organization with more than 220 members from over 30 countries (http://www.dvb.org). In the 1990s, this body drafted a series of digital standards for satellite, cable, multichannel multipoint distribution service, and terrestrial broadcasting that were subsequently approved by the European Telecommunications Standards Institute. Both the U.S. (ATSC DTV) and European (DVB-T) terrestrial television standards use MPEG-2 compression, but they differ from each other in terms of modulation/transmission technology (see Table 7.2). While the ATSC DTV standard relies on 8-VSB to transmit the digital signal, the DVB-T standard uses COFDM that is more resistant to multi-path problems (i.e., complex reception inside buildings). Both organizations are now competing for the world's acceptance of their digital systems, largely based on the merits of their transmission technologies. As of March 2000, at least 28 countries, including the 15 member states of the European Union, have opted for the DVB-T standard, against five for the ATSC DTV standard (United States, Canada, Taiwan, Singapore, and South Korea).

Since the market failure of MAC and HD-MAC (Dupagne & Seel, 1998), HDTV has all but disappeared from the European electronic media landscape and is unlikely to return anytime soon. Instead, over-the-air broadcasters and satellite operators have embraced the lower-resolution SDTV model to deliver a greater quantity of channel options to viewers. As Phil Lavin, European Broadcasting Union's Technical Director, summed up, broadcasters, viewers and manufacturers in Europe "have looked at the business case for high-definition television and they simply don't see it" (European broadcasters, 1999, p. 5). Planning and timing for converting analog platforms (cable, satellite, and terrestrial) to digital vary widely across and within European countries (Lange, 1999). For instance, France became the first country in Europe to offer a full package of digital channels via satellite with the introduction of *CANAL+*'s CSN in April 1996, but digital terrestrial television broadcasting will not be operational before the end of 2001. The most aggressive European country with regard to the deployment of DTV services has been the United Kingdom. In addition to British Sky Broadcasting's digital satellite service, both the British Broadcasting Corporation and ONdigital, a new terrestrial subscription television provider, have been broadcasting digital channels since late 1998. A recent Arthur Andersen report noted that, "In countries such as France, Spain, and the United Kingdom, digital technology is expected to flourish because the new platforms are introducing mass market pay TV services where they have only existed on a limited basis before.... By contrast, countries with rich free-to-air services such as Italy and Germany have seen limited digital activity, because consumers are already adequately served" (Turner, 1999, p. 70).

Japan

Since it reconsidered its commitment to analog MUSE broadcasting in 1994 (Dupagne & Seel, 1998), Japan has worked feverishly to introduce digital satellite and terrestrial broadcasting as soon as possible to catch up with its European and U.S. counterparts. The result has been the formulation of key digital policy initiatives in the last two years. Digital satellite broadcasting will begin on the second BS-4 satellite in December 2000 and will offer seven HDTV channels (or as many as 18 SDTV channels), three SDTV channels, and 23 audio services. While NHK will have the option of delivering one HDTV channel *and* two SDTV channels, commercial broadcasters will have to choose between transmitting one HDTV signal and three SDTV signals within their allotted spectrum (Targeting 10 million households, 1999).

In October 1998, the Advisory Committee on Digital Terrestrial Broadcasting of the Ministry of Posts and Telecommunications announced that digital terrestrial television broadcasting will debut experimentally in the Tokyo metropolitan area in 2000, followed by Osaka and Nagoya in 2003, and finally in the rest of the country by 2006. The technical features of the Japanese ISDB-T (Integrated Services Digital Broadcasting-Terrestrial) standard, a variant of the European DVB-T standard, was summarized in Table 7.2. Analog television broadcasting will cease in 2010, provided that 85% of Japanese households own the necessary equipment to receive digital signals and that digital service covers the same area as the present analog NTSC service (Schedule for introduction, 1999). This provision is not unlike the 1997 U.S. Balanced Budget Act clause mentioned earlier. The Japanese consumer electronics industry plans to introduce 40-inch flat-panel digital television receivers in 2003 at a retail price of ¥400,000 ($3,306) (New televisions, 1999).

Analog HDTV (Hi-Vision/MUSE) broadcasting will continue on the first BS-4 satellite at least until 2007 (Smooth transition, 1999). Since October 1997, NHK and commercial broadcasters have been transmitting 17 hours of HDTV programming a day. From 1996 to 1999, the top categories of HDTV programming were news/documentaries and sports, with a yearly average of 27.4% and 19.6% of total broadcast hours, respectively (see Table 7.4).

Table 7.4
Genres of HDTV Programming in Japan, 1996-1999 (in %)

Year	Sports	Music	Movies/ Dramas	News/ Docu.	Enter-tainmt.	Culture/ Education	Other	Total Broadcast Hours
1996	22.7	15.7	18.9	17.9	10.1	13.6	1.1	4,804
1997	19.4	16.2	18.2	23.8	6.5	14.2	1.7	5,346
1998	23.9	14.8	12.7	26.5	6.3	13.2	2.6	6,365
1999	12.3	13.3	12.9	41.4	6.8	11.7	1.7	6,182

Source: Hi-Vision Promotion Association

Factors To Watch

The global diffusion of DTV technology will evolve over the first decade of the 21st century. Led by the United States, Japan, and the nations of the European Union, digital television will gradually replace analog transmission in technologically advanced nations. It is reasonable to expect that many of these countries will have converted their television transmission standards to digital by 2010. In the process, DTV will influence what people watch, especially as it will offer easy access to the Internet and other forms of entertainment that are more interactive than traditional television.

The global development of digital television broadcasting is entering a crucial stage in the coming decade, as terrestrial and satellite DTV transmitters are turned on and consumers get their first look at HDTV and SDTV programs. Among the issues that are likely to emerge in 2000 and 2001 are the following:

COFDM. The FCC's denial of the Sinclair petition in February 2000 does not signify that COFDM is off the agenda of the Commission and the industry. On the contrary, the Commission will soon conduct field tests to ascertain indoor reception improvements of the latest generation of DTV receivers over the initial models. The FCC is also inviting comment on the status of the 8-VSB transmission technology as part of its first biennial review of the DTV conversion (FCC, 2000b). In a remarkable turnaround, the Advanced Television Systems Committee announced in March 2000 that it was dropping its opposition to revising the DTV *transmission* standard. ATSC Chairman Robert Graves acknowledged that there were significant reception problems with 8-VSB technology, stating that "[t]he problems created by multipath signals are more complex than people imagined when the system was first designed" (McConnell, 2000b, p. 10). An ATSC taskforce will examine the alleged shortcomings of the present transmission system and recommend any changes that should be made. Any significant revisions in the standard (such as the inclusion of a COFDM transmission option) might delay the implementation of digital television broadcasting in the United States by two or three years (FCC, 2000a).

Receiver Prices. How much will prices of DTV sets have to drop before stimulating consumer demand beyond affluent innovators and early adopters? As the Japanese experiment with Hi-Vision demonstrated, high retail prices can thwart the successful diffusion of a new technology despite software availability and consumer interest (Dupagne, in press; Dupagne & Seel, 1998). After more than 10 years of diffusion, sales of HDTV sets in Japan still only numbered 831,000 by January 2000 (http://www.hpa.or.jp). Although prices have fallen since U.S. DTV receivers were first introduced in August 1998, they still remain beyond the financial reach of most of the U.S. population. In January 2000, the cheapest fully-integrated 30-inch HDTV set cost $3,500 (Kerschbaumer, 2000).

Large-Screen Televisions. The advent of DTV broadcasting has not yet affected the sale of large-screen analog sets. Sales of analog large-screen projection sets increased 16.7% in 1998 to cross the million-unit mark for the first time (CEMA, 1999). In fact, the increasing popularity of larger television displays (with their inherently higher costs) may actually smooth the transition to large-screen digital models. The premium the consumer must pay for a DTV model will be less of a disincentive for someone who can afford $2,000 for an analog projection model, compared with $3,000 for a DTV set of similar size. The technological pendulum will increasingly swing toward the DTV side, and sales of large-screen analog sets will decline rapidly in the coming decade.

Compatibility Standards. Even after two years of debate, the cable and consumer electronics industries have yet to finalize the technical specifications for connecting DTV receivers to digital cable set-top boxes. In November 1998, the two industries approved the use of the "Firewire" (IEEE 1394) connector to link digital receivers to digital cable boxes, but they failed to reach an agreement on the issue of copyright protection. Sixteen months elapsed without significant progress. Perturbed by these delays, in January 2000, FCC Chairman William Kennard warned industry executives that he would urge the Commission to adopt compatibility rules if an agreement was not reached by April (Beacham, 2000). As of this writing, the cable and consumer electronics industries have approved a copy protection mechanism called 5C, which would embed a "marker" authorizing or preventing copies of digital content (Brinkley, 2000). However, the major Hollywood studios have yet to endorse this system.

DTV Must-Carry Rule Making: The FCC's decision on DTV must-carry rules by U.S. cable television operators will be a crucial factor in the diffusion of digital broadcast technology. Even if 100% of American commercial broadcast stations are transmitting in DTV by May 1, 2002 (as the FCC requires), it will be a moot issue if the 67 million cable households in the United States cannot receive the digital signal. This impasse must be resolved for DTV to succeed in such a wired nation.

In the coming decade, the merger of the television and the computer will create interesting programming options for viewers. Television programming has the potential to offer viewer interactivity that has been just a pipe dream for the past two decades. DTV telecomputers with high-bandwidth connections to the Internet will have two-way transmission capabilities (Pavlik, 2000). Viewers will have the choice of passively watching what is being telecast, or they may opt to actively participate in the program through a chat room in a program-related Website. Pundits have been predicting the arrival of interactive television for over two decades—perhaps, it is finally here.

Bibliography

Advisory Committee on Advanced Television Service. (1995). *Advisory Committee final report and recommendation.* Washington, DC: ACATS.

Ashworth, S. (1998, March 23). Finding funds for the transition. *TV Technology, 10*, 12.

Balanced Budget Act of 1997, Pub. L. No. 105-33, § 3003, 111 Stat. 251, 265 (1997).

Beacham, F. (2000, February 9). Kennard warns of FCC action on TV/cable standards. *TV Technology, 12*, 22.

Bowser, A. (2000, February 21). Seller's guide: HBO. *Broadcasting & Cable*, 50.

Brinkley, J. (1997). *Defining vision: The battle for the future of television.* New York: Harcourt Brace.

Brinkley, J. (1998, January 12). They're big. They're expensive. They're the first high-definition TV sets. *New York Times,* C3.

Brinkley, J. (2000, February 28). The sound of one foot dragging, again. *New York Times,* p. C9.

Consumer Electronics Manufacturers Association. (1999). *1999 U.S. consumer electronics industry today.* Arlington, VA: CEMA.

Dickson, G. (1999, August 30). HDTV rolling at CBS, ABC. *Broadcasting & Cable,* 10.

Dupagne, M., & Seel, P. B. (1998). *High-definition television: A global perspective.* Ames, IA: Iowa State University Press.

Dupagne, M. (in press). Adoption of high-definition television in the United States: An Edsel in the making? In C. A. Lin & D. J. Atkin (Eds.), *Communication technology and society: Audience adoption and uses.* Cresskill, NJ: Hampton.

European broadcasters continue wait on HDTV rollouts. (1999, October 18). *Communications Daily,* 5.

Fedele, J. (1987, September 25). DTV schedule breeds apprehension. *TV Technology*, 16.

Federal Communications Commission. (1987). *Formation of Advisory Committee on Advanced Television Service and announcement of first meeting*, 52 Fed. Reg. 38523.

Federal Communications Commission. (1990). Advanced television systems and their impact on the existing television broadcast service. *First Report and Order*, 5 FCC Rcd., 5627.

Federal Communications Commission. (1992). Advanced television systems and their impact on the existing television broadcast service. *Second Report and Order/Further Notice of Proposed Rule Making*, 7 FCC Rcd., 3340.

Federal Communications Commission. (1996a). Advanced television systems and their impact upon the existing television broadcast service. *Fifth Further Notice of Proposed Rule Making*, 11 FCC Rcd., 6235.

Federal Communications Commission. (1996b). Advanced television systems and their impact upon the existing television broadcast service. *Fourth Report and Order*, 11 FCC Rcd., 17771.

Federal Communications Commission. (1997a). Advanced television systems and their impact upon the existing television broadcast service. *Fifth Report and Order*, 12 FCC Rcd., 12809.

Federal Communications Commission. (1997b). Advanced television systems and their impact upon the existing television broadcast service. *Sixth Report and Order*, 12 FCC Rcd., 14588.

Federal Communications Commission. (1998a). Advanced television systems and their impact upon the existing television broadcast service. *Memorandum Opinion and Order on Reconsideration of the Sixth Report and Order*, 13 FCC Rcd., 7418.

Federal Communications Commission. (1998b). Carriage of the transmissions of digital television broadcast stations. *Notice of Proposed Rule Making*, 13 FCC Rcd., 15092.

Federal Communications Commission. (2000a, March 3). *Letter denying petition for expedited rule making.* [Online]. Available: http://www.fcc.gov/Bureaus/Engineering_Technology/News_Releases/2000/nret002a.txt.

Federal Communications Commission. (2000b, March 6). Review of the Commission's rules and policies affecting the conversion to digital television. *Notice of Proposed Rule Making*, MM Docket No. 00-39.

Kerschbaumer, K. (2000, January 10). In search of DTV's magic price point. *Broadcasting & Cable,* 52, 54.

Kerschbaumer, K., & McConnell, B. (2000, March 20). LPTV station bumped by DTV. *Broadcasting & Cable,* 44.

Lange, A. (1999). *Developments in digital television in the European Union.* Strasbourg, France: European Audiovisual Observatory.

McConnell, B. (1999, October 11). Sinclair hurls TV gauntlet. *Broadcasting & Cable,* 19-20.

McConnell, B. (2000a, February 28). NAB blasts cable-DTV deal. *Broadcasting & Cable,* 11.

McConnell, B. (2000b, February 28). Smith: Death of 8-VSB is only a matter of time. *Broadcasting & Cable,* 45-46.

New televisions hinge on timing, demand. (1999, July 5). *The Nikkei Weekly*, 1.

Pavlik, J. V. (2000). TV on the Internet: Dawn of a new era? *Television Quarterly, 30* (3), 31-47.

Schedule for introduction of digital terrestrial broadcasting. (1999, February). *NHK Update,* 3.

Smooth transition to BS digital HDTV broadcasting. (1999, August 2). *MPT News,* 1-2.

Targeting 10 million households in 1,000 days. (1999, Autumn). *Broadcasting Culture & Research,* 2.

Telecommunications Act of 1996, Pub. L. No. 104-104, § 336, 110 Stat. 56, 107 (1996).

Turner, M. (1999, January 5). Euro digital "revolution" will take time: Report sees slow rollout for services. *Hollywood Reporter,* 70.

Turner Broadcasting System, Inc. v. F.C.C., 117 S. Ct. 1174 (1997).

8

Streaming Media

Jeffrey S. Wilkinson, Ph.D.*

The technology of "streaming" audio and video signals over the Internet is one of the most exciting multimedia developments at the beginning of the 21st century. Streaming enables an Internet user to take advantage of audio on demand or video on demand capabilities. Streaming is heavily dependent on bandwidth, so, for the time being, we still have to think of it as "low-definition television," although audio streaming can give an experience closer to "the real thing." That perception will change when the necessary improvements have been made in the Internet infrastructure. Even as "low-D TV," tens of millions of users are listening to or watching content provided by countless Websites and portals around the world every day. This chapter will try to capture the essential information about streaming technology, knowing that, by this time tomorrow, there will be another new application, another improvement or upgrade, and another legal and ethical issue that has to be considered.

What Is Streaming?

Streaming allows you to listen to audio or watch video in "real time" on your computer. In a more technical description, Kaye and Medoff (2000) write, "When a signal is streamed, [the computer] starts decoding a signal as it is received and plays it almost immediately. The computer continues to receive the signal, playing it shortly after reception, until the file has ended" (p. 98). An excellent demonstration video by RealNetworks on streaming basics can be found at http://www.IDG.net/.

The main advantages of streaming are speed, control, and flexibility. Playing streamed material is almost instantaneous. Control is maximized because the content always remains on the server—it is not downloaded onto a user's hard drive. Streaming is also extremely flexible. Streamed segments are typically individually placed on a host video server, and may be changed or linked to

* Associate Professor, Hong Kong Baptist University/University of Tennessee-Knoxville (Knoxville, Tennessee).

other clips in any order or fashion the host wishes—without having to make entire new versions or re-releases (a problem with updating CD-ROMs).

Applications

There are primarily two types of streaming. One is the streaming of a "live" broadcast signal, and the other is playing a stored audio or video file "on demand." The first type of streaming is a feature now offered by hundreds of radio stations around the world. A visit to http://www.broadcast.com provides one of the best (but always incomplete) listings of what's available.

The second type of streaming is the application of audio or video on demand. Content (programming) is offered to the user who can simply "click" on the link, launch the player, and watch the program. The number of sites offering this type of streaming is growing every day, and there are currently thousands of hours and countless numbers of audio and video segments available.

The Bandwidth Issue

Bandwidth is a measure of how much data can pass through a network connection per second, and it is affected by network traffic, connection speed, and the size of the file being streamed. A 28.8 Kb/s (kilobits per second, or thousands of bits per second) connection best plays a presentation streamed at the same rate. It is possible for the connection to play a segment streamed at a higher rate (such as 56 Kb/s), but there will be a lot of stopping or stalling because the player is running at twice the speed of the download.

To enable smooth streaming, technicians came up with the idea of creating a *buffer*. The few seconds taken before the content appears on the screen allows seamless, uninterrupted play. This buffering is a key component to streaming and makes it not actually audio or video on demand, but rather near video on demand (NVOD). According to literature from RealNetworks, people generally won't wait for buffering that takes longer than 15 seconds (ideally, it should be under 10 seconds) (RealNetworks, 1998-1999).

At this time, the tradeoff is time over quality. Downloading video files results in higher quality, but that can take a very long time, and the user is left with a large file taking up space on the hard drive. Evidence from millions of users suggests that quality can be sacrificed somewhat to save time. The great unknown at this time is what "quality" for streaming is most acceptable given physical bandwidth/connection speed limitations. For example, normal "full-motion" NTSC video is 30 frames per second (fps). But when applied to streaming video, 15 fps has been considered full motion for lower bandwidth speeds.

Required Components

There are five components necessary for streaming to take place. First, *content must be produced*. This obvious but underappreciated consideration should remain central to the concept of streaming. There still needs to be something interesting to say or show. Think of the Zapruder film of the Kennedy assassination versus the latest Hollywood big-budget flop. So recording, editing, and digitizing the audio and/or video remain primary considerations.

Figure 8.1
Streaming Media Diagram

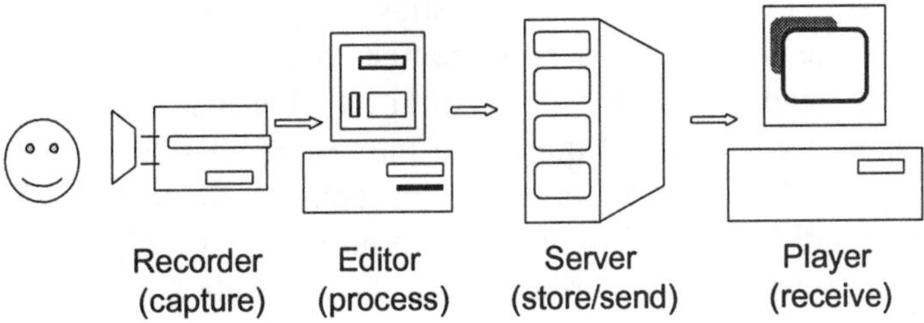

Source: J. S. Wilkinson

Second, *the content must be encoded and stored on a server*. Any video, audio, animation, text, graphic, or other content that has been produced has to be digitized and converted with an encoder that compresses the files and turns them into streaming files. The higher the quality of the source material, the better the resulting streamed images. During the encoding, decisions in terms of optimal rates/space allocated for video and audio need to be made. For example, a musical performance may be encoded so two-thirds of the bandwidth is allocated to the audio portion. A film trailer might use most of the space for the visuals. Knowing whom the audience is and what type of connection they tend to have is crucial.

Third, *special streaming software on the video server* reads the encoded file and sequentially delivers it a little at a time through the Internet connection to the user's computer. Most streaming software is capable of delivering many different streams from many different files at the same time; a typical server can deliver between 10 and tens of thousands of streams at the same time, depending upon the speed of the computer and the bandwidth of the stream.

The fourth component needed is the *Internet connection itself*. Both host and user need to be connected to the Internet at comparable rates. High-quality streamed images demand high bandwidth; the user also needs enough bandwidth to view the streamed content. At present, 80 Kb/s to 150 Kb/s streams result in fairly good images—good enough to argue it looks like low-quality TV. Connections this fast are available in offices with high-speed Internet connections (using T1 lines [1.5 Mb/s] or better) or in homes with cable modems or DSL (digital subscriber line). But the fastest connections over regular telephone lines (56 Kb/s modem) can only handle bit rates of 45 Kb/s and less (20 Kb/s streams are recommended for 28.8 modems). At these lower rates, the streamed video looks more like a sloppy slideshow. Of course, the reader should remember that early television was also criticized, and streaming technology is improving rapidly.

The last component needed is a compatible player. Taking the example from TV, virtually all the dominant streaming formats provide free players. The user has to go to an approved site, download the program, and install it on the user-computer's hard drive. These are generally not large programs, and the steps are quite simple. Any computer can house several types of media players, and we're only now on the verge of interoperability (where one player can play another format).

Background

In 1995, RealNetworks (known then as Progressive Networks) first offered free players to users (Heid, 1998). Taking their cue from broadcasting, they relinquished control over the sets by offering the player as a free download. Users of the first application, RealAudio, were able to take advantage of limited bandwidth and modem speeds (14.4 Kb/s and 28.8 Kb/s were the standards) to offer music and speech of sufficient quality to interest radio stations. It wasn't long before streaming video was also offered, and global interest took off. Since 1995, literally hundreds of radio stations in the United States and overseas have jumped in to offer some form of audio on demand, either through clips of music or streaming their signals "live." The novelty is rapidly becoming an expectation; some recent research (McClung, 1999) found that, among users of college radio sites, streaming was rated as one of the more important components.

Streaming Platforms

There are three primary platforms (companies) providing and playing streamed content: RealPlayer (RealNetworks), Windows Media Player (Microsoft), and QuickTime (Apple). Each has approached streaming from a somewhat different perspective, with somewhat different results. There are merits to each of the platforms, and all three agree in principle on the issue of compatibility. Areas of conflict stem primarily from business concerns, which have helped to make their technological differences seem greater. Further, the fledgling world of streaming is still in its infancy, and undoubtedly there will be other companies and formats that will be tried and tested on the world stage.

RealNetworks/RealPlayer. The early dominance of RealNetworks has been attributed to it being the first on the market, and having the foresight to give away the players rather than sell them. Since 1995, several generations of improvements have been made in its products, resulting in tens of millions of players installed, and countless numbers and hours of streamed segments becoming available on demand. In 1999 alone, 150,000 free RealServers (capable of delivering up to 25 streams each) were downloaded.

As with all the formats, the RealNetworks free player is available (but can be difficult to find). Most Website space is given to advertising the deluxe models. The most recent model as of this writing, the RealPlayer 7 Plus, can be fully-supported for $29.99. The pricing structure has been pretty stable for at least three years.

One of the innovations introduced by RealNetworks is "Surestream" encoding that encodes a stream at many different bit rates at the same time. When a Surestream clip is played, the server attempts to deliver the best quality stream to the player. The player then determines if it is capable of receiving all of the data being sent; if not, it sends a signal to the server telling it to deliver a lower-quality stream. This process continues until the bit rate of the clip is equal to the speed of the connection.

For businesses interested in servers, there have been noticeable improvements and upgrades in the technology. In January 1997, a 20-stream RealAudio server software license was priced at around $2,700 for businesses ($2,170 for educators). Three years later (January 2000), in addition to the free RealServer, a 60-stream server license (including technical support) was listed at $1,995. RealNetworks prices its server software based on number of streams and other factors. For example, the top-of-the-line "RealServer Professional" is priced from $5,995 (100 concurrent streams) to $79,995 (2,000 concurrent streams). The RealNetworks Website also conveniently offers help to ascertain hardware needs. These are useful in conveying server size/requirements.

Memory requirements—For each 20 Kb/s stream (sufficient for audio, but not for video), 240 KB RAM is required. For each video stream at 100 Kb/s, 1,200 KB RAM is needed. With multiple streams, there are some additional memory requirements, so 100 video streams (80 Kb/s) demands 160 MB RAM, and 2,000 audio streams (at 20 Kb/s) needs 544 MB of available memory.

Bandwidth requirements—Clients also have to make sure they have the infrastructure to deliver the content. To provide 100 video streams (80 Kb/s), a business needs to be able to provide 8 Mb/s; 2,000 audio streams (20 Kb/s) similarly demands a T3 (45 Mb/s) connection.

Storage requirements—Storing the content is also an important issue. A three-minute clip encoded with SureStream takes up 900 KB of hard disk space. The same clip encoded at a single rate—say 20 Kb/s—only takes up 450 KB of space. To stream entire libraries of film...well, you get the picture.

Microsoft/Windows Media Player. Shortly after RealNetworks began streaming audio, Microsoft became interested and, for a time in the late 1990s, the two were partners. Shortly after the partnership, Windows Media Player was developed and offered as a free plug-in to users. Since then, Windows Media Player has also been released in stages with various improvements. Microsoft has also continued supporting and investing in WebTV, and seems to be gaining an upper hand by virtue of offering a decent player in addition to being self-compatible (OS, browser, media player, and various tools).

In a keynote address at the Streaming Media West conference in San Francisco (December 7, 1999), Bill Gates and company demonstrated some of the new features for Media Player and Windows 2000. The audience was treated to the prospect of easily skimming through a video stream at various speeds. Stripped-down video editors for the home user are also now being offered, making the idea of "ubiquitous streaming" a possibility for the average user. Other features include movie making, USB connectivity, and voice capability—where the user could tell the computer which multimedia streaming applications to get.

Streaming is just one of the many things Microsoft does, and it's easy to get lost in the vast array of products when visiting the Website of the computer giant. In terms of services and applications, Microsoft does not provide prices to guide you. Also, unlike RealNetworks, you can't simply look up the price of a server, because that's just one part of the larger whole. For example, instead of just listing servers, Microsoft lists the types of opportunities that businesses have to use their services. One such service is Microsoft's Digital Broadcast Manager application.

The literature provided simply states, "With Microsoft Digital Broadcast Manager [DBM] for Windows Media, you can manage the process of generating revenue from your content. Digital Broadcast Manager includes tools that help you create revenue models based on the downloading and streaming of your content." The DBM enables businesses to securely and reliably charge consumers to watch content—"to conduct commerce transactions based on access control methods, including pay-per-view (e.g., pay-per-download or pay-per-stream), registration, subscription, and digital rights management through Microsoft Windows Media Rights Manager." More details regarding price, support, and other conditions are naturally provided, with follow-up communication from interested clients. Microsoft's Digital Broadcast Manager is just one example of a digital rights management (DRM) system that helps keep track of royalties due from distribution of content over the Internet (Microsoft, 2000).

Apple/QuickTime-4. In early 1999, Apple began offering streaming with its QuickTime 4 platform. A few months later, it established a partnership with U.S.-based Akamai Technologies (www.akamai.com) to offer streaming content. Akamai would provide the central storage place for the content, and Apple could concentrate on the streaming.

Apple has an open-source code approach where they allow vendors to configure the QuickTime format for their use, then send the changes back to Apple for incorporation into new versions of the server software. The official news release at the time touted Apple's QuickTime 4 as "the first Internet streaming solution to use non-proprietary industry-standard protocols using Real Time Protocol (RTP) and Real Time Streaming Protocol (RTSP)." Indeed, the QuickTime format (MPEG-4) is highly rated at high streaming rates. The problem with QT4 at this time, however, is some reported instability at the lower rates typically available to consumers. At 120 Kb/s or higher rates, QuickTime 4 provides excellent video quality. Unfortunately, at 14.4, 28.8, and 33.6 Kb/s, a user may experience problems even getting it to run.

One streaming developer has cautioned about the problems in trying to stream across platforms. Streaming QuickTime on PCs can result in some problems, both in installation and operation. Similar problems have been encountered when running a PC-based client on Apple hardware. As for startup costs, the assessment was that there was not much difference, even though RealNetworks has a per stream charge. This was because large numbers of simultaneous streams were not possible without substantial hardware investment. There was also the admission that the quality of QuickTime was very impressive.

Although RealNetworks, Microsoft, and Apple lead the market in streaming technology, many other companies are offering similar tools. Some of these are compatible with one or more of the three discussed above, while others are not. Each company promoting its own player and server software has the same goal—gaining enough market share to become the industry standard.

Recent Developments

Standardization

There have been a number of recent developments that are indicative of the interest and activity in streaming. In 1999, Apple began streaming for the first time. Since then, both Apple and RealNetworks adopted RTSP (Real-Time Streaming Protocol) as a standard for streaming rather than hyptertext transfer protocol (HTTP), enabling improvements in video quality and synchronizing audio and video signals. Both companies (and many academic institutions) advocate MPEG-4 as something of a standard for Internet video.

Multiple Streamed Rates

In 1998, RealNetworks launched its SureStream technology, where each clip has multiple embedded data rates. SureStream works in such a way as to deliver at the optimal rate available to the user's computer. In August 1999, the updated version of Windows Media Player offered Intelligent Streaming that essentially does the same thing—embedding multiple streaming rates for a single clip to ensure an uninterrupted signal for the user.

Information, Entertainment, Education

Media firms and information companies are continuing to migrate to the Internet and take advantage of streaming capabilities. The Web presence of radio, television, and film companies will continue to be fairly high profile. But the lure of "Internet television" is bringing together a range of other businesses as well. For example, Yahoo! announced in March 2000 that it would offer a live-stream Financial News Network. Working in conjunction with MarketWatch, Forbes.com, TheStreet.com, and *Business Week*, Yahoo! plans to offer eight hours of financial news and analysis each day on FinanceVision (http://financevision.yahoo.com), an interactive financial news network broadcast live in streaming video.

New forms of entertainment are being explored for the Internet involving this technology. In October 1999, Hollywood filmmakers Steven Spielberg and Ron Howard announced that their host companies (DreamWorks SKG and Imagine) would be experimenting with new forms of storytelling designed specifically for the Internet. According to the official news release (http://www.pop.com), the content will be "a mix of live action and animation, video on demand and live Web events, and nonlinear interactive features and games. Most features will consist of one- to six-minute episodic streaming video segments, or 'pops,' with an emphasis on comedy, although all genres of entertainment will be represented." In addition, there are a number of sites dedicated to streaming independent short films and animation. Indeed, video streaming offers producers distribution for their projects, which often do not receive traditional distribution opportunities.

New ways of delivering educational materials are also being adapted to streaming. As mentioned before, streaming depends on bandwidth. This consideration is important for the content provider as well as the user. Since normal copper telephone lines (without DSL) cannot physically squeeze the large video signals through them, the content provider cannot encode video at higher rates because users are not able to watch it.

Using streaming media, course lectures can be streamed in real time or stored for on-demand access. Photo courtesy of Professor Jeffrey S. Wilkinson, Hong Kong Baptist University.

At many universities, enough bandwidth is available for experiments to take place with streaming video as educational aids. Thanks to improvements in CODECs (compression/decompression), what used to require a T3 connection (45 Mb/s) is now possible with a T1 (1.54 Mb/s). This allows a university to offer video lectures and course aids at higher rates (80 Kb/s to 150 Kb/s). For example, this author and others are experimenting with streaming lecture material on campus video servers for students to use (note that this uses HTTP rather than RTSP) (see http://hkbulib.hkbu.edu.hk/search/rcomm+2320/1,1,1,B/frameset~1532706&rcomm+2320&1,,0).

Current Status

The battle for platform dominance looks like a repeat of the browser wars, where the smaller company that pioneered the application was overtaken by the larger one. RealNetworks has been the dominant player/platform since 1995, and recent figures note there are over 95 million registered players. In December 1999, however, Media Metrix' SoftUsage Report showed that—for the first time—Windows Media Player was used more than any other multimedia player among U.S. PC households (http://www.microsoft.com/presspass/features/2000/03-14rn.asp). Many of the most popular sites streaming content (such as http://www.broadcast.com) allow users either of the two PC formats. Although Apple has yet to pull in the same numbers (remember, they didn't start streaming QuickTime until 1999), they have an inherent strength in the MPEG-4/QuickTime 4 link.

Coinciding with the release of these figures, a licensing agreement was also announced between RealNetworks and Microsoft, where RealNetworks products would now be compatible with Windows Media Audio technology (Alvear, 2000). This development was viewed as a huge step forward for Windows Media to become the industry's universal format for audio (if not all digital) media. Under this agreement, Windows Media audio will be playable on RealNetworks' software (such as RealJukebox). The biggest benefit for RealNetworks was that Windows Media's secure music format is used on many other software systems and is accepted by some major labels (Alvear, 2000). This type of format also has smaller file sizes than MP3 with the same quality. Industry concerns over pirating was seen as a major factor in the move away from MP3 and toward the Microsoft platform.

Factors to Watch

Compatibility

At the 1999 Streaming Media West conference in San Francisco, a panel with representatives from each of the three platforms agreed that compatibility was an important issue. Currently, each player has to be separately configured, and each server has to provide the same clip in each of the formats. Those in the audience confirmed this was time consuming and inefficient.

As streaming begins to mature, all three format-makers have been stressing "interoperability" and "compatibility." RealNetworks and Apple have touted their adherence to RTSP and RTP standards, but Microsoft has been adamant to its position that it is still "too soon for standards." In light of its increasing market share, this seems like a wise course.

On the academic side, computer engineering/network service units at various universities, for example, have been clamoring for a standard for delivering high-quality video over the Internet. These tend to be the people also working on developing Internet 2, and they argue against giving this lucrative territory to a single commercial business (that they differ with on technical matters anyway). A more favored approach seems to be the open-source code route taken by Apple and QuickTime 4.

Issues related to CODEC, file format, and protocols are unresolved and continue to separate the three platforms. RTSP and RTP standards are viewed by many as a way for all three to come together. As of December 1999, Microsoft was not interested in adopting RTSP/RTP as a standard because it was premature. Instead, its stated intention is to seek the best means of providing the best video to the user.

Ownership and Control of Platform/Content

The factor to watch is whether Microsoft will become the de facto standard for streaming media. Microsoft has so far resisted adopting RTSP and RTP. As Abrahamson (1998) noted, the actions of this computer giant have far-reaching implications for the future development of the Internet. Microsoft's Digital Broadcast Manager seeks to empower businesses and enable them (and Microsoft) to exact revenues by a pay-per-click (PPC) model. Microsoft has championed this approach for a number of years. But rampant piracy due to illegal MP3 downloads and controversy over sites that distribute copyrighted programs without the permission of the copyright holder are changing public attitudes over protecting the rightful owners of content. Whether the PPC model for streaming becomes widely accepted remains to be seen.

Ownership and Control of Infrastructure

Upgrading the Internet infrastructure is seen as necessary for increasing bandwidth, but it is slow in coming. Traditional telephone lines must be replaced/upgraded to ensure high-speed connectivity. The battle for control of the "one line into the home" has been broadly painted as a contest between telephone companies and cable operators. Telephone companies propose DSL technology, and cable companies offer cable modems. Although cable modems have had an early lead, a recent study suggests this may change in a couple of years (Grice, 2000). As more households get wired for high-speed connectivity, streaming services will be in greater demand.

Player Speeds and Screen Size

From the user's perspective, this is a huge factor. Video signals streamed for 56K modems and under simply do not look very good. This limitation is continually being worked on and will change—but slowly. Both Microsoft and RealNetworks admit that 300 Kb/s and higher is needed to offer "a true VHS experience."

In a related issue, high-resolution, full-screen, full-motion video is still a consumer dream, but everyone agrees that making the computer "like television" is the goal. Expect to see more improvements and higher streaming rates available through DSL and/or cable modems.

Of course, the ultimate playback device for such high-quality streams is a television receiver. Advances in home networking and digital television will soon allow the user a choice of whether to watch programming on a computer monitor or television. At that point, Internet television will become a true competitor to broadcast television.

Congestion

Multicasting (sending one signal to many users versus dedicated single streams) is being hailed at this time as a way to ease immediate network congestion caused by streaming. As demand increases and more people want to make video available, servers can become filled up quickly. Also, it is not known how severely the increased traffic in streaming video will stress the system. Several campuses have had to look at curtailing non-academic functions such as ICQ and downloading MP3 files (http://cnn.com/2000/TECH/computing/03/01/napster.ban).

This issue must be kept in perspective. Very few cities have banned the sale of cars because the roads are full. In other words, net congestion is a sign of its popularity and an indicator of demand. In addition, widening our Internet paths is easier and less damaging to the environment.

Bibliography

Abrahamson, D. (1998). The visible hand: Money, markets, and media evolution. *Journalism & Mass Communication Quarterly, 75* (1), 14-18.

Alvear, J. (March 14, 2000). *RealNetworks licenses Microsoft's Windows media audio technology.* [Online]. Available: http://www.streamingmedia.com/news.asp?/news.

Apple Computer. (August 18, 1999). *Apple and Akamai reveal Apple investment to cement strategic agreement.* [Online]. Available: Apple.com/pr/library/1999/aug/18appleakamai.html.

CNN. (March 1, 2000). *Campuses seek compromise over popular bandwidth hog.* [Online]. Available: http://cnn.com/2000/TECH/computing/03/01/napster.ban/.

DreamWorks SKG. (1999). *DreamWorks and Imagine go digital.* [Online]. Available: http://www.pop.com.

Gates, B. (December, 7, 1999). *Keynote address.* Streaming Media West Conference, San Francisco. [Online]. Available: http://www.streamingmedia.com.

Grice, C. (March 1, 2000). *DSL could pull ahead in high-speed race.* [Online]. Available: http://news.cnet.com/news/0-1004-202-1561720.htm/.

Heid, J. (1998), February). "Making waves with streaming audio." *Macworld,* 125-127.

Kaye, B., & Medoff, N. (1999). *The World Wide Web: A mass communication perspective.* Mountain View, CA: Mayfield.

McClung, S. R. (1999). *Uses and gratifications of college radio station Websites: An exploratory study.* Unpublished doctoral dissertation. Knoxville, TN: University of Tennessee.

Microsoft. (March 14, 2000). *RealNetworks licenses Windows Media.* [Online]. Available: http://www.microsoft.com/presspass/features/2000/03-14rn.asp.

RealNetworks, Inc. (1998-1999). *RealSystem G2 release 6.1 production guide.* [Online]. Available: http://www.realnetworks.com.

RealNetworks, Inc. (2000a). *About RealNetworks.* [Online]. Available: http://www.realnetworks.com/company/index.htm/.

RealNetworks, Inc. (2000b). *RealServer 7.0 System Requirements.* [Online]. Available: http://www.realnetworks.com/products/server/.

Wilkinson, J. (2000). *Lecture materials for Communication 2320.* [Online]. Available: http://hkbulib.hkbu.edu.hk/search/rcomm+2320/1,1,1,B/frameset~1532706&rcomm+2320&1,,0.

9

Radio Broadcasting

Gregory Pitts, Ph.D.*

Consumer demand for improved audio fidelity is undeniable. Access to superior digital audio technologies, such as compact discs and—in the near future—satellite digital audio radio service, and the perceived benefits of digitization, generally, fuel such demand. We believe that an important benefit of digital audio broadcasting (DAB) will be enhanced sound quality. DAB technology should permit significant improvements in audio fidelity and robustness over current analog service (FCC, 1999, p. 10).

The switch to digital will reaffirm radio as an essential device of the information society.... [Radio] is perhaps the most underestimated mass medium of the digital age: 90% of the industrial world listen to radio every day (World DAB, 2000b).

The word "radio" fails to deliver the image of a sexy technology capable of competing with the latest digital recording and playback for home audio or video, or to match the innovations of the Internet or broadband or satellite systems capable of delivering billions of data bits. Public attention is not focused on the present analog radio transmission system or on radio's digital future. Technological developments in radio have been described as being mundane and primarily focused on enhancing the current AM/FM radio system.

Although the radio broadcast industry has not abandoned its low-tech approach, a number of factors are starting to redirect the technological path of radio broadcasting:

- Radio station ownership consolidation.
- Improved computer automation of stations and delivery of programming through digital networks.

* Assistant Professor, Electronic Media, Southern Methodist University (Dallas, Texas).

- Pending competition from two digital satellite radio services.
- Technological improvements likely to result in a viable digital terrestrial broadcast system.

Background

The history of radio is rooted in the earliest wired communication, the telegraph and the telephone, although no single person can be credited with inventing radio (Lewis, 1991). The technology may seem mundane today, however, until radio, it was impossible to simultaneously transmit entertainment or information to millions of people. The radio experimenters of 1900 or 1910 were as enthused about their technology as are the employees of the latest Internet-related start-up. Today, the Internet allows us to travel around the world without leaving our seats. For the listener in 1920, 1930, or 1940, radio was the only way to hear live reports from around the world.

Probably the most widely known radio inventor-innovator was Italian Guglielmo Marconi, who recognized its commercial value and improved the operation of early wireless equipment. Another individual who made lasting contributions to radio and electronics technology was Edwin Howard Armstrong.

- He discovered regeneration, the principle behind signal amplification.
- He invented the superheterodyne tuner, leading to a high-performance receiver that could be sold at a moderate price, increasing home penetration of radios.
- In 1933, he was awarded five patents for frequency modulation (FM) technology.

Amplitude modulation (AM) and frequency modulation (FM) provide listeners with information and entertainment. AM varies (modulates) *signal* strength (amplitude), and FM varies the *frequency* of the signal.

The oldest commercial radio station began broadcasting in AM in 1920. Though AM technology had the ability to broadcast over a wide coverage area (an important factor when the number of licensed stations was just a few dozen), the AM signal was of low fidelity and subject to electrical interference. FM, which provides superior sound, is of limited range. Commercial FM took nearly 20 years from the first Amstrong patents in 1933 to begin significant service and did not reach listener parity with AM until 1978 when FM listenership finally exceeded AM listenership.

FM radio's technological add-on of stereo broadcasting, authorized by the Federal Communications Commission (FCC) in 1961, along with the FCC's 1964 mandate to end program simulcasting (airing the same programming on both AM and FM stations), expanded FM listenership (Sterling & Kittross, 1990). Other attempts, such as the 1970s Quad-FM (quadraphonic sound) ended with disappointing results. AM stereo, touted in the early 1980s as the savior in the competitive battle with FM, languished for lack of a technical standard, resulting from the inability of the marketplace and government to quickly adopt an AM stereo system (FCC, undated; Klopfenstein & Sedman, 1990).

Why haven't technological improvements in radio been forthcoming? The obvious answer perhaps is that the marketplace does not want the improvements. Station owners have been unwilling to invest in changes; instead, they shifted music programming from the AM band to the FM band. AM became the home of low-cost talk programming. Listeners perceived present-day AM and FM commercially supported and noncommercial radio programming as satisfying their needs. Government regulators, primarily the FCC, were unable to devise and institute new radio technology approaches. The consumer electronics industry was focused on other technological opportunities, including video recording and computer technology.

The Changing Radio Marketplace

Fueling many of the changes taking place in radio broadcasting has been the FCC elimination of ownership caps. Before ownership limits were eliminated, there were few incentives for broadcasters, equipment manufacturers, or consumer electronics manufacturers to upgrade the technology. Analog radio, within the technical limits of a system developed more than 80 years ago, worked just fine. Station groups did not have the market force to push technological initiatives. (At one time, station groups were limited to seven stations of each service. Later, it was increased to 18 stations of each service, before further deregulation completely removed ownership limits.) The fractured station ownership system ensured station owner opposition to FCC initiatives or manufacturer efforts to pursue technological innovation.

The removal of radio ownership limits mandated by the Telecommunications Act of 1996 rapidly led to ownership consolidation, along with station automation and networking. This change reflects management and operational philosophies that have enabled radio owners to establish station groups consisting of 100 or more stations. As of early 2000, Infinity Broadcasting owned approximately 180 radio stations in some of the largest markets in the United States. The merger between Clear Channel and AMFM will create a radio colossus operating 874 U.S. radio stations. In addition, the group has investment in radio stations or networks around the world and television station ownership in the United States (Yung, 1999).

Group owners have reduced operating expenses by programming the stations with music and announcer comment that can be stored as computer files for playback at the appropriate time. Some owners even feed programming to local stations through networks designed to consolidate on-air production for several stations in a single city. While ownership consolidation has led to significant short-term operation and program delivery changes, the new owners have also realized that they can neither ignore changing technology nor consumer preferences for new improved audio sources.

Recent Developments

There are three areas where technology is affecting radio transmission:

(1) Enhancements to improve the present-day on-air transmission.

(2) Supplements to provide new services within the current AM and FM radio systems.

(3) New transmission delivery modes that are incompatible with existing AM and FM radio.

Enhancements—Stations Install Digital Audio Equipment

Since the introduction of compact discs and digital recording, most radio stations have upgraded on-air and production capabilities to meet the new digital standard. Virtually all portions of the audio chain are digital or are capable of handling a digital signal. Music and commercial playback are digital—either through compact discs, hard drive systems, minidiscs, or other digital media. Many newer consoles and mixing boards are digitally capable. The program signal that is delivered to the station's transmitter is digitally processed and travels to a digital exciter in the station's transmitter where the audio is added to the carrier wave. The only part of the current process that is always analog is the final transmission of the over-the-air signal by AM or FM.

Enhancements—Extending the AM Band

In 1996, the FCC extended the AM band from 1605 KHz to 1705 KHz. The newly authorized portion of the spectrum was intended to alleviate crowding on the existing AM band (535 KHz to 1605 KHz) by migrating stations to the expanded band. New stations on the expanded AM band were given increased daytime power and provided nighttime service, many for the first time. To avoid any economic hardship resulting from the move, stations are allowed to occupy both their previous frequency and the expanded band frequency during the transition period.

Supplements to Existing Services—Radio Data System

FM radio broadcasters have, for many years, utilized part of their spectrum to deliver auxiliary services. In the past, subcarrier frequencies (or subsidiary communication authorizations, SCAs) were used to deliver commercial-free background music marketed to businesses, reading services for the visually impaired, or paging. The latest subcarrier use, RBDS/RDS (Radio Broadcast Data System/Radio Data System), permits specially-equipped radio receivers to display call letters and program information such as station promotional material, music information, or advertising content, and it permits listeners to search for programming based on format (NAB, undated-a).

The system was introduced in the United States in 1993. On April 9, 1998, the National Radio Systems Committee (NRSC), a jointly sponsored committee of the National Association of Broadcasters (NAB) and the Consumer Electronics Manufacturers Association (CEMA), approved a revised version of the U.S. Radio Broadcast Data System Standard. The new system used by these "smart radios" is a compromise between the previous U.S. system and the European system. Of greatest significance was the addition of an "open data application" feature that will allow RBDS to add new features without requiring modification of the existing RBDS standards. The system carries information to FM receivers on a 57 KHz subcarrier utilizing a low bit rate data stream (NAB, undated-a).

The U.S. RBDS standard is based on the European Union RDS system, which results in system compatibility for receiver manufacturers and consumers. In the United States, the Consumer Electronics Association has identified about 700 FM stations that make full or partial use of RDS transmissions (CEA, undated). This is less than 10% of the licensed commercial and noncommercial FM stations in the United States. Receivers for home use are less common and more expensive than models available for automobiles. The most expensive RDS home receiver costs about $4,000, while car radios sell between a low of $300 to a high of about $1,000 (CEA, undated). Although the receiver price has dropped considerably, market penetration has been slow.

An NRSC initiative to create a High Speed FM Subcarrier (HSSC) was suspended in 1998 after a subcommittee reached an impasse on establishing a voluntary FM data subcarrier standard (NAB, undated-b). Recognizing that the RDS system would not sufficiently handle future data transmission needs, the NRSC began evaluating three competing systems. Test results suggested that a single system was not feasible given the variety of anticipated data transmission uses. Ultimately, the International Telecommunications Union (ITU) approved three systems:

- The Data Radio Channel (DARC) technology developed by NHK.
- The Subcarrier Traffic Information Channel system (STIC) developed by MITRE Corporation.
- The High Speed Data System (HSDS) developed by Seiko Communications Systems.

Additional proprietary technology has or is being developed. One factor cited in the NRSC report that inhibited creating a common standard was the consolidation of the radio industry. Broadcasters, especially engineers, were so committed to meeting the engineering needs brought about by consolidation that little attention was available for consideration of the HSSC standard (NAB, undated-b).

New Transmission Modes—Digital Radio

The single biggest change in the free, over-the-air radio broadcasting system in the United States may be the anticipated adoption of terrestrial digital audio broadcasting (DAB): delivering today's radio signals as streams of digital information. Already, radio stations have upgraded their stations to include on-air playback of digital music from CDs, minidiscs, and hard drives. Sometimes called digital audio radio (DAR), digital audio broadcasting will require listeners to purchase a new radio capable of receiving digital programming; the DAB system promises near CD-quality audio delivery. There are questions as to whether consumers would want the new service, given the potential expense of new receivers and the introduction of other technology including digital television, broadband delivery for the home, various wireless communications devices, and the introduction of subscriber-based, satellite-delivered audio services.

So far, the DAB initiative has centered on establishing a hybrid in-band on-channel (IBOC) system that would allow simultaneous broadcast of analog and digital signals by existing AM and FM stations through the use of compression technology without disrupting the existing analog coverage. The desired IBOC system for FM has been described as being capable of delivering CD-quality audio, while the AM IBOC would provide FM stereo-quality signals from AM stations. The two leading companies working on IBOC systems are USA Digital Radio (USADR), which is backed by some of the largest station owner groups, including Clear Channel Communications, CBS/Infinity Radio, Cox Radio, ABC, Bonneville International Corporation, Cumulus Media, Emmis Communications, Hispanic Broadcasting, and Radio One (USADR, 2000a). The second company is Lucent Digital Radio (LDR), a venture owned by Lucent Technologies and Pequot Capital Management (LDR, 2000). Lucent Technologies is the old Bell Labs research unit.

The FCC issued a *Notice of Proposed Rule Making* on November 1, 1999 to begin consideration of DAB technology approaches. The NRSC requested submission of data from interested parties on December 15, 1999. USADR met that deadline; Lucent submitted its report on January 24, 2000. USADR has also made inroads with broadcasters. To begin with, radio broadcaster groups own part of the company. Additionally, USADR has developed a program called EASE (Early Adopter Station Enhancement) to facilitate dialogue with individual station owners, even in small markets, to provide information on digital conversion (USADR, 2000b).

As of mid-2000, IBOC technology, while promising, has not been shown conclusively to work (Rathbun, 2000). The NRSC report noted that both companies failed to submit data for the full range of laboratory and field tests previously deemed necessary by NRSC to effectively evaluate the systems (NRSC, 2000a & 2000b). The NRSC's DAB Subcommittee noted that much work has been done, but that much work is left to do.

One of the most compelling issues facing the two IBOC companies, broadcast owners, and regulators is whether USADR and LDR will agree to work together to develop a common system (Rathbun, 2000). Each company uses a different approach to accomplish its IBOC broadcasts. An alliance, while requiring a split of future profits from an eventual IBOC system, might overcome individual weaknesses in either system and thus shorten the time period to eventual adoption.

DAB service in the United States could probably be launched much sooner if a different technological approach were adopted. This would likely mean the use of new spectrum space out of the existing AM and FM bands. One system that has been tested in laboratory and field settings in this country is the Eureka-147 system, which operates in the L-Band (1452 MHz to 1492 MHz) and another portion of the spectrum referred to as Band III (around 221 MHz) (FCC, 1999). As with an IBOC system, new radios capable of receiving the signals would have to be purchased.

The Eureka-147 system is in operation in Canada, the United Kingdom, Sweden, Germany, and France. The World DAB Forum (http://www.worlddab.org), an international organization to promote the Eureka-147 DAB system, reports that more than 230 million people can receive DAB signals and that there are more than 400 different DAB services available for listeners. Eureka-147 receivers have been on the market since summer 1998, and 16 commercial receivers are currently available that range in price from around $500 to several thousand dollars (World DAB, 2000).

The FCC has emphasized the importance of localism as a "touchstone value" of U.S. terrestrial radio service. An inband DAB service (IBOC) would enable radio listeners to adopt the technology gradually while continuing to receive radio service through locally programmed stations (FCC, 1999). Local or regional out-of-band service could be established to provide localism, but the critical time period necessary to license the stations, develop viable programming, and establish a critical mass of receivers and listeners might produce an economic hardship on the DAB service. Additionally, U.S. commercial broadcasters worry about increased competition from potential out-of-band licensees and the loss of "brand value" from existing station franchises. Unlike the progression of television service from analog to digital, there has not been a process for authorizing out-of-band spectrum for existing broadcaster use. Complicating the U.S. picture even more is the fact that L-Band spectrum is currently used for flight test telemetry, and Band III is used for land, mobile, and amateur use (FCC, 1999).

Old Technology, New Competition—Low-Power FM

The FCC approved the creation of a controversial new classification of noncommercial FM stations on January 20, 2000 (Chen, 2000). The new LPFM (low power FM service) will limit stations to a power level of either 100 watts or 10 watts (FCC, 2000). Community groups (ranging from church groups to community organizations to schools) desiring to provide specific community or organizational programming would be licensed. To make room for the new stations, third adjacent channel separation/protection for current stations is being abolished. Practically speaking, this means that a currently licensed station operating on 95.5 MHz might have a LPFM competitor on a frequency as close as 94.9 MHz or 96.1 MHz. In the past, stations would have received "third adjacent" channel protection from potential interference. The FCC contends there will be no objectionable interference to existing stations, although NAB disputes this claim.

The proposed licensing has proven controversial. At the time of this writing, through lobbying efforts led by NAB, broadcasters were seeking to stop, or at least slow, the licensing procedure set to begin in the second half of 2000. Congressional involvement may result in passage of a bill to kill the service before the first stations are even authorized. Commercial broadcasters fear interference to existing stations, added competition in an increasingly competitive radio market, and possible interference to future DAB signals.

New Delivery Competition—Satellite Digital Audio Radio Service (SDARS)

Another form of out-of-band "radio" service set to launch in the United States during 2000 is subscriber-based satellite radio service. The service was authorized by the FCC in 1995 and, strictly speaking, is not a radio service. Although reception will be over-the-air and electromagnetic spectrum is utilized, the service is national, not local. It requires users to pay a monthly subscription fee of about $10, and it requires a proprietary receiver to decode the transmissions. Two publicly traded companies, Sirius Satellite Radio (previously known as CD Radio) and XM Satellite Radio, plan to offer up to 100 channels of commercial-free and commercial audio channels (Sirius Satellite Radio, 2000; XM Satellite Radio, 2000). After a number of false starts, Sirius Satellite was scheduled to launch its first satellites in late June 2000. Service is scheduled to begin before the end of 2000. XM Satellite Radio is scheduled to launch its satellites by late 2000 and begin its service in early 2001. As of mid-2000, both companies have receiver manufacturing agreements. Many 2001 model automobiles will include factory-installed receivers, and an agreement between Sirius and XM will give second-generation receivers the ability to receive either of the two services (XM Satellite Radio, 2000b).

The question, of course, is whether consumers will pay for audio services they have traditionally received for free. Both companies are developing program content not typically available from over-the-air broadcasters, including show tunes, blues, folk, bluegrass, American standards, and content from Bloomberg, BBC, CNBC, CNN, *Sports Illustrated*, Weather Channel, and USA Network.

Given the development costs of each service, satellite audio faces an uncertain future. From inception through mid-2000, XM Satellite Radio lost nearly $55 million. The company plans to invest $1.1 billion to launch its satellites and begin offering programming services by 2001 (XM Satellite Radio, 2000c). Sirius Radio has lost nearly $108 million since its inception in 1990. Another $1.2 billion will be invested to launch satellites and begin program operation before the end of 2000 (CD Radio, 1999). To put their losses and the additional cost of the SDARS technology into perspective, Clear Channel Communications' merger with AMFM, Inc. put the value of AMFM at $23.5 billion including $6.1 billion of debt. The merged company would be valued at $56 billion (Yung, 1999). Both national SDARS companies, though expensive, will cost less to build than the debt total of AMFM, Inc., even though that company's terrestrial radio stations are already operating and have an established listener base.

New Delivery Competition—Internet Radio

One of the fastest growing uses for the Internet has been audio streaming. Radio stations first used the Web to distribute promotional information and then added programming. Research Firm Cyber Dialogue, Inc. reported that 42% of all Internet users have made use of some music-related content, and about 37% have visited a radio station Website (Clark, 1999). Web-only audio services, sometimes called Internet radio, Net stations, or Net audio, bypass the need for spectrum space. Using RealAudio, Windows Media, or similar software, the Internet makes it possible to create a "radio" station featuring almost any type of programming.

Perhaps the most widely known Internet audio service is Broadcast.com which was purchased by Yahoo! and is now known as Yahoo! Broadcast. Another service, BRS Media, offers Webcasting of Internet-only audio as well as a Web hosting service for radio stations. One of the potential problems users face is heavy traffic and slow download speeds. Internet radio company TuneTo.com has developed a hybrid approach, called metacasting, which involves storing some information on the user's hard drive in advance to improve the signal flow (Clark, 1999). While locating the various stations may be difficult for listeners and the time spent listening to such Net stations may be small, the sheer number of such audio services and the ease of starting such a venture makes them a competitor to radio station listening, especially in workplaces.

Internet delivery of over-the-air content or Internet-only content does have its limitations, chief of which is mobile listening. Even this will change as consumers migrate to wireless communications technology. Downloadable MP3 files that can be saved to portable listening devices present another new use of technology, providing people on the go with listening sources other than the radio. An issue still to be resolved is potential copyright law violations as files are uploaded and downloaded from Internet sites with little consideration for copyright holders (Gomes, 1999).

Factors to Watch

Radio stations have been in the business of delivering music and information to listeners for over 80 years. Listenership, more than any particular technological aspect, has enabled radio to succeed. Stations have been able to sell advertising time based on the number and perceived value of their audience to advertising clients. Technology, when utilized by radio stations, focused on improving the sound of the existing AM or FM signal or reducing operating costs.

The latest technological changes—RDS, DAB, and SDARS—have the potential to shift some radio operators and possibly the entire radio industry into a new mode of operation. Digital broadcasting offers the possibility of streaming data content that might hold added value to consumers. Satellite audio itself holds the promise to create multiple revenue streams—the sale of the audio content, sale of commercial content on some programming channels, and possibly delivery of other forms of data. Price discounting will likely be instituted to establish a subscriber base.

Problems associated with regulatory barriers to these new technologies are not the issue. Properly assessing consumer interests in the technologies, perfecting the technology, and delivering both content and receivers will determine the success of these new radio technologies.

Bibliography

CD Radio, Inc. (November 15, 1999). *CD Radio quarterly report, Securities and Exchange Commission Form 10-Q*. [Online]. Available: http://www.sec.gov/Archives/edgar/data/908937/0000950117-99-002347.txt.

Chen, K. (2000, January 17). FCC is set to open airwaves to low-power radio. *Wall Street Journal*, B12.

Clark, D. (1999, November 15). With Web radio, anyone can be a DJ, but special software confuses users. *Wall Street Journal*, B8.

Clark, D., & Peers, M. (2000, March 1). MP3 chief rocks and roils music. *Wall Street Journal*, B1.

Consumer Electronics Association (CEA). (Undated). *Market overview: Audio division*. [Online]. Available: http://www.ce.org/index.cfm?area=market_overview.

Federal Communications Commission. (1999, November 1). In the matter of digital audio broadcasting systems and their impact on the terrestrial radio broadcast services. *Notice of Proposed Rule Making*, MM Docket No. 99-325. [Online]. Available: http://www.fcc.gov/Bureaus/Mass_Media/Notices/1999/fcc99327.pdf.

Federal Communications Commission. (2000, January 20). *In the matter of creation of low power FM*. MM No. 99-25, [Online]. Available: http://www.fcc.gov/Bureaus/Mass_Media/Orders/2000/fcc00019.doc.

Federal Communications Commission. (Undated). *Audio Services Division, AM stereo broadcasting*. [Online]. Available: http://www.fcc.gov/mmb/asd/bickel/amstereo.html.

Gomes, L. (1999, June 15). Free tunes for everyone! *Wall Street Journal*, B1.

Klopfenstein, B. C., & Sedman, D. (1990). Technical standards and the marketplace: The case of AM stereo. *Journal of Broadcasting & Electronic Media, 34* (2), 171-194.

Lucent Digital Radio. (2000). *Company background*. [Online]. Available: http://www.lucent.com/ldr.

Lewis, T. (1991). *Empire of the air: The men who made radio*. New York: Harper Collins.

National Radio Systems Committee, DAB Subcommittee. (2000a, April 8). *Evaluation of Lucent Digital Radio's submission to the NRSC DAB Subcommittee of selected laboratory and field test results for its AM and FM band IBOC system*. [Online]. Available: http://www.nab.org/SciTech/Dab/Ldrfinalreport_rev1.pdf.

National Radio Systems Committee, DAB Subcommittee. (2000b, April 8). *Evaluation of USA Digital Radio's submission to the NRSC DAB Subcommittee of selected laboratory and field test results for its AM and FM band IBOC system*. [Online]. Available: http://www.nab.org/SciTech/Dab/Usadrfinalreport_rev1.pdf.

National Association of Broadcasters. (Undated-a). *NRSC revises U.S. RBDS Standard*. [Online]. Available: http://www.nab.org/SciTech/Nrscgeneral/rds.asp.

National Association of Broadcasters. (Undated-b). *High-speed FM Subcarrier (HSSC) Committee*. [Online]. Available: http://www.nab.org/SciTech/Nrscgeneral/hsscsub.asp.

Pavlik, J. V. (1998). *New media technology: Cultural and commercial perspectives*. Boston: Allyn & Bacon.

Rathbun, E. A. (2000, April 17). Proceeding with digital radio. *Broadcasting & Cable*, 35.

Sirius Satellite Radio. (2000). *Company background*. [Online]. Available: http://www.siriusradio.com/main.htm.

Sterling, C. H., & Kittross, J. M. (1990). *Stay tuned: A concise history of American broadcasting*. Belmont, CA: Wadsworth Publishing.

USA Digital Radio. (2000a). *Company background*. [Online]. Available: http://www.usadr.com/.

USA Digital Radio. (2000b). *EASE for broadcasters*. [Online]. Available: http://www.usadr.com/broadcasters.html.

Weber, T. E. (1999, July 28). Web radio: No antenna required. *Wall Street Journal*, B1, B4.

World DAB: The World Forum for Digital Audio Broadcasting. (2000a). *DAB receiver archive*. [Online]. Available: http://www.worlddab.org/rarchive/rarch_frame.htm.

World DAB: The World Forum for Digital Audio Broadcasting. (2000). *DAB worldwide*. [Online]. Available: http://www.worlddab.org/.

XM Satellite Radio Homepage. (2000a). *Company background.* [Online]. Available: http://www.xmradio.com/js/xmmenu.htm.

XM Satellite Radio. (2000b, February 16). *Sirius Radio and XM Radio form alliance to develop unified standards for satellite radio.* Company press release. [Online]. Available: http://www.xmradio.com/js/news/pressreleases.asp#.

XM Satellite Radio. (2000c, March 16). *XM Satellite Radio Holdings, Inc., annual report, Securites and Exchange Commission Form 10-K.* [Online]. Available: http://www.sec.gov/Archives/edgar/data/1091530/0000928385-00-000751.txt.

Yung, K. (1999, October 5). Texas merger to create largest radio company. *Dallas Morning News*, A1, A8.

Websites

BRS Media, Inc., http://www.brsmedia.fm/.
Clear Channel Communications, http://www.clearchannel.com/.
Federal Communications Commission, Mass Media Bureau, http://www.fcc.gov/mmb/.
GoGaGa Internet Radio, http://www.gogaga.com/index.html.
Harris Corporation, http://www.broadcast.harris.com/.
Lucent Digital Radio, http://www.lucent.com/ldr/.
MP3.com Homepage, http://www.mp3.com/.
National Association of Broadcasters, Science and Technology, http://www.nab.org/SciTech/.
Radio.Sonicnet, http://radio.sonicnet.com/.
Sirius Satellite Radio, http://www.siriusradio.com/main.htm.
USA Digital Radio, http://www.usadr.com/.
XM Satellite Radio, http://www.xmradio.com/js/xmmenu.htm.

COMPUTERS & CONSUMER ELECTRONICS

If there is one theme underlying the developments discussed in this book, it is "the impact of digital technology." Nowhere is that impact more profound than in the computer industry. This year's computer technology will unquestionably be replaced in less than two years by a technology that has up to twice the performance at almost half the cost (a phenomenon known as "Moore's law"). These advances in computer technology, in turn, lead to advances in almost every other technology, especially those consumer products incorporating microprocessors or other computer components. The next seven chapters illustrate the speed, direction, and impact of the continuous innovation in computer technologies across a wide range of computing and consumer electronics technologies.

The next chapter explores the manner in which computers have moved beyond text to incorporate the melange of video, audio, text, and data known as "multimedia." The following chapter addresses the most significant emerging application of personal computers and explains the Internet and the World Wide Web as an information resource. The development of Internet-based e-commerce has made such a strong impact on individual businesses and the economy in general to warrant detailed discussion of Internet commerce in Chapter 12. The application of computers and other related communication technologies in the office setting (Chapter 13) provides an illustration of the manner in which these technologies are revolutionizing commerce around the world.

The most forward-looking chapter in this section, Chapter 14, discusses a set of technologies that may have the greatest long-term potential to revolutionize the way we live and work—virtual reality. The production and distribution of video and audio programming is the subject of the last two technologies in this section. The home video chapter reports on the incredible popularity of the existing analog video formats, and on new, digital technology that is beginning to challenge the analog incumbents. Finally, the digital audio chapter reports on the early outcome of the battle between competing analog and digital audio technologies, with digital casualties as well as victors.

In reading these chapters, the most common theme is the systematic obsolescence of the technologies discussed. The manufacturers of computers, video games, etc. continue to develop newer and more powerful hardware with new applications that prompt consumers to continually discard two- and three-year old devices that work as well as they did the day they were purchased—but not as well as this year's model. Most software distribution, from movies and music to television and video games, has been based upon the continual introduction of new "messages." The adoption of this marketing technique by hardware manufacturers assures these companies of a continuing outlet for their products, even when the number of users remains nearly static.

An important consideration in comparing these technologies is how long the cycle of planned obsolescence can continue. Is there a computer or piece of consumer electronics so good that it will never be replaced by a "better" one? Will technology continue to advance at the same rate it has over the past two decades? How important is the equipment (hardware) versus the message communicated over that equipment (software)?

Finally, each of these chapters provides some important statistics, including penetration and market size, which can be used to compare the technologies to each other. For example, there is far more attention paid today to the Internet than to almost any other technology, but less than one in three U.S. households has access to the Internet (with even smaller penetration levels in other countries). On the other hand, the VCR is now found in about nine out of 10 U.S. homes. In making these comparisons, it is also important to distinguish between *projections* of sales and penetration, and *actual* sales and penetration. There is no shortage of hyperbole for any new technology, as each new product fights for its share of consumer attention.

10

Multimedia Computers & Video Games

Cassandra Van Buren, Ph.D.*

Interactive multimedia products combine graphics, text, audio, animation, and video in ways that allow the user to make navigational choices that alter the flow of information in terms of form and content. As we continue to progress toward the convergence of media production technologies, distribution systems, conceptualizations of the audience, and types of content (Murray, 1997), interactive multimedia products embody the integration of the communication, computer, and entertainment industries. This chapter describes the background, recent developments, and current status of interactive multimedia hardware and software, including console-based and handheld video games. Relevant industry factors to monitor are included as well.

Background

The multimedia industry revolves around technological developments in authoring and playback systems. Authoring systems are computer-related hardware and software arrays that allow developers to design multimedia titles. These systems include computer languages used to develop proprietary authoring systems and pre-developed and packaged authoring software such as Macromedia Director Studio and Apple Computer's HyperCard. HyperCard was the mainstay of interactive multimedia development in the early years of the business, and eventually became prominent enough to warrant an IBM PC-compatible version in 1990 (Hudson, 1998).

Interactive multimedia products can have many different functions, including education and training, entertainment, sales, and information/reference. These products can also be stand-alone or networked. Networked multimedia encompasses products that allow users to interact with other

* Assistant Professor, Department of Communication, Trinity University (San Antonio, Texas).

people and systems, such as networked multimedia personal computers (MPCs). The Internet is often the network through which these MPCs are connected. Examples of networked MPCs include CD-ROMs and DVDs (digital video, or versatile, disks) that have Internet links programmed into the applications, interactive television, Internet-connected console-based video game systems and handheld video game systems, and most videoconferencing systems. Standalone products include non-networked and non-Internet connected MPCs, CD-ROMs, console-based video game systems, and handheld video game systems.

Interactive multimedia production is similar to motion picture production. The process requires collaborative groups of people to design, prototype, produce, market, and distribute multimedia titles. High-quality graphics and animation are of particular concern to developers, since users typically hold television broadcast-level expectations of image quality. Developers determine the platform, desired user experience, design, and budget. If the title is developed for the video game console platform, developers target the title for playback on a platform such as Sega Dreamcast, Nintendo 64, or Sony PlayStation. If the title is developed for playback on an MPC, developers decide whether to target the Macintosh platform, Windows platform, or both. Once the master (original) of the title is produced, duplicating multiple copies of the title onto CDs or video game cartridges is quite inexpensive (Hudson, 1998).

MPCs

The MPC industry strives to create and distribute hardware standards that developers of peripherals and software can use in creating products for recommended target platforms. Additionally, consumers use MPC standards as a benchmark when making purchasing decisions. In the 1990s, the major developer of MPC standards was the Multimedia PC Council, a group responsible for bringing the industry to consensus and distributing standards. Developing standards helped propel the adoption of the MPC, which simultaneously funded the development of MPC hardware, peripherals, and software. In 1997, Intel introduced a chip developed especially for multimedia applications: the MMX. Other chip companies, including Cyrix and AMD (Advanced Micro Devices), sued for the right to produce MMX-compatible extensions (Hudson, 1998). The lawsuit was settled when AMD agreed to acknowledge Intel's copyright ownership, while AMD maintained the right to use the MMX name in its marketing materials (Reuters, 1997).

By 1998, improved hardware and software, as well as the growing maturity of the industry, allowed increasingly sophisticated high-tech multimedia titles to be created and enjoyed by both the private and public sectors. Computer sales improved, partly due to the MMX processor in Pentium II computers. Simultaneously, superstores that focused on computers, such as Computer City and CompUSA, proved to be the most successful outlets for retail MPC software sales, with 41% of sales. Large retail chains began charging software companies from $10,000 to $20,000 per title per chain for the privilege of shelf space, causing multimedia developers headaches as they competed with each other for access to the retail consumer (Hudson, 1998).

Operating Systems

One of the key differentiating components of computers is the computer's operating system, the set of instructions used to control the manner in which information is read, stored, and manipulated by the computer (see Figure 10.1). Currently, the most popular operating systems are

Microsoft's Windows family, including Windows 98 and ME for consumers and Windows NT and 2000 for business users. The dominance of the Windows operating systems, along with alleged anticompetitive behavior by Microsoft, led the U.S. Department of Justice in mid-2000 to propose that the company be broken into two separate companies. (As of this writing, the case was still under litigation. Indeed, judging by previous such lawsuits, a final resolution is not likely before the next edition of this book is prepared in 2002.)

Figure 10.1

Elements of a Personal Computer System

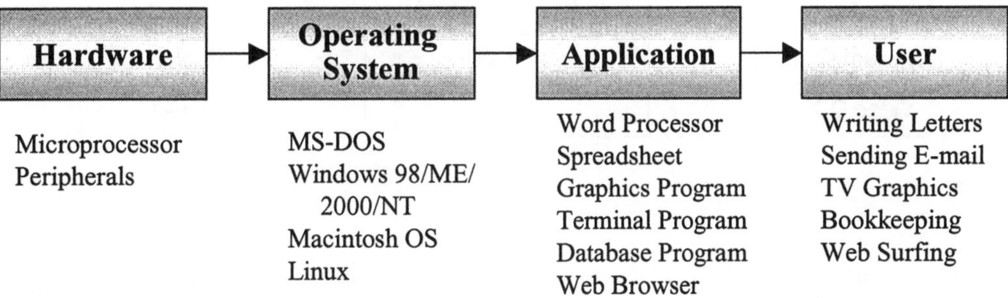

Source: Technology Futures, Inc.

Although one of the four Windows operating systems is installed on the majority of new personal computers, other options are available. All Apple computers use a proprietary operating system identified with the Macintosh computer (Mac OS8, Mac OS9, and Mac OS10). The newest option in operating systems is Linux, which differs from other operating systems because the inventor, a Finnish student named Linus Torvalds, made the source code available through the Internet for anyone to use or modify. The "open" nature of the operating system has yielded significant improvements, although it remains to be seen if such an open model can attract a significant number of users. As of early 1999, about nine million users, primarily corporate computer users and hobbyists, used Linux.

CD-ROMs

Interactive multimedia has evolved over the last 15 years, beginning when Sony and Phillips developed the first audio CD (CD-DA) in 1982. Specifications for CDs were disseminated through "colored books" (see Table 10.1) that covered requirements of disk size and data capacity/type. In 1985, CD technology developed in an expanded format that supported up to 650 MB of a single type of stored data, including audio, text, and graphics. Different formats were required for Macintosh computers and MS-DOS computers, with Mac-compatible CDs using the Hierarchical Filing System (HFS) and MS-DOS-compatible CDs using the IS 9660 directory format (Holsinger, 1995). The standardization of the disk allowed several competing manufacturers to enter the CD-DA player market, thus speeding the adoption of CD-ROM players by consumers. Within 10 years of introduction, digital audiodisc sales outstripped audiocassette player sales. CD-ROMs evolved to become the most heavily used optical storage medium (Hudson, 1998).

Table 10.1
Colored Book Specifications and Formats

Disc Format	Disc Type	CD Book Standard
Audio CD	CD-DA	Red Book
Data CD (CD-ROM)	CD	Yellow Book
Data CD (Interactive)	CD-I	Green Book
Recordable CDs	CD-MO, CD-WO, CD-RW, CD-ROM XA	Orange Book
Laserdisc	CD Extra	Blue Book
Video CD	DVD	White Book

Source: CNet

The integration of audio, text, video, graphics, and animation has become increasingly sophisticated in terms of technology, speed, interactivity, and content. Interactive multimedia titles were developed to fit into genres useful in both the public and private sectors. Typical genres included education, training, entertainment, sales, and reference. Reference titles such as CD-ROM-based encyclopedias were developed for classroom and home use. Small businesses and corporations implemented CD-ROM-based training and multimedia presentation software in an effort to reduce personnel costs. Home consumers purchased CD-ROM-based multimedia for entertainment, productivity, creativity, and education (Hudson, 1998).

Advances in CD-ROM technology developed in terms of the method of reading data; consumers' demands for greater flexibility in data storage options; and greater demand for video playback capability.[1] Previously, the reading of CD-ROM data occurred using constant linear velocity (CLV) mode, which caused inefficient speeding up and slowing down of the motor as the drive head variously read the inner and outer portions of the disk. Engineers later developed constant angular velocity (CAV) mode, in which the disk rotates at a stable rate while the drive head reading rate fluctuates in relation to the location of the data. Some hybrid CD-ROMs use both modes (Hudson, 1998).

The recording capability of CD-ROMs evolved from the CD-R format, which allowed users to record data from another storage medium to a blank CD-R disk. The disadvantage of this WORM (write once, read-many) technology is that the data is unalterable. Multi-session recording capability was developed to allow users to add or "burn" additional data onto a CD-R. The CD-RW (rewritable) was introduced in 1996, allowing users to treat CD-ROMs like large-capacity floppy disks. Users could record, erase, and overwrite data on their CD-RWs thanks to a new phase change technology in which the surface of the disk is not deformed, but rather is changed from a transparent layer to an amorphous visible layer. After a brief compatibility problem (CD-RW

1. CD-ROMs, CD-Rs, and CD-RWs have been surpassed by DVDs as a digital video recording and playback method. Please see Chapter 15 for more information on DVDs.

drives at first could only play back CD-RWs), CD-RWs became well integrated into the MPC technology marketplace. In 1997, prices for blank CD-R discs were down to $2 to $3 each, and prices for CD-R units ranged from $399 to $650 (Hudson, 1998).

The growth of CD-R, in turn, led to an increasing concern over the problem of copyright violations as unauthorized parties made copies of CDs using CD-Rs and CD-RWs. Copy protection technologies were developed for CDs throughout the late 1990s. The technologies added instructions to the CD during manufacturing that were designed to prevent serial copying or to otherwise limit use (Hudson, 1998).

Dedicated Video Game Systems

The development of the dedicated video game market, including home console systems and handheld devices, is marked by a pattern of interdependence between hardware and software specifically related to the coordinated timing of release. Hardware innovations often failed to thrive or even survive if compatible and compelling software was not immediately available for the consumer market. Another general pattern is the intense competition and jockeying for market share and third-party developer relationships between the major players in the industry: Nintendo, Sega, and Sony.

In the early 1970s, the release of the tennis-like video game Pong signaled the beginning of the home-based interactive entertainment industry. Magnavox followed by releasing Odyssey, a video game system that relied on TV sets for display. Atari emulated the TV set model developed by Magnavox by developing a version of Pong that used TV sets. Five years later, Fairchild's Channel F released the first cartridge-based game console that allowed several games to be played on the same hardware. The microprocessors and animation processors used in these early console systems were able to relay megabytes of animation data per second. From early on in the history of the video game console industry, the capability of these proprietary hardware systems to surpass the processor speeds of MPCs was evident. Despite this speed advantage, console manufacturers such as Fairchild, Coleco, and Mattel suffered poor sales due to the unfulfilled promise of an arcade-like gaming experience delivered to the home (Hudson, 1998). In 1998, Coleco filed for bankruptcy and sold most of its product catalog to Parker Brothers and Milton Bradley (Herman, et al., 2000).

It was not until 1977 that home video game sales proved viable when Atari released its Atari 2600 programmable video computer system. Sales of the system and games were moderately successful, building through 1979 with the release of the game *Asteroids*, and taking off in 1980 with the release of the home console version of *Space Invaders.* Video game sales slowed over the next six years until 1986, when Nintendo released its 16-bit cartridge-based system called NES. The release was accompanied by the debut of the *Super Mario* game for home consoles, a stunningly successful version of the arcade game. Nintendo attracted several third-party game developers formerly affiliated with Atari. Sega and Atari attempted to compete with NES by releasing their own systems, but NES outstripped its competitors in U.S. sales by a 10:1 ratio. Atari ceased to exist in 1996 when the company merged with JTS, a hard drive manufacturing company.

In 1989, several video game milestones occurred. NEC released the first system to use CDs for game storage: the TurboGraphx-16. Nintendo released the first black-and-white handheld video game system, known as Game Boy, for $109. Game Boy came bundled with the game *Tetris.*

Super Marioland and *Alleyway,* versions of already popular games, were quickly released for the Game Boy platform. Atari purchased the rights to Epyx's in-color handheld Handy Game and introduced it under the name Lynx. Because of Lynx's $149 price tag, lack of third-party developers, and rumors that Atari would discontinue support for the system, Lynx sales lagged behind Game Boy. In 1990, as Nintendo released *Super Mario 3* for home consoles (which became the best-selling cartridge game to date), NEC released a handheld version of the TurboGrafx-16 called the TurboExpress, marking the first portable system capable of playing games designed for a home console (Herman, et al., 2000).

In the early 1990s, Sega became a strong force in the video game industry. In 1992, the company released the Sega CD system, but despite strong sales of *Sonic the Hedgehog*, lack of openness with third-party developers resulted in Sega games that failed to take advantage of significant hardware-based graphics advances. Sega's market share declined, partially because of Sony's entrance into the market. After ending a development agreement with Nintendo, Sony announced plans to develop its own 32-bit CD system. In 1993, Nintendo and Sega announced the development of 64-bit and 32/64-bit systems. By 1995, Sega discontinued production and support of the Sega CD system as Sony released its PlayStation and Nintendo released Nintendo64 in Japan despite the lack of available compatible games.

Nintendo64 made its American debut in 1996 and sold over 1.7 million units in just three months, prompting formerly skeptical third-party developers to hurriedly crank out titles. Meanwhile, Sony claimed its sales topped $12 million per day during the 1996 holiday season thanks to the number-one position of PlayStation. Japanese, U.S., and European PlayStation sales totaled 11.2 million units by April 1997; this number nearly doubled only four months later (Herman, et al., 2000).

Sony lowered the price of PlayStation to $149 in 1997, causing Nintendo to likewise lower the price of Nintendo64. Sega management chose not to lower the retail price of Saturn. Nintendo emerged as the winner of this price war, while Sega dropped to third place. Generally speaking, 64-bit console development renewed vigor in the industry and allowed retailers to enjoy tripled and quadrupled sales figures. As 64-bit systems gained market share, 32-bit systems lost popularity. In 1998, PlayStation topped the list of number of games developed for its platform with hundreds available. Nintendo64 had only about 40 games available, and Sega trailed a distant third place with roughly 15 games for its platform (Hudson, 1998).

Bandai, a Japanese toy company, released a tiny "virtual pet" game called *Tamagotchi* in Japan in November 1996, resulting in a national craze that traveled to the United States by May 1997. One store alone sold out of its initial stock of 30,000 *Tamagotchi* in only three days, marking the beginning of a new platform and phase of miniaturization for computer-based games.

Content Regulation

Multimedia and video game content captured the notice of government officials in 1993, when U.S. Senators Joseph Leiberman and Herbert Kohl called for and launched an investigation of video game violence. The two games that were the main inspiration for the senators' investigation were *Night Trap* and *Mortal Kombat* (Herman, et al., 2000). As a compromise to the senators' proposal for a ban on violent games, or at least a government-regulated ratings board, the industry agreed to regulate itself via a voluntary rating system established, implemented, and maintained largely by two organizations: the Entertainment Rating Software Board (ERSB) and the Internet Content Rating Association (ICRA).

The ICRA is the result of the 1999 incorporation of the Recreational Software Advisory Council (RSAC), founded in 1994 by the Software Publishers Association and five other trade associations, into the newly formed ICRA. The ICRA rates material, including Websites, according to violence, nudity, language (vulgar and hate-related), and sex. Consumers are able to adjust Web browser and/or blocking device settings using the four-point ICRA scale thanks to agreements between browser developers, Website developers, and ICRA. ICRA claims that, as of August 1999, over 120,000 Websites were rated according to its system (ICRA, 2000).

The ERSB, founded in 1994, rates games according to content related to violence, sex, crude language, tobacco, alcohol, illegal drugs, and gambling. In addition, the ERSB provides ratings for the level of difficulty and genre. Rating symbols such as "E" for "Everyone" and "AO" for "Adults Only" are included on the exterior packaging of game boxes, allowing consumers to make purchasing decisions accordingly. As of March 2000, ESRB claimed to have rated over 4,500 multimedia titles (ERSB, 2000).

In 1997, Senators Kohl and Leiberman, in their annual report card delivered to the gaming industry, stated that, in general, they were happy with the industry's self-regulation efforts, as evidenced by the fact that most packaged software was marked according to ratings systems. Retailers, however, were the targets of the senators' displeasure, since most stores failed to develop or implement policies preventing minors from purchasing inappropriate titles (Herman, et al., 2000).

Recent Developments

MPCs

Packaged software industry revenue, according to International Data Corporation, was $140 billion worldwide in 1998, a 15% increase from the previous year (Packaged software industry, 2000). Revenue rose to $154 billion in 1999. Microsoft was the most successful vendor earning $17.4 billion of the total (IDC, 2000). Worldwide MPC sales totaled roughly 88 million units in 1998, up from 56 million in 1995 (eTForecasts, 2000).

CD-ROMs

CD-RW drives allow the creation of audio and data CDs using CD-R discs. In addition, the drive can be used as a removable storage device or as a second CD-ROM reader to read data and play audio CDs. CD-RW sales grew to 3.447 million units in 1998, a 456% increase over the previous year (Gartner Group's Dataquest says CD-RW, 1999). CD-RW drives gained speed and dropped in price in 1999, with some units dropping to under $200. Manufacturers include Hewlett-Packard, Iomega, Memorex, Plextor, Ricoh, and Yamaha. These drives are available in two interfaces: IDE and SCSI. Technically experienced users usually purchase the SCSI version. Despite the required installation of a host adapter, the benefit of the faster read capability and fewer write errors featured in SCSI units make it worth the effort. Computer novices favor IDE drives because they are simpler to install and less expensive than SCSI units. Unfortunately, these drives suffer from inferior read performance and higher write error rates. CD-RW drives are categorized by three different speed ratings, such as 6x/4x/24x. The notation translates to 6x speed for writing, 4x speed for rewriting, and 24x speed for reading.

A new generation of CD-RW drives emerged in late 1999 that allow portability. Five drives were released, all based on the Matsushita raw CD-RW ATAPI 4X Rewrite 20X read drive with a 2 MB buffer. The drives connect to notebook PCs with PC card slots and are bundled with AC adapters, audio ports, and Adaptec's Easy CD Creator Version 3.5c and DirectCD software. The drives write data to disks at roughly the same speed: 20 MB per minute (Grotta & Grotta, 1999).

Video Games

There has been continued growth in the gaming industry. In 1998, next-generation console system sales were between 6.5 million and 7.5 million units in the holiday shopping period alone (Interactive Digital Software Association, 1999). The electronic game market exceeded $7 billion in 1999, marking a 10% gain over the previous year. For the second year in a row, Americans responding to a 1999 Interactive Digital Software Association (IDSA) survey indicated that they preferred PC and video games (34%) to watching TV (18%) or going to the movies (16%) for entertainment purposes (Americans overwhelmingly, 1999).

The export of *Pokemon*—the extremely popular Japanese characters and game—to the United States, combined with the release of Game Boy Color, spurred sales of Game Boy sets as *Pokemon* quickly became Nintendo's best-selling game ever. The release of Pocket Pikachu (a Pokemon character), a miniature game device resembling *Tamagotchi*, also raised Nintendo's status in the market (Herman, et al., 2000).

Sega released its new console system, Dreamcast, which is reliant upon the Microsoft CE operating system, to the Japanese market in November 1998. The first 150,000 available sets, priced at $250 each, sold instantly. Dreamcast was then released in the United States in 1999, where it retailed for $199. Sega reported earnings of $98 million on the first day of U.S. sales.

In 1999, Nintendo announced plans to co-develop a new console, code-named "Dolphin," with IBM using a 400 MHz copper microchip. The company also announced the Game Boy Advance, which, when combined with a cellular phone, gives users Internet access. The device features 32-bit color and compatibility with Game Boy and Game Boy Color software. Meanwhile, in March, Sony released specifications for PlayStation 2, including a new Toshiba/Sony 250 MHz microprocessor. U.S. introduction is expected in fall 2000. Since 1998, Sony has dominated the console market, with Nintendo coming in second and Sega third (Lemos, 2000).

Despite successful growth in the video game market, the industry has some troubles. The rosy sales figures indicate a strong industry, but analysis of the software side indicates a slightly different situation. Of the 1,000 to 1,500 video games released every year, only about half are commercially successful (Game programmers getting, 2000). In addition, pirated software in 55 countries (excluding the United States, Mexico, and Europe) during 1999 cost the computer and video game industry an estimated $3 billion (IDSA, 2000).

The industry continued to show concern over the issue of content regulation, taking new steps to ensure that government officials would be appeased by efforts at self-regulation. In September

1999, the three major console manufacturers announced the inclusion of ESRB ratings information in all hardware packaging, just in time for the 1999 holiday shopping season. Industry leaders and ESRB executive director Arthur Pober lauded the plan, claiming it demonstrated the industry's continuing commitment to providing valuable educational information about video game ratings to consumers, especially parents (IDSA, 1999).

Current Status

MPCs

The MPC-1 (1990) and MPC-2 (1994) standards are now considered out-of-date. MPC-3 (1996) standards are a barely-acceptable minimum for companies such as Macromedia (Macromedia, 2000). Since Microsoft is the dominant force in the software industry, however, the system specifications developed jointly by Intel and Microsoft are the standard to which most consumers and developers adapt (Venezia, 1999).

The current standard is the PC 99 System Design Guide (see Table 10.2). The *2001 System Design Guide* is available in draft form for review and comment, and is scheduled for implementation on July 1, 2001 (Intel Corporation and Microsoft Corporation, 2000).

Table 10.2
PC System Design Guide Specifications, Non-Mobile PCs

System Requirements	300 MHz processor, 128K L2 cache, 32 MB RAM
System Buses	2 USB ports, no ISA add-on devices or expansion slots
I/O Devices	Keyboard & pointing device, connections for serial & parallel devices
Graphics and Video	640 × 480 to1024 × 768 resolution, 8- to 32-bit color, hardware support for 3-D acceleration
Audio	16-bit stereo, 44.1 KHz and 48 KHz
Storage	8x CD, 2x DVD (4x recommended)
Communications	Internal 56-Kb/s V.90 data/fax modem

Source: Intel & Microsoft

PC sales in 1999 were marked by tumbling prices and falling profit margins, although the overall number of units sold continued to increase. Worldwide sales of PCs topped 113.5 million units, an increase of 21.7% over 1998. Nearly 45 million MPCs were shipped in the United States in 1999, creating $78.1 billion worth of sales. Compaq ranked first in sales, with Dell in second and IBM ranking third (Gartner Group's Dataquest says worldwide, 2000; Volpe, 2000). In January 2000, the industry was surprised by PC prices climbing to $100 to $200 more than the same model cost four months earlier. The price increases were expected to cease within the next quarter, however (McWilliams, 2000). Microsoft controls 95% of the worldwide market for PC operating systems. When Apple computers are included with PCs for consideration, Microsoft controls 80% of the worldwide OS market (Excerpts from, 2000).

It is significant to note that Apple Computer exhibited a major turnaround in 1999, propelled primarily by the introduction of the iMac computer. The iMac continued Apple's tradition of innovative design and proprietary operating systems and generated significant profits for the company. Apple's financial success was significant, but the company's U.S. market share remained relatively low, with less than one out of 20 computers sold during 1999 using any of Apple's operating systems.

Table 10.3

Top Five U.S. PC Shipments (in Thousands)

Manufacturer	1999 Shipments	1999 Market Share (%)	1998 Shipments	1998 Market Share (%)	Growth (%)
Compaq	15,035	13.2	12,785	13.7	17.6
Dell	11,123	9.8	7,361	7.9	51.1
IBM	8,932	7.9	7,613	8.2	17.3
Hewlett-Packard	7,242	6.4	5,388	5.8	34.4
Packard Bell	5,936	5.2	5,914	6.3	0.4

Source: Gartner Group

Advanced Micro Devices and Intel announced the latest advance in chip technology in March 2000. Both companies introduced 1-gigahertz processors within days of each other. AMD's Athlon processor was announced on March 6, and Intel followed two days later with an announcement about the Pentium III processor. AMD's processor is based on its 0.18-micron manufacturing process, with an external level-2 cache of 512 KB and 200 MHz system bus support. MPC manufacturers including Gateway and Compaq quickly announced 1 GHz systems using the new AMD chip (Spooner, 2000), while Dell announced plans for a Pentium III-based MPC release (Intel unveils, 2000). As impressive as these speeds seemed when they were announced, they are sure to be eclipsed by even higher-speed processors in 2001 and 2002.

Games

Internet games, including Internet connectivity embedded within console games, proved a strong factor in 1999. Game players accessed free games on the Web (in exchange for exposure to advertising), and multiplayer game networks, which usually charge subscription fees, showed signs of growth among the "hard-core" gaming audience. The 128-bit console systems, including Sony's PlayStation 2, Sega's Dreamcast, and Nintendo's Dolphin, will integrate Internet connections. Connectivity will require additional peripherals to be released later, which will allow Web browsing, multiplayer game play, and e-mail. Always hungry for new markets and increased marketing advantages despite the almost 2:1 advantage over the computer-based gaming market, the console manufacturers view Internet connectivity as a way to increase sales and open new revenue streams (Berst, 1999; Broersma, 1999).

Sony's PlayStation 2 debuted in Japan in March 2000, breaking industry sales records by selling 980,000 units in the first two days of sales (Sony, 2000). The console has the capability of playing movie DVDs and audio CDs, and eventually will have Internet connectivity. While PlayStation I sales accounted for an impressive 40% of Sony's profits in 1998, PlayStation 2 is expected to prove even more lucrative for the company (Levy, 2000).

In early 2000, Microsoft announced plans to enter the video game console market, marking the company's first venture into large-scale hardware development. Code-named the X-Box, this product is now scheduled for release in late 2001 (the system was originally scheduled for release in fall 2000) (Microsoft unveils, 2000). The system will use Pentium III technology and will be designed to compete directly with console industry leaders Sony, Nintendo, and Sega, just as these companies plan to hybridize their console technology to compete with Microsoft in the Internet access and browser markets. At the 2000 Game Developers Conference, Microsoft chairman Bill Gates asked game developers to make their best effort "to take what's in this platform and create amazing games around it" (Chang & Ward, 2000).

Factors to Watch

Multimedia developers that target the MPC platform continue to be highly sensitive to changes in platform standards. Rapid advances in the development of authoring and playback technology exacerbate the already contentious nature of standards development. This, in turn, delays software releases and revenue flow for developers.

As computer users, especially those who are new and inexperienced, grow weary of dizzying hardware and software advances and changes in standards, watch for the rise of alternatives to full-scale MPCs. Alternatives might include video game consoles and "thin client" operating systems such as Citrix MetaFrame. Thin client systems allow users to pay a monthly fee in exchange for the use of a Winterm (Webpad) device (instead of a PC) that plugs into a high-speed, high-bandwidth Internet connection. Upon connection, users will see their personalized desktop and enjoy access to a myriad of applications, including e-mail, without having to install any software or worry about upgrades (Arquette, 1999).

CD-Rs and CD-RWs will gradually fade out as the higher-capacity DVD-video and DVD-ROM standards coalesce and consumers increase adoption. Also, copyright protection technology will continue to improve, easing the fears of Hollywood and the recording industry about rampant illegal copying.

Interactive games using both computers and dedicated game consoles will soon offer enhanced interactivity, as manufacturers of both hardware and software work to allow users to connect over the Internet and other networks (including home networks). This will allow users to play games against or in cooperation with other players across the room or across the world. This increased level of interactivity will allow new capabilities, including new types of collaborative gaming. Ultimately, video games may evolve into home-based virtual reality systems that allow players to wield "light sabres" and other game peripherals as their images are captured and melded in real time into the game interface via tiny video cameras.

Bibliography

Americans overwhelmingly rate video/PC games as most fun entertainment activity for second straight year. (1999, May 13). *Interactive Digital Software Association.* [Online]. Available: http://www.idsa.com/releases/consumerusage99.html.

Arquette, B. (1999, October 10). The looming demise of the PC. *PC Week.* [Online]. Available: http://www.zdnet.com/filters/printerfriendly/0,6061,2347754-54,00.html.

Berst, J. (1999, December 10). Five reasons your next PC could be a Nintendo. *ZDNet Anchordesk.* [Online]. Available: http://chkpt.zdnet.com/chkpt/adem2fpf/www.anchordesk.com/story/story_4219.html.

Broersma, M. (1999, May 9). Games get dot-commed. *ZDNN.* [Online]. Available: http://www.zdnet.com/filters/printer-friendly/0,6061,2254921-2,00.html.

Chang, G., & Ward, D. (2000, March 11). Playing for keeps. *San Antonio Express News*, D-1.

C|Net. (2000, March 8). *Glossary.* [Online]. Available: http://coverage.cnet.com/Resources/Info/Glossary/.

Entertainment Rating Software Board. (2000 March 25). [Online]. Available: http://www.esrb.org.

eTForecasts. (2000, February 25). *Worldwide PC sales will surpass 200M units in 2005.* [Online]. Available: http://www.etforecasts.com/pr/pr200.htm.

Excerpts from the ruling that Microsoft violated antitrust laws. (2000, April 4). *New York Times,* C14.

Game programmers getting a piece of the pie. (2000, February 22). *Business Wire.* [Online]. Available: http://www.businesswire.com/webbox.bw.022200/200531928.htm.

Gartner Group's Dataquest says CD-RW shipments will nearly double in 2000. (1999, December 20*). Gartner Group Dataquest.* [Online]. Available: http://www.gartner4gartnerweb.com/dq/static/about/press/pr-b9973.html.

Gartner Group's Dataquest says worldwide PC market topped 21% growth in 1999. (2000, January 24*). Gartner Group Dataquest.* [Online]. Available: http://www.gartner4gartnerweb.com/dq/static/about/press/pr-b200004.html.

Grotta, D., & Grotta, S. (1999, November 3). CD-RW hits the road. *PC Magazine.* [Online]. Available: http://www.zdnet.com/filters/printerfriendly/0,6061,2387133-3,00.html.

Herman, L., Horwitz, J., & Kent, S. (2000). *History of video games.* [Online]. Available: http://videogames.gamespot.com/features/universal/hov/index.html.

Holsinger, E. (1995). *How multimedia works.* Emeryville, CA: Ziff-Davis.

Hudson, T. (1998). Multimedia computers and video games. In A. Grant & J. Meadows (Eds.). *Communication technology update, 6th ed.* Boston: Focal Press, 89-105.

Intel Corporation and Microsoft Corporation. (1999). *PC 99 System Design Guide.* [Online]. Available: http://www.pcdesguide.org/pc99/default.htm.

Intel Corporation and Microsoft Corporation. (2000). *PC2001 System Design Guide.* [Online]. Available: http://www.pcdesguide.org/pc99/default.htm.

Intel unveils its one gigahertz chip. (2000, March 8). *MSNBC Technology.* [Online]. Available: http://www.msnbc.com/news/379363.asp.

Interactive Digital Software Association. (1999, September 21). *Video game console manufacturers to provide ratings information with purchase of all hardware.* [Online]. Available: http://www.idsa.com/releases/ESRB-Ratings.html.

Interactive Digital Software Association. (2000, February 18). *U.S. computer and video game publishers lost nearly $3 billion worldwide to software piracy in 1999.* [Online]. Available: http://www.idsa.com/releases/021800.html.

International Data Corporation. (2000, January 25). *IDC puts the worldwide packaged software market at $154 billion in 1999.* [Online]. Available: http://206.35.113.28:8080/Data/Software/content/SW012500PR.htm.

Internet Content Rating Association. (2000, March 25). [Online]. Available: http://www.icra.org.

Leemon, S. (1999, November 14). CD-RW: The write stuff. *Computer Shopper.* [Online]. Available: http://www.zdnet.com/filters/printerfriendly/0,6061,2388413-3,00.html.

Lemos, R. (2000, March 7). Gaming's battle for your living room. *ZDNet News.* [Online]. Available: http://www.zdnet.com/zdnn/stories/news/0,4586,2457243,00.html.

Levy, S. (2000, March 6). Here comes PlayStation 2. *Newsweek U.S. Edition.* [Online]. Available: http://www.newsweek.com/nw-srv/printed/us/st/a16816-2000feb27.htm.

Macromedia, Inc. (2000, March 9). The evolution of man and machine. *Macromedia Director: Determining the Appropriate minimum hardware playback platform.* [Online]. Available: http://www.macromedia.com/support/director/how/expert/playback/playback07.html.

McWilliams, G. (2000, January 13). Surprise! PC prices climb. *ZDNet Business.* [Online]. Available: http://www.zdnet.com/zdnn/stories/news/0,4586,2421803,00.html.

Microsoft unveils plans for X Box. (2000, March 10). *X Box.* [Online]. Available: http://www.xbox.com/press.htm.

Morris, J. (2000, March 6). 1 GHz: A new standard for desktops. *MSNBC Technology.* [Online]. Available: http://www.msnbc.com/news/378734.asp.

Murray, J. (1997). *Hamlet on the holodeck: The future of narrative in cyberspace.* New York: Free Press.

The NPD Group. (2000, March 6). *TRSTS video games top-sellers ranked on units sold month of January 2000.* [Online]. Available: http://www.npd.com/corp/videogames/videogames/c_vg0001.htm.

Packaged software industry revenue and growth. (2000, March 10). *Software and Information Industry Association.* [Online]. Available: http://www.siia.net/pubs/research/softwareoverview.htm.

Poor, A. (2000, March 6). DVD RAM standards update. *ZDNet.* [Online]. Available: http://www.zdnet.com/pcmag/stories/reviews/0,6755,2444616,00.html.

Reuters. (1997, April 22). Intel and AMD settle MMX lawsuit. *Communications Media Center at New York Law School.* [Online]. Available: http://cmcnyls.edu/Public/Bulletins/InAMDSet.HTM.

Sony Computer Entertainment America, Inc. (2000, March 6). *PlayStation sales reach 980,000 units during opening weekend in Japan.* [Online]. Available: http://www.scea.com/news/press_example.asp?ReleaseID=9558.

Spooner, J. (2000, March 6). It's official: AMD first to 1,000 MHz. *MSNBC Technology.* [Online]. Available: http://www.msnbc.com/news/378383.asp.

Venezia, C. (1999, May 25). Inside next year's PC. *PC Magazine.* [Online]. Available: http://www.zdnet.com/pcmag/stories/reviews/0,6755,2161864,00.html.

Volpe, N. (2000, February 28). PC makers move beyond the box as growth slides. *CBS MarketWatch.* [Online]. Available: http://www.marketwatch.newsalert.com/bin/story?StoryId=ColOa0b9DtJi4mtaXmtC4&FQ.

11

The Internet & the World Wide Web

Philip J. Auter, Ph.D.*

hen this chapter was first published in 1996, the Internet and the World Wide Web were just beginning to enter the global consumer arena. The Internet had advanced from an experimental to an elite stage of communication—and was on its way to becoming a form of mass communication.

Today, just four years later, the "Net" has become a major influence for people young and old around the globe. Used so regularly for so many diverse purposes, it is safe to say that the Internet has truly become a new channel for all forms of human communication. Emphasis is now on stabilizing this constantly-changing medium and finding ways to penetrate new and ever expanding markets.

The Internet is a way of linking many small, local area networks (LANs) into a global database of amazing proportions. The interconnections allow instant access to a world of information. Each individual network is administered, maintained, and paid for separately by individual commercial, government, private, and educational institutions or individuals (Eddings, 1996; Gaffin, 1996). Initially, Internet access was free, but of limited availability. Various ways to pay for the growth of this medium have been attempted, including pay Websites, Internet service provider (ISP) subscription fees, and advertising. Today, most (but not all) sites are "free" in that they are supported almost exclusively by advertising. Standard 56K modem ISPs range from free (supported by advertising bars) to $19.95 per month. Cable and DSL (digital subscriber line) ISPs provide a much faster service, but charge quite a bit more.

Although terms like Internet and Web have become synonymous for all forms of computer-mediated communication, a more accurate way to define the Internet is as a channel for a variety of new forms of communication. Some of the most important services provided by the Internet

* Assistant Professor, Department of Communication, University of West Florida (Pensacola, Florida).

include electronic mail, chat, file transfer, access to remote computer systems, newsgroups and mailing lists, search capabilities, and the World Wide Web.

Electronic Mail

Electronic mail, or e-mail, is a way of sending an electronic letter, nearly instantly, between any two people in the world who are linked via LAN, computer, modem and phone, or even fax machine. Although basic concepts behind e-mail parallel those of regular mail, e-mail offers the speed of a telephone call with the detail of a letter. Messages can be one-on-one, like regular mail, or can be used for mass distribution (Eddings, 1996; Gaffin, 1996).

Chat

Unlike e-mail, chat is the utilization of computers and the Internet for real-time communication between two or more individuals. Chatting can be done privately using one-to-one devices such as Yahoo's "messenger" by text (Yahoo, 2000) or in one of the thousands of chat "rooms" dedicated to discussion of a variety of specific topics.

A more recent addition to online chatting is voice chat. Several of the messengers, including Yahoo's, are capable of doing quasi-real-time voice chat if both people involved have PCs equipped with speakers, a microphone, and a sound card. Other options include PC-to-PC "phone calls" and PC-to-telephone calls (e.g., www.dialpad.com). All of these services are "free," meaning they are supported by advertising. However, network congestion on traditional 56K connections has limited their viability as of mid-2000.

File Transfer

File transfer protocol (FTP) allows computer users to download and upload shareware programs, documents, and pictures to and from databases that are stored in archives at hundreds of sites around the world. FTP allows the user to log-in to distant computers to access and transfer files. Files are typically compressed to save space and speed transfer and must be decompressed before they can be used (Eddings, 1996; Gaffin, 1996). Many PC users take advantage of programs like WSFTP to upload changes to their Web pages. Downloading software, etc. can often be done through a Web browser with no additional program required.

Access to Remote Computer Systems

Telnet is a terminal emulation program for TCP/IP networks that allow a user to access a remote terminal from a local desktop. Once connected, the user enters commands locally to control the remote host, providing access to remote and otherwise closed Intranets containing databases and services (Gaffin, 1996; Webopedia, 1998b). With the exception of some Unix-based corporate and educational work environments, most telnet services have been shifted to the graphical Web browser environment. In fact, if you do a search for telnet information, you'll find that most of the links have expired.

A gopher was a type of client/server software using telnet and other Internet applications to send information back-and-forth between the gopher server and a distant client. Gopher servers once existed on almost every large, publicly accessible computer on the Internet, allowing access to a system's resources, including FTP, search programs, newsgroups, and much more. Today, most gopher databases have been converted to Web systems. Gopher was originally created at the University of Minnesota (the "Golden Gophers"), hence the name. The name also refers to the fact that the software "burrowed" through the Internet to find information (Eddings, 1996; Gaffin, 1996; Webopedia, 1997a).

Today, most access to remote hosts is gained directly through a Web browser. In instances where this is not the case, a proxy server may be utilized. Proxy servers sit between a client application and a real server, intercepting all requests, filtering them, improving performance, and modifying information to fit more basic connection appliances (Webopedia, 1999). An example of the latter system is Puma Technologies' ProxiWeb (www.proxinet.com). Users of the PalmPilot PDA (personal data assistant; see Chapter 23) can install this software and surf the Web with a rich, graphical experience, even though PDAs have very limited memory. The ProxiWeb software submits all requests to a server, which obtains Web pages, but then distributes the data before shipping the information to the PDA via modem.

Mailing Lists and Newsgroups

A mailing list is an Internet database of people interested in a particular topic. Anyone can post to a list, but only subscribers will receive the posts. When a message is sent to the list, it automatically goes to *all* list members.

A moderated list is screened by an administrator for duplicate messages or unacceptable content. Unmoderated lists have an administrator, but she or he does no censoring of messages whatsoever. List server software is even more automated and can subscribe, unsubscribe, and perform several other commands without the aid of an administrator (Eddings, 1996; Gaffin, 1996).

Newsgroups work much like mailing lists and may be moderated or unmoderated. Unlike mailing lists, however, newsgroup data is stored on all local sites that subscribe to the newsgroup. Because of this, only the most recent portion (thread) of the discussion may be available. Some sites do archive data, however (Eddings, 1996; Gaffin, 1996).

The World Wide Web

The World Wide Web is the most recent and exciting development on the Internet. A client/server-distributed system, the Web creates a rich graphical environment, incorporating enhanced text, graphics, sound, and moving visuals. Creative use of newer technologies—including cgi scripts, Java applets, and cookies—have allowed the Web to evolve into a powerfully interactive communication environment. For individuals whose computers, PDAs, cell phones, or other Internet devices cannot handle the information transfer load of the Web, a text-based equivalent known as "Lynx" is an option.

A Web browser, such as Microsoft Internet Explorer or Netscape, creates a unique, hypermedia-based menu on your computer screen. "Hypermedia" is the foundation of the Web and is a different, nonlinear way of linking data. One can jump around in hypermedia documents, clicking on highlighted topics and switching to new documents, which are often in entirely different locations on the globe. Web servers maintain pointers or links to data spread out over the entire Internet (Eddings, 1996; Gaffin, 1996).

The *best* Web pages are true hypertext documents, taking advantage of the concept of nonlinear information linking by allowing users to jump from concept to concept at many logical points. The *worst* hypertext documents fall into two extremes: traditional text works simply "dumped" into the Net with few or no links, and documents where the author has appeared to have gone "hyperlink crazy."

The Web browser handles all connections and switches. Netscape was once the dominant Internet browser because of its easy-to-use interface, and because Netscape Communications Corporation (http://home.netscape.com/) offered the software free as a download from their Website. But Microsoft's Internet Explorer (http://www.microsoft.com/ie/) has stolen much of Netscape's thunder. Although some services such as America Online have proprietary browsers, many are legally modified versions of Explorer or Netscape.

Bundled with the Windows operating system—a standard on most PCs purchased in America beginning in the mid-1990s—Explorer quickly landed on more desktops than Netscape could ever hope to. The U.S. government looked into claims of monopolistic tactics by Microsoft, even as Netscape's market share and stock value dropped. As of this writing, the courts are considering the breakup of Microsoft. If such an event occurs, Windows would most certainly be separated from Microsoft's Internet browser software.

Hypermedia documents are written in hypertext markup language (HTML), a series of codes that define the graphical layout and links on a page.[1] HTML is the key to the hypermedia format of the Web. Although already a powerful system, additional functionality has been added over the years with Java applets, JavaScript, and VRML (virtual reality modeling language).

Java applets are linked or imbedded mini-programs written in Java (originally developed by Sun Microsystems in 1995). Web browsers equipped with Java virtual machines can interpret applets from Web servers. Because applets are small in file size, cross-platform compatible, and highly secure, they are ideal for small Internet applications accessible from a browser such as a loan or mortgage calculator (Webopedia, 2000). JavaScript is a scripting language developed by Netscape to allow Web designers to create interactive sites. Despite the fact that it shares many of the attributes of Java language (the foundation of Java applets), it was developed independently (Webopedia, 1997b). Web authors have found additional ways to enhance their sites with plug-ins from other companies including RealMedia, Cold Fusion, and Macromedia, which allows for rich, multi-dimensional Websites.

Developed in the late 1990s, VRML is a three-dimensional version of HTML, serving up interactive 3-D worlds on the Web. VRML was originally expected to replace HTML and offer a fully-interactive 3-D Net environment. But use of the language has been slow to catch on due to current limitations in data transfer rates.

---

1. See *A Beginner's Guide to HTML* at http://www.ncsa.uiuc.edu/General/Internet/WWW/HTMLPrimerAll.html.

Philip J. Auter's Home Page

Copyright 2000, Philip J. Auter

Table 11.1
Partial HTML Source Code for Phil Auter's Home Page

```
<html><head>
<meta http-equiv="Content-Type" content="text/html; charset=iso-8859-1">
<meta name="GENERATOR" content="Microsoft FrontPage 3.0">
<title>UWF Faculty - Philip J. Auter</title></head>
<body bgcolor="#FFFFFF">
<table border="0" cellpadding="0" cellspacing="3" width="551" height="224">
  <tr>
    <td colspan="3" width="547"><!--webbot bot="ImageMap" startspan
    rectangle=" (400,2) (481, 32)  http://www.uwf.edu/"
    rectangle=" (262,0) (389, 32)  http://www.uwf.edu/~commarts/"
    rectangle=" (189,3) (252, 32)  http://www.uwf.edu/~commarts/faculty/index.html"
    src="http://uwf.edu/commarts/images/uwfbar.gif" align="right" border="0" hspace="0" --><MAP
NAME="FrontPageMap0"><AREA SHAPE="RECT" COORDS="400, 2, 481, 32" HREF="http://www.uwf.edu/
"><AREA SHAPE="RECT" COORDS="262, 0, 389, 32" HREF="http://www.uwf.edu/~commarts/"><AREA
SHAPE="RECT" COORDS="189, 3, 252, 32" HREF="http://www.uwf.edu/~commarts/faculty/index.html"></
MAP><img src="http://uwf.edu/commarts/images/uwfbar.gif" align="right" border="0" hspace="0"
usemap="#FrontPageMap0"><!--webbot
    bot="ImageMap" i-checksum="7800" endspan --></td>
  </tr>
```

Source: P. J. Auter

WYSIWYG Page Development Software

Until the mid-1990s, Web page authors were forced to work directly with arcane HTML code. Rather than focus on creativity, interactivity, and graphic design, home page managers had to learn to decipher the less-than-user-friendly HTML tag system. Finally, in the mid-1990s, a number of software producers began producing WYSIWYG (what you see is what you get) software that allows home page producers to focus on the look of their pages—laying out text and graphics in a manner similar to traditional desktop publishing applications. One popular WYSIWYG Web authoring program is Adobe PageMill (http://www.adobe.com/products/pagemill/main.html).

Like other WYSIWYG Web authoring software, PageMill was designed for the non-technical user. Pages are written and designed in a word processor-style environment. Results appear as they will when accessed by Web browsers such as Netscape and Internet Explorer. Styles can be applied and images resized, and PageMill checks and corrects URL (uniform resource locator) links as they are copied and pasted throughout Website documents (Adobe, 2000). Newer WISI-WYG Web page software such as Dreamweaver allows advanced features (such as forms, cgi scripts, and Java applets) to be previewed. For those who can't afford the cost of WYSIWYG Web authoring software, Microsoft includes a "lite" version of Microsoft Front Page with the free download of Internet Explorer (see http://www.microsoft.com/catalog/display.asp?site=808&subid=22&pg=2). The only problem is that inexperienced users may not know how to debug even the most basic HTML code when it is badly formed.

The Domain Name System

All of these applications rely on the assumption that each computer in the "network of networks" can find any other computer linked in cyberspace. The domain name system (DNS) establishes a numeric Internet protocol address (e.g., 123.45.67.8) for every single computer account with Internet access. Major domains maintain lists and addresses of other domains at the next level down, and so on to the end computer user. Servers located all over the Internet maintain databases of domain names and their numeric equivalents. These servers perform the translation between cryptic numbers and easy-to-remember names, vastly simplifying navigation throughout cyberspace.

Most U.S. domains use three-letter identifiers and are divided by application or theme. The primary U.S. domains are gov (government), org (other organizations on the Internet), edu (educational institutions), com (commercial companies connected to the net), mil (military installations online), and net (companies and groups concerned with Internet administration). The rest of the world has been using two-letter country codes as their top domains from the beginning (Eddings, 1996).

In reading an address, domain hierarchy goes from right to left. In pauter@uwf.edu, for example, the major domain is "edu" (for educational institution). The next domain is "uwf" which represents the University of West Florida in Pensacola; "pauter" is the address on that server for Professor Phil Auter's e-mail account. With this information, anyone in the world with Internet access can e-mail Phil Auter a letter.

Web pages have similar addresses, known as URLs (uniform resource locators). They are based on the same system, but can sometimes be quite a bit longer. Take, for example, Phil Auter's academic vita Web page address: http://uwf.edu/pauter/Auter/autervita.html. In this case, "http://" is the standard direction to a Web browser that it will be searching for (and retrieving) a document using hypertext transfer protocol, done when searching for all Web pages. Since http:// is the standard beginning of a Web page URL, newer browsers are configured so that you do not need to type in a direction unless you are planning to use the software for something unobvious (e.g., "telnet://" or "ftp://"). The next section, "uwf.edu," is the domain of the computer at the University of West Florida on which campus Web pages exist. "pauter" and "Auter" are two levels of subdirectories on the uwf system. Finally, "autervita.html" is the actual page. The ".html" (sometimes only ".htm") suffix is a common, but not mandatory, appendage to many Web page files. Other common suffixes are ".cfm" (Cold Fusion) and ".wml" (VRML pages).

Although some computer literacy is required, the Internet and Web continue to become more user-friendly, more visually appealing, more globally accessible, and more necessary in our daily lives. Much has changed about this new medium in its relatively short history.

Background

The history of the Internet and World Wide Web dates back to Cold War tensions between the United States and the former Union of Soviet Socialist Republics. The U.S. government formed the Advanced Research Projects Agency (ARPA) to establish a lead in science and technology applicable to the military in the 1960s. ARPA worked with the RAND Corporation to create a successful way to communicate *after* a nuclear war. RAND came up with the idea of a "network" with no central authority that could operate even if a number of its major nodes lay in ruin (Gaffin, 1996).

During the 1960s, researchers experimented with linking computers to each other and to people through telephone hook-ups using funds from ARPA. They developed a packet switching technology that allowed multiple users to share the same data lines, unlike any telephone lines to at the time (Gaffin, 1996).

The first node of the new network was at UCLA. Stanford, the University of California at San Bernadino, and the University of Utah completed the original four-node system known as ARPANET by December 1969 (Zakon, 2000). Shortly thereafter, individuals with access to the Net developed electronic mail, remote log-in to distant computers, and the ability to transfer files via data lines. The world now had data communication (and data transfer) at the speed of a phone call. ARPANET began online conferences with the help of several college (and one high school) students. Elite scientific discussions later changed to more general, mass appeal topics (Gaffin, 1996).

During the 1970s, ARPA supported the development of internetworking protocols for transferring data between different types of computer networks (Gaffin, 1996). Many LANs being developed in both academia and commercial industries began establishing gateways to the Internet to allow for electronic mail transfer. Links developed between ARPANET and counterparts in other countries (Gaffin, 1996).

In 1983, this network of networks collectively became known as the Internet. At about this time, ARPANET was split into ARPANET and MILNET (the Military Network). The Defense

Communication Association required the two subgroups to use TCP/IP protocols at all stations (allowing internetwork communication). This development also made it possible for additional nodes to come online. The domain name server addressing system was established in 1984 (Zakon, 2000).

In the mid- to late-1980s, the Internet expanded at a phenomenal rate. Hundreds, then thousands, of colleges, research companies, and government agencies began to connect their computers to this worldwide network. The National Science Foundation (NSFNET) began providing backbone service to U.S. supercomputers by 1986, leading to an even faster dissemination of the Net from purely government into university and commercial arenas (Zakon, 2000). NSFNET linked mid-level Nets that, in turn, connected universities, LANs, etc. Also in 1986, the first freenet—a network designed to provide an Internet link to individuals with limited access to traditional avenues—was formed in Cleveland, Ohio and administered by Case Western University (Eddings, 1996).

In 1989, Tim Berners-Lee created the World Wide Web while working at CERN, the European Particle Physics Laboratory (Berners-Lee, 2000). However, the Web was not released to the public until 1991 (Cailliau, 1995; Zakon, 2000). This text-only Web, although a powerful research tool, did not take the public by storm. It would take a more graphical and user-friendly Mosaic-enhanced Web to begin doing that in 1993 (Zakon, 2000).

In 1990, the Electronic Frontier Foundation (EFF) was formed by the founder of the Lotus Development Corporation, Mitchell Kapor, and John Perry Barlow to "address social and legal issues arising from the impact on society of the increasingly pervasive use of computers as a means of communication and information distribution." EFF (http://www.eff.org/) began its mission of encouraging public debate on telecommunications and society issues, as well as supporting litigation that extends First Amendment rights to Internet-published work (Electronic Frontier Foundation, 1990). The same year, ARPANET was decommissioned by the Defense Communications Agency because NSFNET and mid-level nets superseded it (Zakon, 2000).

Until the early 1990s, the Internet community was strongly against going commercial. Because the Net had been primarily supported by government and educational institutions, its primary purpose was considered to be research and education, although it was becoming more and more of an entertainment medium. Attempts to post advertisements to listservs and discussion groups were violently opposed by members. Later, some usenet newsgroups were specifically developed as "classified ad" services, where individuals could post information about second-hand items for sale.

Commercial companies did not really begin to participate in the Internet until the emergence of the World Wide Web. With the introduction of the graphically-oriented Web, the growing number of commercial online services, and the dissemination of the Net to wider areas of society, the definition of the Internet changed dramatically. Such alien concepts as subscriptions, ads, leasing arrangements, and product sales were not only becoming accepted, it became more and more apparent that they would shoulder much of the burden of financially supporting the rapidly growing medium (Vincent, 1995).

Yahoo! (http://www.yahoo.com/), one of the first Web search engines, was released to the Internet public in 1994. More recently, Digital Equipment Corporation introduced AltaVista (http://

www.altavista.com/), a service that makes heavy use of multi-threading 64-bit addressing capabilities to search for any word in any document published on the Web (including "hidden" text) or in Usenet discussion groups (AltaVista, 2000). Led by pornography sites, many Web page developers soon realized that more people would find their sites if dozens of key search words were hidden within the HTML source code.

Another way to drive traffic to sites was the registration of multiple domains that all redirect the surfer to the same site. Quite a few search engines have sprung up since then, many of which also offer "people finders," mapping programs, and even Web page language translation. More recently, "meta-search engines" like Dogpile.com have been developed that allow you to enter search criteria once, then survey the results of multiple search engines. (An excellent database of search engines can be found at http://hawking.NHGS.Tec.VA.US/ internet_resources_index.html.)

According to Zakon (2000), Radio HK, the first 24-hour, Internet-only radio station began offering online programming in 1995. (It ceased operation in 1996.) Leaders in audio streaming since 1995, Real Networks (http://www.real.com/) announced RealVideo software for real-time streaming of video across the Net in 1997 (Real Networks, 1997). Although many people still do not have computers or Internet connections powerful enough to take full advantage of video streaming, the merging of the TV and the PC had begun. (For more on streaming media, see Chapter 8.)

Web pages, originally useful because of their hypertext links, began evolving into rich, textured multimedia documents in the mid-1990s. CGI scripts and Java applets (mini-applications) allowed the Net to become interactive with forms, hit-counters, calculators, and other scripts, making the Web an even more powerful tool for business, education, and entertainment (Sun Microsystems, Inc.). Image maps—pictures that include several imbedded hyperlinks—offer users a more visual method of navigating through a database of information. Animated icons, streamed audio and video, and plug-ins such as Macromedia's Shockwave Flash (http://www.shockwave.com/) now enhance the graphical interface with sound and motion.

In 1996, America Online initiated a revolution in Internet service pricing by offering unlimited access for only $19.95 per month. Demand swelled as users switched from other Internet service providers, and as many new users entered the market. Even though AOL spent millions of dollars upgrading their service, and other providers lowered their prices to match AOL, there was still a brief bottleneck as demand far outstripped access. While the original Internet experienced growing pains, the government began to fund "Internet 2" (http://www.internet2.edu). The next-generation Internet (NGI), Internet 2 is a new backbone that could be up to 1,000 times faster than the existing system (Cronise, 1998).

Commercialization of the Internet allowed for an exponential growth in for-profit and subsidized sites, which generated greater demand from a larger user-base. Once the Internet went commercial, profitability became an issue. One attempt to stimulate online purchasing was "cybercash" accounts at Net banks like First Virtual. The need for these "middlemen" in online purchasing has almost vanished, however, due to the relative safety of a secure socket layer (SSL) of communication developed by Netscape and now widely adopted (Microsoft Corporation, 1999).

The late 1990s saw the Internet wielding a significant impact on people and businesses of all types—and many were concerned. In 1996, U.S. telecommunications companies asked Congress to ban Internet telephone technology, fearing that free long distance would put them out of business. The controversial U.S. Communications Decency Act (CDA) became law, prohibiting distribution of indecent materials over the Net. One year later, the Supreme Court unanimously ruled most of it unconstitutional (Reardon, 1997). Around the world, many countries put restrictions on ISPs and Internet users, including China, Saudi Arabia, Germany, and New Zealand. Sites all over the world, including Internic, NASA, and the Indian government, are hacked for a variety of reasons. More and more e-mail viruses and worms are introduced, with three of the most infamous being "Melissa" and "ExploreZip" in 1999, and "ILOVEYOU" in 2000. Interestingly, though, the much-feared Y2K bug had almost no effect on the Internet, e-mail, or most of the world's computers (Zakon, 2000).

Even as some individuals and organizations grew concerned about the Internet's potential negative impact, "dot com" businesses had an explosive impact on the stock market. New online businesses and cheaper, Net-based investing has brought "Main Street" closer to "Wall Street," and both streets appeared to be somewhere in Las Vegas. As secure socket and fill-out form technologies improved, a number of Internet-based brokerage houses sprung up, offering deep discounts on commissions by minimizing traditional costs. Simultaneously, the longest running bull market in history was further fueled by an explosive tech and Internet stock sector. Unheard of gains occurred over and over again as new Internet stock initial public offerings (IPOs) caught investors' fancies and exploded by hundreds of percent—sometimes overnight. As of this writing, the Dow and NASDAQ are off their stellar highs of 11,000 and 5,000 respectively, but they are still significantly above their levels of a few years ago, and thousands of people who have never invested before are now in the stock game, most through online investing sites. This trend concerns some traditional professionals; they are legitimately worried that some investors lack the knowledge necessary to make these trades, yet have become addicted to the experience. "Gambler's Anonymous" has started getting calls from "average individuals" who have become hooked on day trading. Traditional investment professionals are also worried at the narrowing profit margins in the stock trading business as investors cut out one layer of middleman.

As of mid-2000, mergers, acquisitions, and IPOs continued to be the theme of big business—especially Internet businesses—even as the Justice Department continued a long, heated battle to break up Microsoft.

The gambling theme also carried over into another of the latest Internet crazes—auction sites. From "ebay" to Amazon.com, auction sites sprang up all over, turning the Net into a giant flea market. Ebay came under fire for allowing the sale of firearms on its site, a practice that it soon abolished. Additionally, some sellers had given ebay and other sites a bad name by reneging on deals after buyers sent their checks, or by misrepresenting the items they were selling.

Over the last few years, the Net and the Web have grown at exponential rates. In response, government and other users have tried to expand the Net itself and increase the speed of data transfer. The Internet backbone has been upgraded again and again. Proposals have been made for more new generic top-level domains (e.g., "dot com") for the United States, as more and more country domains are being registered. The U.S. Department of Commerce began making plans to privatize the DNS in order to increase access and global competition in the area. The government has begun testing Internet2, a system that should work at approximately 100 times the speed of the fastest traditional T1 connection. At such a phenomenal rate of change, recent social and technological developments are quickly becoming outmoded. With that in mind, here are some of the most recent developments involving the Internet and WWW.

Recent Developments

Events occur in cyberspace at an alarming speed and in an incredible quantity. Recent developments involving the Internet have occurred in the areas of technology, litigation, and commerce, just to name a few. What follows is a short summary of just a few of the important developments that were occurring in mid-2000. They may be ancient history by the time you read this chapter. (If you would like a regular, summarized update on events in information technology, consider subscribing to the free *Edupage* e-mail newsletter at http://www.educause.edu/pub/edupage/edupage.html.)

Internet access options have changed while cable companies roll out digital modem services for about $50 per month (http://www.cox.com/CoxatHome/?c=faq.asp&), and telephone companies are racing to release DSL (http://www.dsl.com), which allows for Internet and the telephone to be simultaneously used over one line for about $25 per month. But if you're satisfied with 56K and on a budget, you may want to consider free Internet service. A number of national services provide free Internet service on the model offered by one of the earliest national companies in the field, Netzero (http://www.netzero.net). Subscribers download software that displays a 1-inch by 3-inch advertisement window that appears whenever they are connected to the Internet. The company makes its money solely through advertising profits. A new twist to this concept has been developed by 1stUp (http://www.1stup.com), a CMGi company, which co-brands its browsers/ad bar with other organizations such as Fox's *The Simpsons* (http://www.simpsons.com). 1stUp is also one of the first to display local as well as national ads in the ad bar window, presumably based on the local dialup used to connect to the service.

If free isn't good enough—and your monitor and patience are large enough to support multiple ad windows and still allow you to see what you're doing—you may want to consider "pay to surf" companies. These businesses pay you about $0.50 for every logged hour you are surfing on the Internet and have their ad window open. You are also paid a smaller amount for every referral you bring into the fold, a borderline pyramid scheme. The granddaddy of all of these organizations—and the one with the best reputation for actually paying—is AllAdvantage (http://www.alladvantage.com). Some of the other companies have had trouble with software and payouts. All of them have had to continually modify and update their software because ingenious PC users have developed "fake surfing" software.

Fraud continues to be close on the heels of new technology development. The MP3 format, which allows for high compression ratios and CD-quality sound, has permitted the posting, access, and often theft of copyrighted music on a global scale. Much like the cassette tape many years ago, most of the music taken by MP3 format is being done by individuals who feel it is within their rights to copy and distribute music to their friends and also to complete strangers. The recording industry feels differently, believing that each individual incident of a song being distributed this way without royalties being paid constitutes theft. The American Society of Composers, Authors & Publishers (ASCAP, http://www.ascap.com) licenses music use and offers Website owners four different type of agreements from which to choose in order to legally distribute music via the Internet (Kimmel, 1997).

The individual "thefts" add up to a large projected revenue loss. Sites such as MP3.com and Napster.com have inflamed the process, with the latter offering software and links that easily allow anyone to access, take, and redistribute music. But the industry has recently taken some countermeasures. An online music service invented by MP3.com was found to violate copyright law by a U.S. District judge. MP3.com has already reached a licensing agreement with Broadcast Music International (BMI), the other major agent of songwriters and composers. The agreement lets MP3.com play BMI's 4.5 million compositions via their Website for a fee. Presumably, any individual recording these songs off the MP3.com site would be considered the same as someone making a tape of a song off the radio (Bloomberg News, 2000). Perhaps even more seriously, the band Metallica fought back, suing Naptster.com and providing them with documentation of over 300,000 individuals, mostly high school and college students, that the group says have copied their music illegally (CNET News, 2000).

Net surfers may soon have to pay for copying music, but they may not have to pay sales tax on online purchases, at least for the next five years and maybe never. In May 2000, the Republican-dominated House Judiciary Committee approved legislation that extends the Internet tax moratorium until 2006. At the time of this writing, the legislation still had to be taken up by the full House, the Senate, and the President. Other bills, including a permanent ban on Internet access taxes, were pending (Shiver & Goldman, 2000). (The ban on state taxes on Internet purchases applies only to interstate transactions; state tax may still be required on transactions that don't cross state lines.)

Current Status

Determining the number of users and user demographics of the Internet and WWW as growing media is challenging at best. Surveys almost always consist of convenience samples and frequently cannot screen out multiple responses made by the same individual. It is often impossible to differentiate between households, individuals, and even complete strangers because, despite potential security issues, people often let others use their accounts. Additionally, there has seldom been a way to corroborate responses given to Internet-based surveys: A respondent claiming to be a 45-year-old female may actually be a 12-year-old male. Finally, even quasi-accurate results become quickly outdated as the Net community continues to grow at exponential rates.

A number of online surveys have been performed under a variety of conditions and with varying results—many are unscientific and seldom can they claim to be representative. (One listing of some of the Internet usage surveys is available at http://www.gvu.gatech.edu/user_surveys/others.) With that caveat in mind, what follows are excerpts from research reports produced by two respected institutions: the Graphics, Visualization & Usability (GVU) Center at Georgia Tech University in Atlanta; and CommerceNet in conjunction with Nielsen Media Research. Change occurs so rapidly that it is important to note that this information may already be out of date as of this printing.

The GVU Tenth Survey

The Graphics, Visualization & Usability Center was established at Georgia Tech in 1991 and has been performing biannual surveys of the Internet community for the last five years. Their most recently posted survey, the tenth, ran from October 10, 1998 through December 15, 1998, and collected over 5,000 responses worldwide. Participants are gathered through non-probabilistic methods including banner ads and cash incentives, and the researchers are aware of the limitations of the sample selection method (http://www.gvu.gatech.edu/user_surveys/survey-1998-10/#methodology). Survey results focused on three categories: technology issues, user demographics, and electronic commerce (Georgia Tech, 1999).

Technology Issues. Connection speed was a crucial issue in this sample, with 34% of the U.S. respondents in 1998 using a 56K modem, twice the number of Europeans using that speed modem. Three-fourths of the sample said they had recently upgraded their modem or intended to very soon. Of those who did not intend to upgrade, the majority were relatively new users of the Internet. As of this survey, Netscape Communicator was the Web browser of choice, far outstripping Microsoft Internet Explorer. E-mail and Web browsing for information and entertainment were far and away the primary uses of the Internet. Over 89% of the users sampled in 1998 knew that they had used Java or JavaScript programs, and 74% had worked with audio over the Internet. Chat use remained constant to earlier surveys at 61%. Newer technologies such as Web phone and Web fax all fell below 20%.

Most respondents preferred using the Web to a variety of daily tasks and diversions, including watching television. U.S. users are more inclined to use the Web than European users. Respondents under 21 and over 50 use the Web more on a daily basis than do those between 21 and 50 years of age (Georgia Tech, 1999).

User Demographics. The percentage of female users declined slightly to 33% in the 10th survey, but researchers were not sure if this was a trend or an artifact of the sample gathering method. New and novice users remained split almost equally between men and women in 1998, with the majority of the gender difference coming from respondents who had been working with the Internet for quite some time. The respondents, as in all the previous surveys, were overwhelmingly white (87%). However, younger respondents and those newer to the Net were more racially diverse. Still, African-Americans made up less than 5% of the respondents overall. The average age of respondents was 37. Most (47%) listed themselves as married, while the second largest group (32%) checked themselves as single. Most were quite educated, with 88% having at least some college education and 59% having obtained at least one degree.

In 1998, 49% of the respondents said they lived in suburban areas, while 37% lived in urban areas and 13% in rural communities. The vast majority of the sample (85%) was from the United States. Most of the respondents (37%) have been on the Web for four to six years. Seventy-nine percent access the Web from home on a regular basis, and 69% access from work instead of, or in addition to, their home access. Only 28% said they had accessed the Web from public terminals in 1998, although this percentage was higher for respondents between 11 and 20 years old (Georgia Tech, 1999).

Electronic Commerce. In 1998, the average respondent made a purchase from Web-based vendors less than once a month, but did make purchase decisions from Web information once or twice every month. Sixty-seven percent of the sample had paid for a personal purchase via the Web, and 59% for a professional purchase. Security was not the most important concern regarding Web shopping and purchases. Quality information was the most important consideration, followed by easy ordering, reliability, and finally security. Women were more concerned about security issues than men in this sample. Most respondents browse far more than buy. Still, 98% of the sample had made online purchases, although the percentage for novices was somewhat lower (85%). Software and hardware were the items most often purchased online, followed by books, music, travel, electronics, video, magazines, flowers, apparel, and banking services. In 1998, the average amount spent online during the previous six months was $500. Convenience was far and away the primary reason for people shopping and purchasing items online (Georgia Tech, 1999).

The CommerceNet/Nielsen Data

Unlike most surveys of Internet/WWW usage, results of the CommerceNet/Nielsen Internet Demographics Surveys are based on completed telephone-based interviews of U.S. and Canadian citizens. Nielsen Media Research attempts to employ rigorous scientific methodology in order to obtain a sample that accurately and proportionally reflected the population of Internet users and non-users in both countries (CommerceNet/Nielsen, 2000). Although their surveys have been widely debated, CommerceNet and Nielsen claim that their surveys can be generalized to the population with an acceptable margin of error. In addition to the data synthesized by their own rigorous sampling techniques, CommerceNet/Nielsen now offers summarized data of Internet usage from other sources as well.

CommerceNet/Nielsen Surveys. Some of the key findings of the most recent (April 1999) survey were that, in 1999, there were approximately 92 million Internet users in the United States and Canada that were 16 years of age or older, up from 79 million just nine months earlier. This number is 20 million below their projected growth figure of 110 million, a number based on a 10% per year growth estimate (http://www.commerce.net/research/stats/wwwpop.html#IPOP). The number of women purchasing online increased 80% from the last survey. The number one category of items shopped for online was cars and car parts.

Other Summaries. CommerceNet/Nielsen has combined data from their North American and new United Kingdom surveys with information from other sources to project current global Internet usage trends (http://www.commerce.net/research/stats/wwstats.html). It is estimated that in 1999 there were 242 million Internet users worldwide, with almost half (120 million) residing in North America. Europe was in second place with 70 million, followed by Asia and the Pacific Rim (including Australia), which may have had 40 million users. South America, Africa, and the Middle East fell well below that with 8 million, 2.1 million, and 1.9 million users, respectively.

Computer Industry Almanac projected that there would be 259 million Internet users worldwide by 1999, a number similar to CommerceNet/Nielsen's. This group also forecast that this number would jump 90 million the following year, and be over 765 million by 2005. They also suggested that the percentage of all Internet users living in North America would decline from half to about one-quarter by 2005 as growth in usage explodes in other regions of the globe. Significantly, more conservative estimates put global Internet use at around 350 million by 2005 (CommerceNet/Nielsen, 2000).

CommerceNet/Nielsen has also produced a summary of Internet industry statistics compiled from press releases and other online sites (http://www.commerce.net/research/stats/advert.html and http://www.commerce.net/research/stats/indust.html). It is estimated that 1997 online transactions totaled US $9 billion, with business-to-business transactions accounting for $7.5 billion. It has been projected that the value of business-to-business transactions will increase five-fold by 2002. It has also been suggested that online investing will swell to 60% of the discount brokerage industry by 2004. The Internet Advertising Bureau announced that Internet advertising revenue had increased 322% in the first half of 1997, compared to the first half of 1996. The largest segment of Internet advertising in 1997 (30%) was consumer advertising, followed by financial services (22%), computing (21%), telecom (7%), and new media (7%).

Factors to Watch

If keeping track of recent developments is challenging, deciding what the future holds for the Internet is almost impossible. Future prognostications on this topic often lead to blind alleys and deserted "dot com" storefronts. Still, there are a few important, and not so important, items of interest on the horizon.

One of the most interesting and important stories to follow is the Justice Department's ongoing battle to break up Microsoft. The debate has been going on for years now, but the federal government, the courts, and many states all now seem to support some sort of dismantling of this technology giant. One of the most popular scenarios is that the company be divided into two competing firms: one that controls the Windows operating system and another that manages applications such as MS Office and the company's Internet presence. Microsoft has vowed to fight any decree all the way to the U.S. Supreme Court (Wilcox & Junnarkar, 2000).

Another area of Internet litigation to watch is government efforts to crack down on cyberstalking. California was the first state to put a cyberstalking law on the books in 1999. Now, Congress is debating a similar law. The Stalking Prevention and Victim Protection Act of 2000 has been approved by the House and, as of this writing, was under review by the Senate (Dean, 2000).

Law is also at issue concerning university involvement in distance learning, academic use of materials, and intellectual property. As federal grant dollars decrease, many schools are looking for ways to increase revenue and reduce the impact of global competition for students via the Web. Virtual universities, and their older "bricks-and-mortar" sisters are trying hard to figure out how to allow distant students to transfer credits taken at their schools, without allowing distant semester hours to be transferred in. It has been a long-standing situation in corporate America that anything you create "on the clock" is usually owned by the business, not you. However, because much of what academics created (outside of granted work) has produced three cents for every dollar invested, most universities have allowed employees to retain the rights to work created on university time using university facilities. With many potential online markets for this work, especially Web-based curricular material, some universities are rethinking their policies regarding intellectual property. (For more information on intellectual property issues in higher education, see http://www.nea.org/he/abouthe/intelprop.html.)

Another important trend to watch is how Internet access will spread through an already saturated U.S. market. "Web Without Wires" (http://www.gx-2.net/wwow/overview.html) is a term used for all the various ways the Internet may be accessible away from the traditional wired, high-end PC. Wireless PDAs, cell phones, and new Internet-only devices are some of the newer ways people might go online. Researchers are also coming up with ways to access the Web via traditional technologies including fax-to-Internet and phone-to-Internet via voice commands.

And if all this isn't good enough, the Web may soon be talking back to you. Motorola will soon release "Mya" (http://www.gx-2.net/wwow/mya_flash.html), an intelligent agent that will allow you to listen to the Web as she reads it to you over any telephone connection. In early 2000, the British Press Association introduced a digital newscaster named Ananova that reads the news and looks better than the cyber-newscasters featured in the popoular WB cartoon, *Batman Beyond*. Ananova can be heard and seen at www.ananova.com.

Finally, if the Web hasn't fully assaulted your senses yet, brace yourself—another one or two are about to be attacked. If you're an early adopter willing to spend about $400, you can now download tastes and smells from the Net using a device released in late April by TriSenx (http://www.trisenx.com). Users click on an image that has a digital scent or taste programmed into the Web page, and the system mixes water-based chemicals to create smells and tastes that are printed on a fiber card-stock paper. (For a light hearted, fictional look at this technology, check out the site www.realaroma.com. Published in 1996, this site purports new Web scent technology and RATML, the Real Aroma Text Markup Language.) The technology has attracted the immediate attention of fragrance and food companies, and additional applications are only limited by one's imagination. Despite the faddish sound of this development (remember how odd streaming video sounded at one time), this doesn't appear to be a one shot deal. Hot on the heals of TriSenx are DigiScents (www.digiscents.com) and AromaJet.com. These new devices may bring new meaning to the phrase "the Internet stinks."

Will the excitement never end? The ever-growing Internet may one day stop expanding, making way for a newer, more exciting form of communication. Perhaps one day it will—maybe sooner than you think. The *last* page of the Internet has already been published from Dayton, Ohio and can be found at http://home.att.net/~cecw/lastpage.htm. In case, for some typical technological reason, you follow this link and it's lost, broken, or outdated, here is what the page once read: "Attention Please: You have reached the very last page of the Internet. We hope you have enjoyed your browsing. Enjoy the rest of your life."

Bibliography

Note: The Web is an active environment, and, unfortunately, page addresses often change. Although links to Websites referenced in this chapter were accurate at the time of publication, pages may have since moved. To find a "lost" link, enter key words from the references into a search engine such as Snap (http://www.snap.com/).

Adobe Systems, Incorporated. (2000). *Adobe PageMill: Building and enhancing Web pages.* [Online]. Available: http://www.adobe.com/products/pagemill/feature1.html.

AltaVista Company. (2000). *AltaVista Company background.* [Online]. Available: http://doc.altavista.com/company_info/about_av/background.shtml.

Berners-Lee, T. (2000). *Bio.* [Online]. Available: http://www.w3.org/People/Berners-Lee/.

Bloomberg News. (2000, May 5). *MP3.com, BMI in music licensing deal.* [Online]. Available: http://news.cnet.com/news/0-1005-200-1823327.html?tag=st.cn.1.lthdne.

Cailliau, R. (1995). *A little history.* [Online]. Available: http://www.w3.org/History.html.

CNET News.com Staff. (May 4, 2000). *Lawyers explain copyright rules as bands point fingers.* [Online]. Available: http://news.cnet.com/news/0-1005-201-1816837-0.html?tag=st.

CommerceNet /Nielsen. (2000). *Internet statistics.* [Online]. Available: http://www.commerce.net/research/stats/stats.html.

Cronise, E. (1998, August 22). *Next Generation Internet (NGI) Initiative.* [Online]. Available: http://www.jhu.edu/~hac_ns/ngi/.

Dean, K. (2000, May 1). *The epidemic of cyberstalking. Wired News.* [Online]. Available: http://www.wired.com/news/politics/0,1283,35728,00.html.

Eddings, J. (1996). *How the Internet works,* 2nd Ed. Emeryville, CA: Ziff-Davis.

Electronic Frontier Foundation. (1990). *New foundation established to encourage computer-based communications policies.* [Online]. Available: http://www.eff.org/pub/EFF/Historical/eff_founded.announce.

Gaffin, A. (1996, December 11). *EFF's guide to the Internet, v. 3.20* (formerly *The big dummy's guide to the Internet).* Electronic Frontier Foundation. [Online]. Available: http://www.eff.org/pub/Net_info/EFF_Net_Guide/netguide.eff; also from online archives at ftp.eff.org, gopher.eff.org, http://www.eff.org/ and elsewhere. (Published in hardcopy by MIT Press as *Everybody's guide to the Internet.*)

Georgia Tech Graphics, Visualization & Usability (GVU) Center. (1999). *GVU's 10th WWW User Survey.* [Online]. Available: http://www.gvu.gatech.edu/user_surveys/survey-1998-10/.

Kimmel, B. (1997). *Distributing music over the Internet.* [Online]. Available: http://www.duke.edu/~bdk3/music.html.

Microsoft Corporation. (1999). *Secure sockets layer/transport layer security.* [Online]. Available: http://www.microsoft.com/security/tech/ssl/default.asp.

Real Networks, Inc. (1997, February 10). *Progressive Networks announces RealVideo, the first feature-complete, cross-platform video broadcast solution for the Web.* [Online]. Available: http://www.realnetworks.com/company/pressroom/pr/97/realvideo.html.

Reardon, S. (1997). *Supreme Court declares Communications Decency Act unconstitutional: Analysis.* [Online]. Available: http://www.ema.org/html/at_work/cdauncon.htm.

Shiver, J., Jr., & Goldman, A. (2000, May 5). *House* committee OKs 5-year ban on new Internet taxes. *Los Angeles Times.* [Online]. Available: http://www.latimes.com/news/front/20000505/t000042336.html.

Sun Microsystems, Inc. *The Java language: An overview.* [Online] Available: http://java.sun.com/docs/overviews/java/java-overview-1.html.

Vincent, C. (1995, October 25). *A cultural transition: The commercialization of the Internet.* [Online]. Available: http://swissnet.ai.mit.edu/6095/student-papers/fall95-papers/vincent-culture.html#introduction.

Webopedia. (1997a). *Gopher* (definition). [Online]. Available: http://webopedia.internet.com/TERM/g/gopher.html.

Webopedia. (1997b). *JavaScript* (definition). [Online]. Available: http://webopedia.internet.com/TERM/J/JavaScript.html.

Webopedia. (1997c). *VRML* (definition). [Online]. Available: http://webopedia.internet.com/TERM/V/VRML.html.

Webopedia. (1998a). *Chat* (definition). [Online]. Available: http://webopedia.internet.com/TERM/c/chat.html.

Webopedia. (1998b). *Telnet* (definition). [Online]. Available: http://webopedia.internet.com/TERM/T/Telnet.html.

Webopedia. (1999). *Proxy server* (definition). [Online]. Available: http://webopedia.internet.com/TERM/p/proxy_server.html.

Webopedia. (2000). *Applet* (definition). [Online]. Available: http://webopedia.internet.com/TERM/a/applet.html.

Wilcox, J., & Junnarkar, S. (2000, April 28). *Government to judge: Break up Microsoft.* [Online]. Available: http://news.cnet.com/news/0-1003-200-1777348.html?tag=st.ne.ni.rnbot.rn.1003-200-1777348.

Yahoo. (2000). *Yahoo messenger.* [Online]. Available: http://messenger.yahoo.com/.

Zakon, R. (2000). *Hobbes' Internet timeline v5.0.* [Online]. Available: http://www.isoc.org/guest/zakon/Internet/History/HIT.html.

12

Internet Commerce

Julian A. Kilker, Ph.D.*

In the popular media, electronic commerce (e-commerce) is associated with creating a booming "new economy," particularly in the United States, that is based on digital rather then physical commercial transactions. Fueled by the tremendous recent growth of Internet access and home computer ownership, e-commerce has increased substantially and is predicted to continue to do so for the next few years. Nevertheless, e-commerce is at an early stage, in which regulations, business models, formal and informal standards, and impacts on related business sectors are still emerging.

In 1999, the U.S. Bureau of the Census released a working definition of e-commerce as "any transaction completed over a computer-mediated network that involves the transfer of ownership or rights to use goods or services" (Mesenbourg, 1999). This definition does not take into account transactions conducted using other communication technologies, "free" transactions (such as downloading trial software), or barter and in-kind transactions (such as exchanging Web advertising placement for services). Most current definitions, and this chapter, emphasize commerce using the Web, although there are electronic predecessors to Web-based commerce. Business-to-customer (B2C) e-commerce has precedents in television shopping (which grossed approximately US $4 billion in 1998) and in the use of communication and information technologies for catalog shopping (which grossed over US $150 billion in 1998) (Grant & Meadows, 2000). Business-to-business (B2B) e-commerce, which has recently received media and corporate attention, is related to the earlier electronic data interchange (EDI), in which businesses electronically exchanged business documents in standardized formats over private rather than Internet data networks.

E-commerce relies on a wide range of hardware, software, and communications technologies. Thus, the major players in e-commerce combine established companies, such as General Electric, IBM, SAP, Oracle, Cisco, and AOL, with more recent Web-centric companies, such as Network Appliance, Commerce One, Ariba, and Verisign, to name a few. Virtually every business sector is—or will soon be, if current trends continue—using e-commerce technology. These businesses

* Assistant Professor, Hank Greenspun School of Communication, University of Nevada (Las Vegas, Nevada).

range from relatively new companies such as Amazon.com (a company founded in 1996 as an online bookstore, but has since branched into several product categories) to more established "bricks-and-mortar" companies with a physical presence such as General Motors. The latter have recently realized the value of e-commerce and are establishing "*clicks*-and-mortar" operations that join the efficiencies of e-commerce with traditional services and distribution.

Tracking the rapid pace of e-commerce development requires close monitoring of current business and technology sources. At present, e-commerce is a challenging research topic because of its rapidly changing nature because, by the time academic publications on the subject are published, the field has changed, and because competitive concerns limit detailed or critical coverage (Lohse & Spiller, 1999).

Background

The non-commercial roots of the Internet led many early proponents of the technology to extol its virtues as a medium unencumbered by commercial content. But the same technology that offered netizens an easy and efficient way to share messages, data, and entertainment has proven to be just as efficient at connecting businesses with their customers.

Indeed, the commercial online services that many consider to be predecessors to today's Internet, such as Compuserve, Prodigy, and AOL, all considered commercial transactions over their networks to be an important part of their business model. The pioneers who chose to take a chance on the nascent Internet such as Amazon.com and Travelocity were rewarded for being first by capturing incredible market share in the growing online market.

Because e-commerce replaces physical with digital transactions, it takes advantage of the time—and distance—collapsing nature of the Internet, as well as the data processing capabilities of computers to manage billing and order fulfillment. Ideally, e-commerce provides an efficient means for consumers and merchants to interact:

- Consumers can rapidly locate and compare product specifications and pricing.
- Merchants can reduce their physical presence, have automated electronic storefronts open around the clock, achieve economies of scale, and reduce payroll expenses.

E-commerce has been influential in three ways. First, it adds value to existing services or makes existing services more accessible; for example, banking and shipping companies provide online account or tracking information. Second, it substantially alters the nature of existing businesses:

- Online bookstores provide automated searches and reader comments.
- Online brokerages provide rapid, low-priced trades and immediate access to research documents.
- Online music stores can create custom compact discs and play audio samples.

Third, Internet commerce reduces barriers to the implementation of new business models; for example, eBay.com provides a forum for online trading, Priceline.com provides a "demand collection system" in which individuals make offers that are then aggregated to obtain low prices, and portal sites provide free services subsidized by advertising and tracking customer behaviors.

Recent Developments

Economic

The challenges of defining e-commerce and its rapid development make it difficult, but increasingly important to quantify its impact in the domestic and global economy. The Census Bureau's first quarterly report of retail e-commerce sales, using techniques similar to its Monthly Retail Trade Survey, estimated sales of US $5.3 billion, or about 0.64% of total U.S. retail sales for that quarter (U.S. Department of Commerce, 2000b). This is in accord with Forrester Research's estimate of US $20.3 billion for all of 1999 (see Figure 12.1), up from an estimated US $400 million for 1997 (Grant & Meadows, 2000). Note, however, that e-commerce estimates and forecasts should be viewed with particular care: eMarketer found that 1999 estimates for e-commerce varied from a low of US $3.9 billion to a high of US $36 billion, and none of these estimates take into account B2B e-commerce, which, according to the Gartner Group, is estimated to be US $145 million for 1999 and is projected to increase rapidly (see Figure 12.2).

Ready access to venture capital has enabled e-commerce "dot coms" to proliferate, but because of high start-up expenses (including establishing a brand identity), high ongoing costs (such as delivery, promotions, and technology upgrades), tight margins due to high price sensitivity of online retailing (because customers can easily compare prices), and intense competition, it is likely that many of these companies will fail or merge. In the online toy market, for example, eToys (launched in 1997) purchased its competitor Toys.com in 1998. In contrast, Toysrus.com, affiliated with the traditional retailer, has done poorly since 1998.

Because e-commerce spans geographic, and hence regulatory, boundaries, its regulation has received increased U.S. and international attention. The Anticybersquatting Consumer Protection Act was signed into U.S. law in November 1999 in response to corporate concerns about the unauthorized registration of domain names using trademarks, in some cases in order to hold them hostage or resell them for a profit. Fair taxation has also been a serious concern: To support then-fledgling e-commerce, the U.S. Internet Tax Freedom Act of 1998 declared that "the Internet should be a tariff-free zone," and placed a three-year moratorium on "multiple and discriminatory taxes on electronic commerce." This moratorium has applied to transactions across the United States. The ongoing controversy is whether a state should be able to tax interstate sales initiated within its boundaries, as well as tax revenues for companies having a physical presence within its boundaries regardless of customer location. Internationally, the European Commission recently proposed that companies collect a value-added tax (VAT) on products such as software and music sold and distributed over the Internet (Andrews, 2000).

Figure 12.1
Predicted Online U.S. Retail Revenues (US$ Billions)

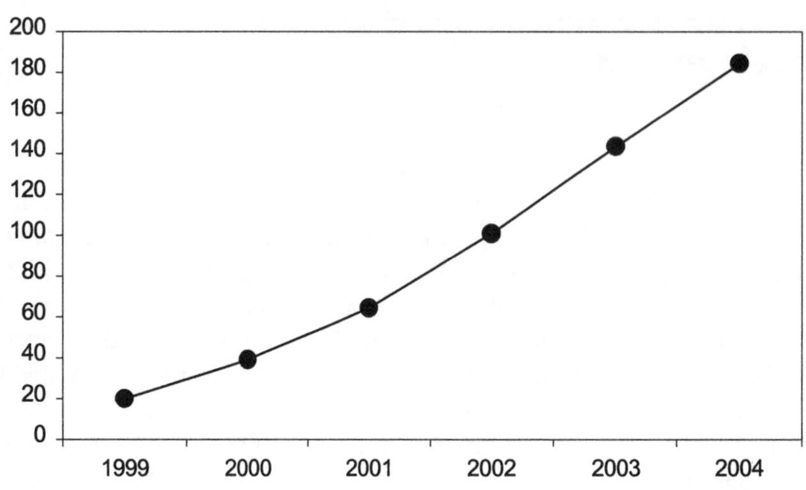

Source: Forrester Research, Inc.

Figure 12.2
Predicted Online Worldwide Business-to-Business Market (US$ Billions)

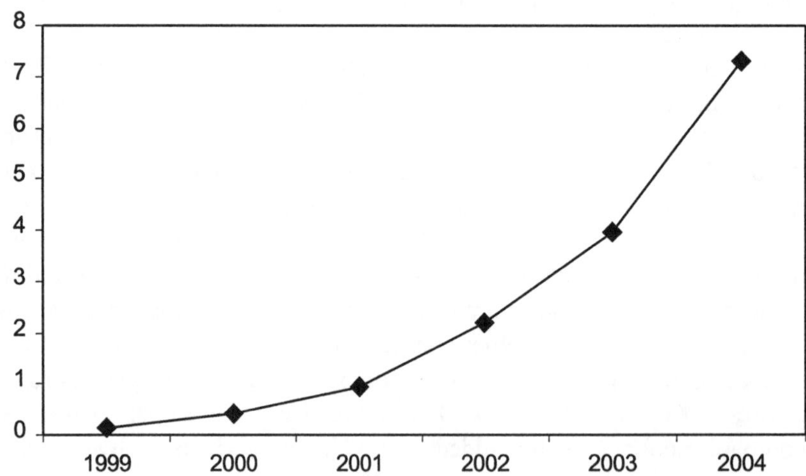

Source: Gartner Group

Developing and maintaining online privacy standards has been an ongoing concern because data gathered online can be misused (at least from the perspective of consumers). In October 1998, the U.S. Congress passed the Children's Online Privacy Protection Act (COPPA), effective April 2000, to protect the collection and use of personal information from children up to the age of 13. In March 2000, the U.S. Department of Commerce and the European Commission agreed on a "safe harbor" arrangement that handles the gap between the two systems for protecting individuals' private information (U.S. Department of Commerce, 2000a).

E-commerce technology's ability to act as an intermediary in transactions has hit some established bricks-and-mortar businesses particularly hard (booksellers and travel agents are notable examples), leading them to seek legal protection. Some companies, such as Levi Strauss, have removed e-commerce capabilities from their sites in order to avoid alienating their distributors.

Users

As Internet usage has grown spectacularly in recent years, so too has e-commerce. According to 1999 estimates by eMarketer, 3.2% of Americans aged 14 or over were online buyers in 1997, and by 2002, 30% are predicted to be online buyers. An important barrier to e-commerce is that the Internet is not yet as accessible as traditional communication technologies: In 1999, far fewer households had access to the Internet (approximately 38% of households) (Nielsen Media Research, 1999) than had access to other communication technologies used for commerce—the television (approximately 99% of households) and the telephone (approximately 94% of households). In addition, access to the Internet has been linked to higher education and income categories, as well as some ethnic groups (the "digital divide"), with implications for the market segments served by e-commerce (Hoffman, Novak & Schlosser, 2000).

As with traditional businesses, key challenges in e-commerce are attracting and retaining customers. The Internet Advertising Bureau reported revenues for online advertising of nearly US $1 billion in the second quarter of 1999, over twice the revenues for the same quarter in 1998. A key player is the Doubleclick network (founded in 1996), which sells advertising space online. In February 2000, Doubleclick reported serving an average of 1.5 billion banner ads per day on 11,000 sites. But because "click-through" rates for online banner ads are very low—approximately 1% to 2%—e-commerce sites have recently begun to spend heavily on traditional mass media such as billboards, radio, and television to attract customers. E-commerce advertising on the Super Bowl broadcast in 2000 led to impressive increases in visitors to the advertisers' Websites (Media Metrix, 2000).

Holiday periods with traditionally high purchasing volumes have strained the ability of e-commerce companies to rapidly scale their businesses, maintain customer satisfaction, and retain customers (Hansell, 1999). E-commerce sites are especially concerned about alienating customers because of the ease with which they can compare, critique, and discuss products and Websites.

Technology

At its simplest, e-commerce relies on a basic client/server model. A customer uses a standard Web browser (the client) to view product information on a merchant's Web server. If the customer decides to purchase a product, encrypted delivery and billing information is entered by the customer and sent to the merchant. After payment verification, the product is then transmitted if it is data (such as software, an electronic report, or a music file) or packed and shipped using delivery services if it is a physical product.

"Fulfillment" technologies, including inventory management, warehousing, logistics, and shipping, are critical but often overlooked components of e-commerce, and they proved to be a serious bottleneck for some companies during the 1999 holiday season. eToys.com, for example, gave $100 gift certificates to each customer whose promised delivery date was missed because of limitations in fulfillment planning. Just as managing physical inventories is difficult, so is predicting and managing the use of communication resources. This is because of the cost of network technologies, including bandwidth and servers, and because of the challenges of balancing loads across multiple systems and planning for the scale-up of capacity. For example, when Britannica provided free access to its encyclopedia in October 1999, it dramatically underestimated the load its servers would receive from the onrush of visitors. Similarly, Stephen King's release of an online novella in March 2000—one of the first such releases by a prominent author—produced 400,000 orders in 24 hours and exceeded the resources of several online booksellers (Carvajal, 2000).

Software is crucial to managing e-commerce transactions, maintaining product and customer databases, and serving Web pages. Key challenges for software include linking with legacy systems (older systems with different standards), coordinating multiple communications channels for marketing and customer service (Web, print catalogs, e-mail, and telephone), responding in real time to customer requests, maintaining general system usability (balancing server loads and minimizing downtime), and unobtrusively archiving transaction data. E-commerce packages for small to medium-sized businesses that include server, Web design, and billing and customer service applications are provided by a wide range of companies as different as Earthlink (an Internet service provider), Network Solutions (originally the only domain name registrar), and Broadvision (an e-commerce software developer). Enterprise resource planning (ERP) applications integrate a company's entire range of software tools into a single application that often includes electronic data information communications with other companies (Kumar & Hillegersberg, 2000). Traditional ERP firms such as SAP AG (founded in 1972) and Oracle (founded in 1977) have incorporated Internet e-commerce technologies into their latest products. The recently developed extensible markup language (XML) standard allows new markup elements to be defined and simplifies document viewing on multiple platforms; these characteristics make XML particularly useful for both B2B and B2C e-commerce.

Early e-commerce implementations used simple Websites, whose design often imitated existing brochures and catalogs with the addition of an online order form. More recent sites use program scripts in CGI, PERL, C++, and Java, and use specialized production tools (such as Dreamweaver, Shockwave, Flash, and RealProducer) to create content. Overall, retail site design is still in its infancy (Lohse & Spiller, 1999), and there are key problems with downtime, access speed, security, and usability (the last due to confusing navigation and interfaces and slow responses). It is also possible that design innovation has been stunted by patent disputes: To the consternation of e-commerce retailers, Amazon.com patented its "one-click" improvement to the popular shopping cart metaphor, and then took legal action against competitor Barnes & Noble for using a similar system (Gleick, 2000). Priceline.com has been similarly aggressive in patenting e-business concepts.

Online credit card fraud and the theft of billing databases are serious concerns (Brunker, 2000). One survey found that merchants (rather than customers) must often bear the costs of fraud, and approximately two-thirds of merchants polled viewed fraud as a serious problem (Helft, 2000). Secure Socket Layer (SSL) encryption is used by most e-commerce Websites to keep transaction information secure, and digital certificates and secure e-mail (such as Sendmail's Secure Switch) provide additional security. In response to concerns about the online "pirating" of audio, video, and software files, encryption and authentication technologies are also being used by producers to control access to purchases delivered online.

Online advertising is becoming increasingly sophisticated through the use of "cookie" files that store data on customers' computers, registration databases, click-stream tracking by which e-commerce sites trace the patterns users follow in moving from page to page in a Website, and personalization tools. The appearance of advertising has also changed: Images were replaced with banner ads with standard dimensions, which have now been replaced by animated and interactive ads. E-commerce sites have begun to use e-mail as a sophisticated marketing and customer service tool rather than just for mass mailings. Using appropriate software, e-mail content can be personalized, triggered by customer- or merchant-defined alerts, and programmed to confirm orders.

Current Status

Forrester Research reported that, in 1999, the average U.S. household that shopped online spent US $1,167. The five top e-retailers for February 2000 are shown in Table 12.1.

Table 12.1
Top Five E-Retailers

Company	# of Customers
CDnow.com	1,000,000
Amazon.com	984,000
Americangreetings.com	851,000
Ticketmaster.com	385,000
Barnesandnoble.com	250,000

Source: PC Data Online

Despite the current enthusiasm over e-commerce, the viability of many e-commerce businesses remains unclear. In early 2000, online retailers spent unprecedented amounts of money attracting customers by routinely providing free shipping, rebates, and substantial discounts on purchases. For example, CDNow.com gave US $10 off each order of at least US $25, Petstore.com gave a US $20 rebate and free shipping to orders of US $30 or more, and customers were able to purchase, after discounts, a best-selling hardcover book from Amazon.com for one-third its cover price (Streitfeld, 1999). The Industry Standard reported that the subsection of consumer-oriented

Web companies it studied spent an "eye-popping" average of 69% of their revenues on sales and marketing efforts in 1999 (Mowrey, 2000).

With rare exceptions, consumers have resisted paying for online content. Slate.com and *New York Times on the Web* have dropped access fees (the latter charged for non-U.S. Web access) and are now subsidized by online advertising and their parent companies. Similarly, other sites provide "free" services such as e-mail (Hotmail.com) and Website hosting (Tripod.com), and portal sites such as Yahoo.com, Excite.com, and Altavista.com, among many others, continue to compete to provide a growing array of personalized services (such as tailored weather, news, stock, and sports information). These sites use demographic data, personalization choices, and customer searches to develop marketing profiles and display targeted advertising.

Recently, Yahoo.com and Aol.com have begun to court the small business B2B market: Yahoo aggregates information about products from other B2B sites, while Aol.com and PurchasePro.com have created a strategic alliance to allow companies to easily interact with their suppliers. B2B e-commerce, estimated by the Gartner Group at US $145 million in 1999, has begun to receive attention from "old economy" companies: GM, Ford, and DaimlerChrysler announced plans to collaborate on a US $250 billion B2B parts purchasing system, and the Grocery Manufacturers of America has explored the creation of an online B2B marketplace for its members' goods and services.

As DSL and cable modems have become more popular, e-commerce sites are beginning to use more bandwidth-consuming multimedia content such as streaming audio and video, interactive views of products, and real-time online customer support. Eddie Bauer's site features a "virtual dressing room," for example, and the March 2000 version of Nike's site suggests that users install the latest versions of Shockwave, Flash, or QuickTime multimedia players on their computers. Such advanced installations require extra effort on the part of potential customers and are slow when using a standard modem, but they are not unreasonable given the technical sophistication of some companies' target markets.

The weaknesses of online security were emphasized by reports of a massive 1999 theft of online credit card records (Brunker, 2000), and the February 2000 malicious "distributed denial of service" attacks that flooded several e-commerce and government Websites with spurious requests, resulting in outages typically lasting several hours (Abreu, 2000). Shortly thereafter, a U.S. government working group released a report examining whether "existing federal laws are sufficient to address unlawful conduct involving the use of the Internet" (President's Working Group on Unlawful Conduct on the Internet, 2000); the report hints at increased government interest in regulating and protecting e-commerce.

Factors to Watch

Mergers and business failures will continue to remove the less-sustainable e-commerce companies—particularly those in the mid-range of each business sector—and create alliances among companies with e-commerce experience, content, and customer databases. Interest in B2B will increase as various industries collaborate to market their goods and services, and it will likely soon dwarf B2C revenues. As e-commerce matures and its revenues increase, maintaining a tax moratorium on Internet purchases is likely to lose support.

Differences in access to information technologies and the Internet—and the reliance on credit cards for B2C billing—suggests that e-commerce growth will vary by demographics and socio-economic status and that e-commerce providers may need to develop new strategies to maintain growth. Forrester Research predicts that the number of new online shopping households in the United States will drop from a high of 11 million in 2000 to 2.6 million in 2004, but that the amount spent by each household will rise as they begin to buy routine or frequent purchases (such as groceries) online. The globalization of e-commerce will increase as more international users, particularly those in Europe and Asia, adopt Web technologies. However, high telecommunications costs and the lower popularity of the credit card as a form of payment, remain limiting factors in some countries.

New e-commerce software and services are likely to emphasize:

- Integration of multiple tools and systems.
- Real-time analysis of purchasing and browsing behaviors.
- More sophisticated and personalized customer assistance.
- Robust scalability and data archiving.

The use of personal digital assistants and cellular telephones for wireless access to Web, including e-commerce, is an area to watch. Much depends on whether customers accept the restrictions on the devices: potentially high communication charges, limited bandwidth, and small screen displays. If wireless e-commerce is successful, it may influence the design of e-commerce sites, force highly differentiated wireless and wireline versions, or provide opportunities for additional intermediaries.

As very high-capacity network bandwidth and storage devices become affordable to consumers, e-commerce retailing is likely to develop into a more personalized and richer multimedia version of television shopping, and more bandwidth-intensive products such as videos will be regularly delivered online. Adequately securing such files, however, is a serious concern for content providers.

Forecasts and press reports are generally quite optimistic about the future of e-commerce, although it is important to keep in mind that the physical reality of many commercial transactions remains a limiting factor. While purchases that involve digital content such as software, music, images, tickets, and e-books can be ordered and delivered online, the physical nature of some purchases makes warehouse and delivery technologies critical to e-commerce success. In addition, some customers and certain product categories will continue to require hands-on interaction. The opportunities for success and failure as we sort out these various distinctions among products, services, and customers will keep e-commerce an interesting topic of study for many years.

Bibliography

Abreu, E. (2000, February 21). The hack attack. *The Industry Standard,* 66-67.

Andrews, E. L. (2000, March 2). Europe plans to collect tax on some Internet transactions. *New York Times,* C4.

Brunker, M. (2000, March 17). Major online credit card theft exposed. *MSNBC*. [Online]. Available: http:// www.zdnet.com/zdnn/stories/news/ 0,4586,2469820,00.html.

Carvajal, D. (2000, March 16). Long line online for Stephen King e-novella. *New York Times,* A1, C10.

Gleick, J. (2000, March 12). Patently absurd. *New York Times Magazine,* 44-49.

Grant, A. E., & Meadows, J. H. (2000, In press). Electronic commerce: Going shopping with QVC and AOL. In C. A. Lin & D. J. Atkins (Eds.). *Communication technology and society: Audience adoption and uses*. Cresskill, NJ: Hampton.

Hansell, S. (1999, December 17). As sales boom online, some customers boo. *New York Times,* C1, C10.

Helft, M. (2000, March 6). The real victims of fraud. *The Industry Standard,* 152-165.

Hoffman, D. L., Novak, T. P., & Schlosser, A. E. (2000). The evolution of the digital divide: How gaps in Internet access may impact electronic commerce. *Journal of Computer-Mediated Communication, 5* (3), unpaginated.

Kumar, K., & Hillegersberg, J. (2000). Enterprise research planning (ERP): Experiences and evolution [Special Issue]. *Communications of the ACM, 43* (4), 22-69.

Lohse, G. L., & Spiller, P. (1999). Internet retail store design: How the user interface influences traffic and sales. *Journal of Computer-Mediated Communication, 5* (2), unpaginated.

Media Metrix. (2000, February 2). *Dot.com Super Bowl television ads drive major increases in unique visitors according to Media Metrix*. [Online]. Available: http:// www.mediametrix.com/PressRoom/Press_Releases/ 02_02_00.html.

Mesenbourg, T. L. (1999). *Measuring electronic business: Definitions, underlying concepts, and measurement plans*. U.S. Census Bureau. [Online]. Available: http://www.census.gov/epcd/www/ebusines.htm.

Mowrey, M. A. (2000, March 6). Financial spotlight: Wall Street impatient with runaway dot-com marketing spending. *The Industry Standard*. [Online]. Available: http://www.thestandard.com/research/metrics/display/ 0,2799,12553,00.html.

Nielsen Media Research. (1999, May). *TV viewing in Internet households*. [Online]. Available: http://www.nielsen-media.com/reports/TVinInternetHomes.pdf.

President's Working Group on Unlawful Conduct on the Internet. (2000). *The electronic frontier: The challenge of unlawful conduct involving the use of the Internet*. [Online]. Available: http://cybercrime.gov/unlawful.htm.

Streitfeld, D. (1999, December 19). Online specials bring in buyers, ring up losses. *Washington Post,* A1.

U.S. Department of Commerce. (2000a, March 17). *Documents regarding Department of Commerce's work to develop a "safe harbor" that would help U.S. organizations comply with the European Union's Directive on Data Protection*. [Online]. Available: http://www.ita.doc.gov/td/ecom/menu1.html.

U.S. Department of Commerce. (2000b, March 2). *Retail e-commerce sales for the fourth quarter 1999 reach $5.3 billion, Census Bureau reports* [Press Release]. [Online]. Available: http://www.census.gov/Press-Release/www/ 2000/cb00-40.html.

Websites

Academic & Business

- Vanderbilt's e-lab: http://ecommerce.vanderbilt.edu/index.html
- *Journal of Computer-Mediated Communication*: http://www.ascusc.org/jcmc/
- www.redherring.com
- www.thestandard.com
- news.cnet.com/news/0-1007.html
- www.nytimes.com/library/tech/reference/indexcommerce.html

Metrics

- www.emarketer.com
- www.forrester.com
- www.ecommerce.gov
- www.mediametrix.com
- www.pcdataonline.com
- www.nielsen-netratings.com

13

Office Technologies

Mark J. Banks, Ph.D. & Robert E. Fidoten, Ph.D.*

any of the technologies used in office settings and for office functions are described in other chapters. This chapter looks at some specific technologies, but also explores the larger picture of the use and impact of information and communication technologies in the workplace.

In addition to the well-established technologies such as the typewriter, word processor, telephone system, and the personal computer, there are a number of other important office technologies and applications that are discussed in this chapter:

Desktop publishing (DTP) and *multimedia* allow a single worker to use a computer, printer, and appropriate software to incorporate research, art, photography, charts, graphs, writing, layout and design, and printing into documents such as newsletters, notices, and reports on a professional-quality level. These technologies have also been expanded in recent years to dovetail with mediated presentations and Web-oriented documents and media.

The *facsimile* (fax) machine transfers documents and images electronically over ordinary telephone lines to another similar machine. Fax functions have also become incorporated into personal computers, mobile and handheld technologies, and the Internet.

Local area networks (LANs), *wide area networks* (WANs), and *external networks* such as the Internet and "Extranets" connect office devices, primarily computers, over appropriate network configurations. The *PBX* (private branch exchange) is another type of network used to connect telephones within an office.

Multifunction products (MFPs) are automated devices that combine several functions into one unit, such as printing, scanning, fax, word processing, and the telephone.

* Dr. Banks is Professor and Dr. Fidoten is Associate Professor, Department of Communication, Slippery Rock University (Slippery Rock, Pennsylvania).

Telecommuting and *mobile offices* make use of several portable devices, most notably the notebook computer, the personal communications device, and, increasingly, the personal electronic "organizer." Several of these are also multifunction devices.

Teleconferencing, e-mail, voice mail, and *presentation programs* are among the technologies that augment the communication functions of the office.

Background

The old office technologies often seem to hang on forever. Although writing documents by hand gave way to the typewriter in the early 1900s, that technology, along with the telephone, dictating machine, and hand-delivered mail, dominated the office environment through the first seven decades of the 20th century. The copy machine, which was added in the 1960s, represented one of the first major "modern" additions to the office. Although purported to be a major labor-saving device, the copy machine led almost immediately to an increase in the use of paper and file cabinets, and, in effect, an increase in workload.

Writing Instruments

During the late 1970s and throughout the 1980s, a convergence of several technologies led to a mini-revolution in office technologies—innovations that would significantly change the nature of the office and its workers. These technologies included the personal computer. The change was so significant that Smith-Corona, vendor of over 70% of typewriters in the United States, filed Chapter 11 bankruptcy in mid-1995, and, to survive, the company had to venture into other electronic products. Sales of typewriters fell to less than 800,000 in 1995, and the company's stock declined to one-eighth its value from 1997 to March 2000. Although the typewriter is still part of most offices, it has been relegated to those few chores that are difficult to do on a computer, such as single mailing labels, filling out forms, and writing postcards (Elsberry, 1995; Deutsch, 1998, Smith Corona considers, 2000; Thomas, 1997). According to Cullen (1999a), today, only about half a million typewriters are sold each year.

Mailing Functions

In the late 1970s and 1980s, the nature of mailing changed. Companies increased their need for faster delivery of mail, and overnight carriers such as FedEx emerged.

Fax technology has been around since the mid-1800s and was used for specialized functions such as transmitting photos for news services during the first half of this century. During the 1970s and 1980s, fax technology went through several improvements including standardization, leading to the proliferation of business and personal use after 1980. Even higher-speed fax technologies continue to evolve. Since the 1990s, fax technologies are being incorporated into desktop and portable computers and multifunction machines, moving from analog to digital systems, and finding increased use through the Internet (Harper, 1998; Wetzel, 1997; Cullen, 1999b).

Probably the most significant evolution of mailing technologies has been the expansion of electronic mail. E-mail began as a proprietary mainframe technology, but has achieved wide use through the convergence of personal computer technology and networking, both within organizations as LANs and among organizations through networks such as the Internet. Although long-distance carriers such as MCI offered early e-mail services, the Internet has played a significant role in the tremendous growth of e-mail technology in the 1990s. Cullen (1999a) reports that, in a 1999 PricewaterhouseCoopers survey, 48% of respondents said that electronic mail was the main reason for Internet subscription.

Telephony

Few things were as unchangeable over the decades as plain old telephone service (POTS). Although developments such as direct dialing, easier access to international calling, and switched networks progressed significantly throughout the century, the end user saw little change in the way the telephone was used. In the United States, one telephone company, AT&T, owned virtually all aspects of the telephone network, and it wasn't until the breakup of the AT&Ts monopoly in the early 1980s that telephone service providers and equipment manufacturers were able to introduce their own equipment and to vary the functions of telephone service. This change led to a host of add-on technologies, again centered on the desktop or portable computer. Among these technologies were fax machines, PBX, voice mail, and automated call routing. On the wireless front, portable and cellular telephones proliferated, as well as pagers and personal communication devices. Videoconferencing also grew out of telephony and satellite communication. (Videoconferencing is explored in more detail in Chapter 24. For more on telephony, see Chapter 17, and for more on wireless telephony see Chapter 22.)

Recent Developments

The technologies named above are not unique to the office environment, and several of them are described elsewhere in this book. The converging application of office technologies in the workplace have led to at least three major developments:

- The so-called "paperless office."
- The compression of office activities.
- The "virtual office."

The Paperless Office

The "paperless office" is a misnomer. Few offices, if any, will never use paper. In fact, one estimate shows that the number of pages of paper generated per worker doubled in the 1990s, abetted by such technologies as desktop publishing, color copying, home offices, and inexpensive computer printers (Comeau-Kirschner, 1999). Many of the functions and activities that relied on the printed form in the past, however, are being replaced by office technologies that allow them to be put into electronic form.

An important component of the paperless office is the development of "Intranets," private networks that have the look and feel of Internet Websites. Much of the software used for Intranets is the same as that used for the Internet. A 1999 study by International Data Corporation (IDC) shows that more than 75% of employees will have access to the Internet, and more than half of the employees in small, medium, and large businesses will have access to Intranets by the end of 2000 (Gantz, 2000).

IDC also found that the top four Intranet uses are "information sharing, information publishing, email, and document management," with data conferencing a growing use as well (IDC, 1999).

A selection of other uses includes:

- Regional or international sales forces can examine new products.
- Employees can get updated company news or check the lunch menus for the week.
- Personnel offices can make applicants' résumés available to members of a work team or department.
- Work groups can collaborate, often at great distances.
- Production can become more automated, using electronic inventories for just-in-time ordering.
- Management can share its decision-making processes by making information about decision factors available online.
- Employees can communicate via e-mail, voice mail, and internal audio- and video-conferencing.
- Files of information can be shared widely and made immediately accessible.
- Archives can be stored and accessed by everyone.
- Training and development multimedia programs can be accessed from individual workstations.
- Expert systems can be tapped for assistance with work problems.
- Students can take courses through distance learning.

A comparable application is the "Extranet," which provides special links between companies and customers, suppliers, or clients. Instead of just using the Internet for such links, Extranets are more secure, while still allowing external browsers access to internal information systems (Moody, 1997).

Indeed, the very definition of "office" is changing rapidly. Because of the proliferation of not only wired technologies such as Intranets but also wireless technologies such as cellular, personal communications systems, handheld personal digital assistants (PDAs), and portable computers, flexibility in the configuration of the office has led to several developments.

Compression of Activities

Early office computers were used almost exclusively by secretaries, and executives avoided them because of this clerical identity. As the technology evolved through the 1980s and 1990s, however, computers in the office became used by more and more people at all levels of work, including executives. This led to some job compression. For example, memos can be conceived, written, and printed or mailed in one basic operation by the originator. Telephone calls reach the desk of the recipient because automated voice call routing has eliminated most or all of the intervening human steps. The same is true with voice mail messages, which no longer need to be written. In some specialties, such as desktop publishing, what used to take several steps in several places for writing, artwork, photography, typesetting, layout, and printing is now compressed into a single workstation where the job can be done by a single person in one place. Employees continue to learn more skills, including communicating through LANs and WANs and creating information materials that grow ever more versatile, such as streaming media, multimedia, and presentation materials.

One of the significant office developments in recent years has been the appearance of multifunction peripherals. These are automated devices that combine several functions into one unit. Such functions may include printing, scanning, faxing, word processing, telephone answering machines, and other computer functions such as data processing, networking, and CD-ROMs. Although researchers predict the gradual replacement of individual office machines by the convergence of several technologies into one MFP (PC use, 1997), Cullen (1999) argues that they are more useful for small or home offices than for larger companies, and that the advantages of combining several functions into one is often outweighed by the multiple user advantages of individual peripherals.

The Virtual Office

The "virtual office" has emerged as a feasible solution for contemporary and future work environments. Many types of traditional office work that required a fixed physical setting can be relocated to a wide variety of alternative sites. The employee's home, automobile, client/customer location, or even temporary hourly/daily space can substitute for traditional centralized office space (Weston, 1997). The Employment Policy Foundation predicted in late 1999 that one-fourth of the workforce will telecommute from home by 2005. Among the advantages the report listed are "increased productivity, lower real estate and travel costs, reduced employee absenteeism and turnover, increased work/life balance, and improved morale and access to additional labor pools, including disabled workers, to ease shortage skills" (Hrisak, 1999).

Telecommuting technologies also permit freedom of location, instantaneous interaction, and fast response and spontaneity. From a business perspective, the "virtual office" provides substantial economic benefits. Enterprises are partially relieved of relatively high-cost real estate investment or rental. "When your file cabinets are electronic, you don't need a building to put them in" (The virtual office, 1997, p. 56). The burden of providing office space is often shifted to the employee or independent contractor. Thus, the home takes on a new multi-faceted role. Capital investment shifts from fixed physical space to communication facilities, computing and associated equipment, and communication lines and related costs.

A major implication of this approach is the need for homes or telecommuting sites to be equipped with high-speed communication lines, as well as wireless technology. The speed and quality of residential communication lines shifts from primarily providing voice-oriented facilities to one that provides rapid data and image transport, as well as videoconferencing capability.

Employers must often provide up-to-date computers with organizationally standardized and compatible software, fast modems, communications services, network access, and other related facilities. Further, it is essential that security be given additional emphasis, since there is markedly increased difficulty in maintaining control and limiting unauthorized access to proprietary information.

Since the office can also move into mobile virtual locations, employers may also provide laptop portable computers, modems, PDAs, and fax facilities so that office workers can have almost infinite flexibility in reaching clients, colleagues, and "headquarters."

A major office activity is the ever-present need to conduct meetings. Traditionally, meetings required physical presence, and telephone conference technology is commonly used. Despite repeated introductions, however, video telephony has not been widely accepted or deployed. Videoconference facilities have been introduced in various manifestations over the past two decades, but these, too, have had limited acceptance. Probably the largest growth in this area is the increased capability of audio and video "streaming" over the Internet, especially when faster and increased broadband functions evolve.

A development that combines telecommuting and teleconferencing is the emergence of "virtual teams" of workers who collaborate using the Internet and other communication technologies. In a Ceridian Employer Services survey, 40% of small companies allow telecommuting and about one-third have virtual teams, while 90% of larger companies allow telecommuting and more than half have virtual teams (De Lisser, 1999).

Despite the many economic and practical advantages posed by virtual offices, many new sets of problems emerge. The social fabric of an organization is significantly changed. Traditional authority structures must be completely reworked, since hierarchical organization is less appropriate. Even the modern "flatter" organization may not be effective as new sets of work relationships materialize. Projects and activities can be built around the expertise required, wherever that expertise may be geographically located. Reporting structures may be based upon areas of specialized knowledge rather than traditional levels of authority. However, lack of employee interaction and managing distractions can pose new threats. Training and development programs may be needed to help employees understand their new independence and heightened individual responsibilities (Avoiding a "virtual disaster," 1998).

Current Status

As much of the previous discussion shows, office technologies are in a constant state of change. Because so many technologies are involved, they do not change at the same rate, and their changes are seldom coordinated. The changes occur at three levels: internal independent office technologies, wired services, and wireless services.

Independent office technologies—computer workstations, multifunction stations, and desktop publishing—continue to be a large sector of office technology. As the storage capacity and versatility of information management capability of these increase, their use has become all the more embedded in the definition of the office of the 21st century. In addition, all these technologies dovetail more and more with external technologies, the Internet being the most prominent.

The growth of Internet usage is probably the largest business application of wired services being used for research, communication, and, increasingly, as a means of promotion and advertising. Faster access through ISDN, T1 and T3 lines, DSL, and other high-speed connections makes this medium even more convenient and versatile enough to also accommodate videoconferencing. Wireless networks are proliferating more slowly, mostly because of the high cost. Although there are some companies that use wireless data interconnection regularly such as United Parcel Service, most have adopted a wait-and-see approach pending more facile technology and lower costs (Girard, 1998).

Factors to Watch

Major trends in office technologies include the gradual transition to wireless devices, the convergence of electronic media and devices, the continuous streaming of information, the emergence of a feasible 24-hour workday, and the growing practice of outsourcing work.

Communication devices are concurrently moving into digital and wireless mode, providing substantial flexibility with respect to the need for fixed office locations. Wireless technology provides the organization and worker with almost infinite possibilities with respect to location, office organization, work systems, and almost any aspect of traditional and contemporary work. Although basic messaging among coworkers is simplified, opportunities for "meetings on the run" from disparate locations have become the norm, not an exception. The time required to contact and organize meeting participants is reduced from days to minutes, providing the possibility for almost instantaneous decision making.

Beyond the issue of meeting and conferencing speed, these wireless technologies provide a decided economic advantage relative to the cost of office facilities. Office space selection no longer needs to be governed by the requirement of a prestigious address, employee commute time, or client/customer access. Wireless technologies permit any office facilities to be situated in lower-cost locations and be designed for less than the full complement of staff since many staff members will be mobile and working from home offices, vehicles, temporary offices, or any other nontraditional locations.

The streaming of information is also an important development. It is no longer necessary to organize information by calendar or clock constraints. Information that is financial, statistical, event, news, or continuous stream in nature may have a substantial impact on an organization's operations, tactics, and strategies. Traditionally, such information was reviewed and assessed on a periodic basis, and only after such a process were decisions generated that would result in significant changes in operations. The continuous streaming of information, perhaps targeted at computer-based decision support systems or artificial intelligence models, permits organizations to initiate change in a more dynamic mode. As conditions change, analysis can be initiated, and recommended changes in operations can be made. Weekly or daily review meetings can literally be compressed into automated decision making, or an almost minute-by-minute change process.

Continuous streaming of information plays a major role in changing the way office work is managed. It is no longer necessary to have structured layers of employees that traditionally have accessed and interpreted information. This traditional process of decision making required a hierarchical structure, many discussions and conferences, and delays. Streaming brings the information directly from its source to the decision maker, obviating the need for intermediate layers. Thus, information availability and accessibility become the criteria for organizational structure.

Until recently our society has maintained an allegiance to the traditional five-day, eight-hour-per-day work schedule. Although this may have been nominal for many office situations, it has nevertheless been the formal work system. The explosion of communication technologies has all but eliminated that work system. With online accessibility, 24-hour communication availability from anywhere, information streaming, and related technologies a new office work culture has been created. Further, business structures continue to grow on a global basis with little or no recognition of national boundaries or time zones. Thus, the workday becomes a manifestation of the global demands of the business and, in turn, its office functions.

Office workers become available wherever and whenever needed. Weekends are no longer privacy protected. Since various cultures have varying work-week structures, it is essential to many organizations that employee availability not be bound by local national traditions. However, there needs to be some attention to the problems of the "24-hour work day." Employee burnout, never having free time, and always being "on call" can interfere with family life and private time.

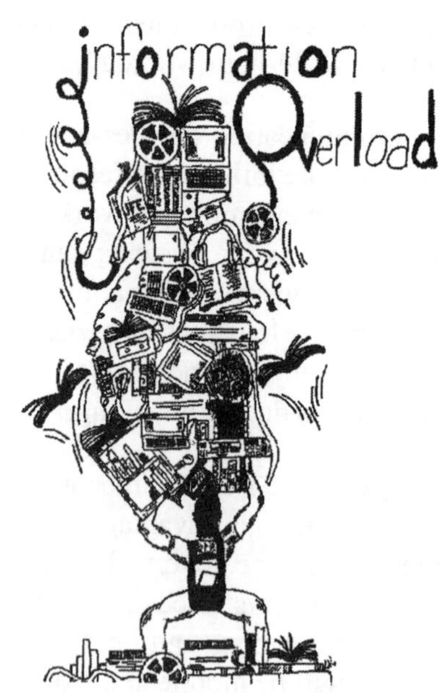

The outsourcing of work has become a growing practice, one that is growing exponentially. Work can be shipped via telecommunications to any corner of the earth where the requisite skills are available at the right price. This practice allows organizations to ferret out the required skills, employ personnel on a contract basis without longer-term commitment, handle overload situations, and shop for the most advantageous labor rates. Tasks can be assigned, reviewed, and submitted regardless of the worker's location. The economy has grown at such an accelerated basis that labor shortages have become commonplace. Recruiting, employing, and training a sufficient staff to meet workload requirements has become a major organizational problem. The outsourcing practice alleviates this problem since work can be shifted to underutilized workers in areas with less booming economies. Communication technologies provide the engine to facilitate this manner of organizing work.

The ability of office organizations to manage their information resources and sources has reached a critical juncture. In the very recent past, the typical office organization could access manageable chunks of information on demand, essentially as required to perform required work. Now, there is a glut of information available, making it not only more difficult to manage, but also more difficult to determine its reliability. Data flood into organizations in exponentially increasing amounts, and it is an almost insurmountable task to filter and convert it into useful information. The task of making decisions is hampered by this "information overload." There is, however, the potential for the development of artificial intelligence technologies to assist with that filtering process. There may also be an increased role for the so-called "information manager" whose job it is to find, organize, evaluate, and synthesize information for decision makers.

This discussion ends with some attention to paper trails and information security. The communication revolution of the past decades, and the changes anticipated for the next decades, demand that the traditional "paper trail" concept be rethought and redefined. As office records continuously shift from paper to electronic media, the nature of how to organize, maintain, and secure an information trail is being redesigned.

Traditionally, paper trails were specified and defined in organizational manuals, government and tax regulations, and operational procedures. Today's movement of information, however, has accelerated with respect to time and technologies. No longer can a reviewer anticipate locating a clear, clean, logical trail. Office information may be located in computer files, e-mail messages, paper memos and letters, voice mail, or file cabinets—in effect, ever-widening geographic domains. The problem calls for a new set of generally-accepted procedures that can be incorporated into office practices, particularly as electronic forms of information carry increasing status as legal documents.

Related to the electronic paper trail is the issue of security. As records and information move more and more through cyberspace, the future will bring more challenges and solutions to the need for security, confidentiality, and protection of sensitive or proprietary information. While modern communication methods have brought enormous flexibility in how information is generated, stored, and distributed, it has brought a corresponding need for ways to protect that information.

Bibliography

Avoiding a "virtual disaster." (1998, February). *HR Focus,* 11.
Comeau-Kirschner, C. (1999, June). Paper, paper everywhere. *Management Review, 88* (6), 1-5.
Cullen, S. (1999, November). The truth about MFPs. *Office Systems, 16* (11), 26-32.
Cullen, S. (1999a, December). Telecommunications in the office. *Office Systems, 16* (12), 24-28.
Cullen, S. (1999b, December). Tools of the trade. *Office Systems, 16* (12), 32-36.
De Lisser, E. (1999, October 5). Update on small business: Firms with virtual environments appeal to workers. *Wall Street Journal.*

Deutsch, C. H. (1998, March 23). Using a key still works; Smith Corona's future rests in putting its name on other products. *New York Times,* Cl.

Elsberry, R. (1995, December). Test of time. *Office Systems,* 95.

Gantz, J. (2000, March 6). Controlling the coming chaos of Intranets. *Computerworld, 34* (10), 33.

Girard, K. (1998, February 23). Wireless revolution fizzles. *Computerworld, 32* (8), 6.

Harper, D. (1998, February). Telephony and faxing on the Net; New technology allows low-cost communications over the Internet. *Industrial Distribution, 87* (2), 94.

Hrisak, D. M. (1999, December). Millions move to the home office. *Strategic Finance, 81* (6), 54-57.

International Data Corporation. (1999, August 31). *IDC survey reveals Intranets are becoming backbone of companies' IT infrastructures.* [Online]. Available: http://www.IDC.com.

Moody, G. (1997, May 29). Get a little more from your Extranet. *Computer Weekly,* 54.

PC use drives sales of multifunction products. (1997, December 11). *Purchasing, 123* (9), 66.

Rybczynski, T. (1999, October). Distributed workplace networking: Making the right choice. *Business Communications Review,* 7-14.

Shellenbarger, S. (1998, February 18). Forget juggling and forget walls; Now, it's integration. *Wall Street Journal,* B1.

Smith Corona considers strategic alternatives. (2000, February 14). *Business Wire.*

Taking virtual office public. (1998, January 12). *PC Week,* 66.

Thomas, P. (1997, December 19). Clack, clack. The typewriter's back? Smith Corona hammers away with a machine still useful for labels and forms. *Wall Street Journal.*

The virtual office; Groupware creates consultants offices without walls. (1997, December 19). *San Francisco Business Times,* S6.

Weston, C. (1997, December 30). Taking offices to a "hire" level. *The Guardian,* 17.

Wetzel, R. (1997, November). The argument for Internet-based facsimile. *Telecommunications, 31* (11), 32.

Websites

www.businesswire.com. This is a Website primarily for businesses to post news on the Internet and send news to the media. It is a good source for up-to-date developments, although the news is primarily generated by the companies themselves. A service also provided by Business Wire is Newstream.com, "multimedia news for the 21st century newsroom." There are several other related business news sites, two of which are www.medialink.com and www.prnewswire.com.

www.newmedia.com. Recently, the magazine *NewMedia* converted to an all-electronic format. The site provides updates on recent developments, including technologies affecting businesses.

www.office.com. According to the Website itself, "Office.com is the new way to work—complete online business resource with industry-specific news and analysis, tools and access to relevant services, business news, business search, business research, business needs assessment, online business, and Internet business...."

14

Virtual & Augmented Reality

Karen Gustafson*

Virtual reality is not a single technology. Rather, it is a collection of technologies that are increasingly implemented in a variety of industries including medicine, entertainment, defense, transportation, and architecture. Virtual reality (VR) allows the user to interact with three-dimensional representations of information and, in this way, is a sort of "ultimate interface" (Larijani, 1993). While the term "virtual reality" has been used to refer to certain technologies or industries, VR can generally be considered as an "emerging communication system" (Biocca & Levy, 1995, p. 15).

Some VR systems are highly immersive, making the user feel almost entirely within an alternate, computer-generated reality (Slater & Usoh, 1995). In the virtual environment (VE), graphics change according to the user's apparent point of view and movement, so that the user feels encompassed within a dynamic, interactive virtual space (Heudin, 1999). Several speculative films—*Lawnmower Man* and, a few years later, *Virtuosity* and *Existenz*—were made in the 1990s about the potentials of these immersive systems. Many books addressing the potentials and perils of virtual reality were written during this time as well. *Snow Crash*, written in 1992 by Neal Stephenson, is one of the best-known forays into VR in fiction, and describes a common virtual reality called the Metaverse. In 1991, cyberspace pundit Howard Rheingold predicted in his book *Virtual Reality* that "in the coming years, we will be able to put on a headset, or walk into a media room, and surround ourselves in a responsive simulation of startling verisimilitude" (Rheingold, 1991, p. 388).

Virtual reality became possible due to advances in several areas of computing, including graphics generation, processing, and new input and output interfaces (Loeffler & Anderson, 1994). Immersive examples of VR systems may include head-mounted displays (HMDs), treadmills, audio input, and haptic force-feedback devices, which lend virtual objects palpable qualities and allow the user to feel as if he or she is physically manipulating 3-D solid objects. These systems can offer a convincing sense of presence, so that the user genuinely feels part of the virtual environment (Biocca & Delaney, 1995). Engineering teams can experience immersive VR as a group

* Graduate Student, Department of Radio-TV-Film, University of Texas at Austin (Austin, Texas).

in a CAVE (Cave Automated Virtual Environment). With this technology, groups of people can collaborate on virtual automobile or building designs without having to build costly models.

Augmented reality (AR) uses technologies similar to VR, but focuses on overlaying material reality with information or graphics. Augmented reality does not completely immerse the user. Rather, it superimposes information over the "real world" and allows the user to combine real and virtual views (Slater & Usoh, 1995). This technology has possible applications in medicine, where information could overlay the patient's body to assist the surgeon, or in the production plant, where a worker could see instructions superimposed over assembly pieces (Mahoney, 1999b).

Background

History

Although VR is often considered to be a phenomenon of the very late 20th century, VR technologies can actually be traced to flight simulator research in the 1930s (Hillis, 1999). Flight and military simulation continue to be a primary interest in VR research. The film industry likewise has historically been involved with VR development. In 1960, Morton Heilig developed a stereoscopic display, hoping this would catch on as an entertainment device. While the stereoscopic display did not immediately become popular, it has become an integral part of contemporary virtual reality interfaces.

In 1965, Ivan Sutherland suggested that computer graphics could create "windows to virtual worlds," and this vision has framed subsequent virtual reality research (Sutherland, 1965; Brooks, 1999). Sutherland's prescient paper, "The Ultimate Display," described an ideal concept of virtual reality in which users could be surrounded with 3-D displays of information, and today Sutherland is considered one of the pioneers of VR (Vince, 1998; Larijani, 1993; Hillis, 1999). Sutherland focused on VR development in flight simulation, and created the first head-mounted display in 1969 (Hillis, 1999). Sutherland also developed early image generators that were forerunners of contemporary graphics accelerators. These generators could produce very simple scenes at a rate of 20 frames per second (fps), but more complex animation suffered at these speeds (Burdea & Coiffet, 1994).

Later, the computer graphics industry became a powerful partner in VR development, as much of VR depends on the production of convincing, immersive stereoscopic imagery. Most computer graphics generated for VR are constructed from polygons, a flat shape frequently used in 3-D modeling. Because polygons are very simple, they are an ideal building block for the more complex 3-D images one might encounter in VR, such as desks or houses (Vince, 1998). Also, these shapes can be generated quickly, and time is of the essence in the real-time, dynamic imagery.

Military spending has traditionally been a primary income stream for VR development, with the hope of creating more efficient flight simulators (Burdea & Coiffet, 1994). During the 1970s, however, research interests in VR broadened, although the National Aeronautical & Space Administration (NASA) continues to be a driving force (Hillis, 1999). The United States was the primary nation involved in VR research until the 1980s, when Japan, Germany, Canada, and France began to show serious interest (Larijani, 1993; Burdea & Coiffet, 1994). Although different components of VR research have been developing for decades, it was not until 1989 that another VR pioneer, Jaron Lanier, coined the term "virtual reality" (Heudin, 1999).

VR Hardware Components

Today, virtual reality configurations consist of three primary hardware elements. Either liquid crystal displays (LCDs) or CRT (cathode ray tube) displays are used to provide the visual component of a virtual environment. Another component is the tracking system, which relays the user's orientation information to the computer, so that, if the user moves, the virtual images will be altered according to the user's new perspective. Trackers detect head and hand movements with optical or electromagnetic technologies, or with ultrasonic and gyroscopic devices that detect the user's physical orientation.

Finally, high-speed image rendering systems are needed to generate changing graphics in close to real time to produce a convincing experience of virtuality (Larijani, 1993). If image processor speeds are unable to produce graphics at a speed of at least 20 fps to 30 fps, the system latency disrupts the immersion of the user. Latency refers to the delay between user movement and the readjustment of the graphics system (Loeffler & Anderson, 1994). Ideally, the computer running the graphics would adjust to the user's movement in real time. If the delay is too long, such as in a highly-detailed environment, the user may notice and feel disoriented or even become affected by motion sickness (Regan, 1995).

As mentioned above, there are several levels of immersion in virtual reality, and these correspond to the various presentation modes. High-immersion modes of VR presentation include:

- CAVEs (Cave Automated Virtual Environments) or dome displays.
- Cylindrical displays.
- Head-mounted displays.

Less immersive modes of VR include:

- Panoramic displays.
- Workbenches.
- BOOMs (binocular omni-orientation monitors).
- Traditional computer screens.

Group viewing technologies such as CAVEs and dome displays give a surrounding view of the virtual environment, and several people can experience it at once. However, it is costly to obtain multiple image-generation systems, and color saturation and contrast are not perfect. Panoramic displays are also useful for groups, but these do not surround the users. Also, only one person may control the viewpoint in a panoramic display, so all other users' visual perspectives will be slightly off-kilter. Workbenches allow one or two viewers to be tracked so that each perceives customized point-of-view images; this hardware is sometimes configured to look like a drafting table (Brooks, 1999).

Individual VR technologies include the HMD and the BOOM, an alternative to HMDs, which consists of a viewing box suspended on a rod. The HMD is one of the most common components of a VR system. The visor displays on these headsets consist of either LCDs or CRTs, and each display mode has benefits and drawbacks. LCDs were formerly used quite frequently, but these tended to offer grainy images with low resolution. CRTs had much higher resolution, but tended to be prohibitively bulky and more expensive (Biocca & Delany, 1995). In the 1990s, LCDs improved in image quality, and CRTs became smaller. An emerging technology that may eventually supersede the HMD is the virtual retinal display (VRD). The VRD projects images directly to the back of the eye, eliminating cumbersome head-mounted screens; it has been in experimental development since the 1980s (Rheingold, 1991).

Other interface devices for virtual and augmented reality include reality gloves, such as the DataGlove, which was commercially introduced in 1987 (Burdea & Coiffet, 1994). Fiber optics run along each of the hand joints of the glove, relaying movements of the user's hand back to the computer generating the virtual environment (Sun, 1995). With this interface, the user can manipulate virtual 3-D objects, rearranging a room or redesigning an automobile engine. Other VR systems include force-feedback devices such as joysticks, which generate varying amounts of force according to the user's interaction with the virtual environment. Technologies like these make it possible for virtual objects to offer sensations of weight or resistance, further adding to the sense of immersion.

Current Uses of Virtual Reality

Entertainment

Vehicle simulation is one of the oldest uses of VR, and is one of the most advanced areas of application. In the entertainment industry, this application has translated into the development of virtual rides. In 1999, Disney opened its second virtual reality theme park in the United States and plans to open 20 more parks around the world. These parks are cheaper to produce than traditional amusement parks, and use less real estate—for example, the new DisneyQuest Park in Chicago is housed in a five-floor building (Hill, 1999).

Construction and Design

In the construction industry, CAD (computer-assisted design) developers have embraced VR. Users can experience fully immersive environments wearing HMDs, and building designs can be altered in virtual reality with immediately visible results. These technologies are useful for sales and marketing applications in the construction industry, and even allow a building's future inhabitants to participate in the building's actual design (Making an appearance, 1999).

Medicine

Surgical simulation is one of the most common uses of virtual reality in the field of medicine. Images are becoming increasingly photorealistic, and haptic devices even allow users to feel a scalpel cutting into virtual flesh. These technologies are useful in medical training, certification, and diagnosis (Hill & Jensen, 1998; Satava, 1995). Medical VR can also simulate needle, catheter, and scope procedures (Satava & Jones, 1998).

Another exciting use of VR is telemedicine, which enables doctors to perform procedures remotely. Telemedicine combines robotics, haptic technologies, and stereographic imaging to allow surgeons to operate on patients in other rooms or even other states. The surgeon sits at a virtual workstation, the "master" site, and manipulates surgical instruments whose movements correspond to those of instruments at the remote "slave" site, which is video and audio monitored (Moukheiber, 2000). Unfortunately, however, liability issues may continue to inhibit the widespread adoption of this technology.

Aerospace and Defense

VR and AR technologies are being utilized in a variety of ways by government and military institutions. These include warfare simulations, enhanced information systems for pilots in low-visibility conditions, and interactive laboratory models. The European Space Agency has constructed a virtual reality model of the Columbus Orbital Laboratory, which is scheduled to dock with the International Space Station in 2002. The virtual reality model in The Netherlands provides a 3-D, interactive simulation of the space facility (Smith, 1999).

The Synthetic Environment Tactical Integration (SETI) Project of the U.S. Navy is focused on developing a distributed undersea warfare simulation designed to run in real time. The Navy is investigating how simulated weapons tests can save money with less risk, while still providing valuable military training (Corless, 1999).

Recent Developments

Although virtual and augmented reality tools are clearly useful to a variety of industries, these emerging technologies are still just beginning to diffuse into everyday use. Indeed, just a few years ago, VR was referred to as a "zero billion dollar industry" (Loeffler & Anderson, 1994, p. xxiii). With the exception of entertainment and vehicle simulators, there are probably fewer than 100 examples of routine VR installations currently in use (Brooks, 1999). In the meantime, however, many advances continue to be made in nearly all aspects of VR and AR technologies.

Image rendering engines have grown exponentially more powerful, while HMDs have improved in resolution and decreased in size. Tracking and haptic technologies are also becoming more precise, increasing VR's usefulness for delicate procedures such as microsurgery. Finally, new standards are being sought for virtual reality modeling language (VRML).

Image Generation

Rendering engines have improved dramatically, so that many more polygons can be generated per second. In 1994, the fastest rendering engines could only produce 30,000 polygons in a 1/20 second frame. Now, the SGI RealityMonster can produce 180,000 polygons in a 1/20 second frame. Because fast CPUs (central processing units) are getting much cheaper, it is possible to configure a good VR system from mass-market image generation engines (Brooks, 1999).

These faster and cheaper chips have led to the rise of "immersive videogames." Sony Playstation 2, released in March 2000, should be able to produce real-time, 3-D animation similar to that of the film *Toy Story*. The graphics chip in the Playstation 2 also uses curve-based modeling, rather than the angular polygons used by most 3-D computer graphics programs (In the eye, 1999). (See Chapter 10 for more on videogame technology.)

Displays

There has been considerable improvement in HMD displays as well, and current LCDs have good resolution and color saturation (Brooks, 1999). Meanwhile, the virtual retinal display may eventually obviate the need for head-mounted screens. Tom Furness, a researcher at the University of Washington's Human Interface Technology Laboratory (HITlab) has designed a visual technology that beams images directly into the viewer's eyes. VRD will provide better image brightness, contrast, and color fidelity, and can also display stereoscopic images (In the eye, 1999).

Tracking

Developments in tracking technologies have addressed the disorienting lag time in VR. InterSense, a company producing precision motion tracking technologies, introduced the lightweight headset component that tracks the user's head position for simulators and virtual reality systems. Eventually, these researchers hope to track many points on a single VR user's body and provide wireless operation, enabling the entire body of the user to interact with virtual environments (Stowell, 1998; Nordwall, 1999).

Haptic Technologies

Haptic (relating to the sense of touch) technologies provide a sense of physical force in VR, offering resistance in joysticks, levers, or other tactile interfaces. Traditionally, haptic technology has relied on mechanical pulleys and gears to produce sensory feedback, but a new form of haptic technology has been developed at Carnegie Mellon University's Robotics Institute. This new form uses magnetic levitation technology to better simulate forces in virtual environments. The technology can be applied to medical training as well as to design tasks (Galuszka, 1999).

VRML and the Web3D Consortia

The VRML Consortium set standards in the 1990s for versions of VRML, which is used to enable Internet browsers to interact with 3-D virtual worlds (Vince, 1998). Because VRML programs are too large for easy online use, the Consortium is attempting to find a new language standard that will be better integrated with the Web and allow increased interactivity. The new Web3D Consortium is a non-profit organization of about 50 companies working toward a new VRML browser standard via a community source program (Web3D Consortium, 1999).

Current Status

Although much of the research on virtual reality was originally linked to government interests, current research on VR and AR technologies includes work within "a complex and intertwined hybrid of profit-driven private consortia such as Autodesk, Apple, Division Corporation, and Silicon Graphics" (Hillis, 1999, p. 14). Also, there are several major academic centers of research including the HITlab at the University of Washington, MINDlab at the University of Michigan, the Electronic Visualization Laboratory at the University of Illinois, Massachusetts Institute of Technology, Brown University, UC Berkeley, Stanford, the University of Southern California, and Carnegie-Mellon University. Finally, NASA continues to pursue VR research at the Ames Human Factors Research Division in Mountain View (California).

Market and Current Uses

VR and AR are now being adopted in various fields requiring precision, increased productivity, and communication between remote teams. These areas include vehicle and architectural design, aerospace training at NASA, and medicine, especially in the work of probe microscopy. These fields are primary drivers for ongoing VR research and development (Brooks, 1999). Other markets with a strong potential to incorporate virtual reality technologies include games, retail commerce, military applications, and industrial uses such as factory design and large-scale engineering processes (Astheimer & Rosenblum, 1999). The market for VR and AR is growing, as more and more industries adopt these technologies. CyberEdge Information Services, a VR industry analyst, measured the total value of the VR and visual simulation market at $13.6 billion in 1998 and $17.7 billion in 1999, with an estimated growth rate of 35%. At this rate, the market is predicted to top $23 billion in 2000 (CEIS, 1999; Third year, 2000).

Companies in VR

While there are several companies currently in the VR and AR markets, major commercial virtual reality developers and vendors include Silicon Graphics, FakeSpace, Alternate Realities, Division, Inc., and Sense8. SGI's Onyx2 graphics engines can produce imagery at rates of about 60 frames per second. The Unix-based computer processes geometry, imagery, and video simultaneously, creating simulations at a constant frame rate.

FakeSpace, which recently merged with Pyramid Systems, markets immersive rooms, workbenches, BOOM displays, other stereoscopic displays, and interactive devices. At the 1999 SIGGRAPH (Special Interest Group in Computer Graphics) conference, FakeSpace demonstrated a "magical electric room" that displays 3-D data on its walls and ceiling, completely surrounding the users (Dixon, 1999). The company sells a commercial version of its CAVE for about $400,000, and has sold about 26 around the world (Hill, 1999). Pyramid Systems has introduced Immersadesk, an immersive virtual office space. Another company, Alternate Realities, markets a portable 3-D environment called the VisionDome, which can accommodate multiple users. Panoram Technologies offers a panoramic display also intended for collaborative virtual reality applications. In software, Division, Inc., makes tools for building virtual product designs, and Sense8 makes programming kits for coding 3-D graphic and VR applications.

Factors to Watch

Virtual reality is increasingly implemented as a collaborative tool, for both remote users and groups interacting in the same space. VR has not yet reached the mass market, and it is still largely used by specialized industries. An exception to this is virtual Web environments, such as Activeworlds.com. Web VEs allow anyone with Web access and the appropriate plug-in applications to enter a 3-D world as an interactive character or avatar (see Figure 14.1). Finally, wearable computing, such as that demonstrated at SIGGRAPH in 1999, may influence the adoption of VR technologies on a mass scale.

Figure 14.1
Screen Avatar

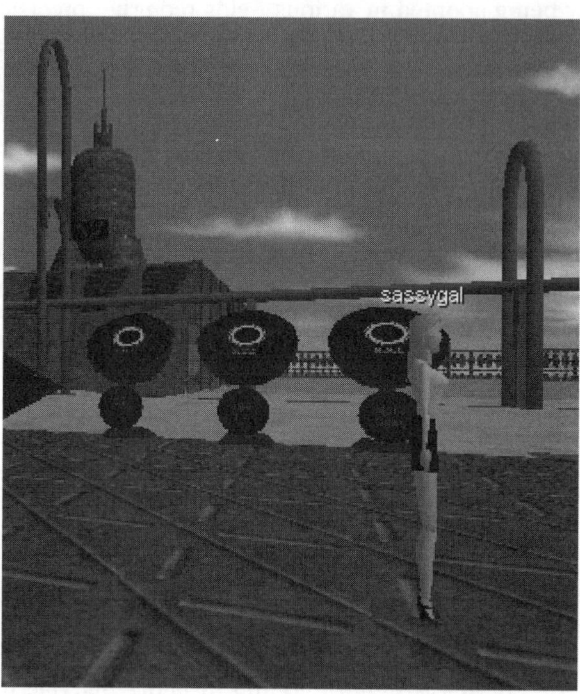

Source: Activeworlds.com

Collaborative Virtual Environments

VR allows groups to interact with 3-D representations of data, and this makes collaborative virtual environments (CVEs) an important factor in the development of virtual technologies. Collaboration can occur between users with CAVE interfaces or with the virtual workbench. Groups congregate in the same space, and can individually manipulate these 3-D objects, "interact[ing] with the data while maintaining their real-world connections" (Mahoney, 1999b, p. 15).

Also, locomotion in VEs may become more natural, so participants can walk long distances in virtual environments instead of simply "flying." The University of North Carolina at Chapel Hill, the University of Tokyo, and the University of Utah have all experimented with various ways to simulate walking. One proposal to enhance this interface is a torus-shaped treadmill, which uses 12 treadmills side by side to generate an apparently infinite walking surface. This hardware does not require a safety harness and does not cause instability for user. One possible application of this new locomotion interface will be as an evacuation simulator, to help ship personnel practice emergency procedures (Iwata, 1999).

Augmented Reality

AR is frequently viewed as a more practical alternative to full-scale immersive VR, but visual AR displays are still being perfected. Currently, there is a choice between optical or video display technologies. Optical viewers consist of transparent displays that also allow users to see their material surroundings, but images generated in optical displays tend to be confusing in their transparency (Biocca & Levy, 1995).

Optical displays project computer-generated images onto partly silvered mirrors angled over each of the viewer's eyes. The other option, video display technology, superimposes the generated images on a video display of what the user would normally be seeing, but this use of video display limits the user's field of view substantially (Mahoney, 1999b). Also, AR applications continue to suffer from registration problems. This refers to the accurate relation of virtual and material objects viewed by the user. Virtual and material objects need to be convincingly integrated, especially in precise applications such as surgery (Tatham, 1999).

Web-Based Virtual Worlds

Another less immersive form of the CVE is the Web-based environment. Users interact via monitor displays, text, and sometimes voice in these virtual worlds. Larger commercial virtual world sites, such as Activeworlds.com, host about 250,000 visitors per day, and about 20% of Web users currently have the 3-D capability for these online applications (Hill, 1999). These interactive sites may become more popular as wider bandwidth for "full-sense interpersonal communication" becomes increasingly available to the typical recreational user (Astheimer & Rosenblum, 1999). A major driver for this technology, not often discussed in the literature (understandably) may well be "pornography on the Web," which represents a large market pushing the technology envelope. The high-speed, fiber optic backbone of the Internet 2 project, which involves academic and government interests as well as commercial interests, such as Cisco Systems and IBM, may be a new venue for networked VR (Higgins, 1999).

Virtual reality has come a long way in recent years, and is emerging from the realm of science fiction into everyday applications. This technology is integrating into a variety of fields such as medicine and engineering, and has great potential in many more, such as education, psychotherapy, and theater. In the future, input devices may use brainwave signals or neural impulses to direct the computer, creating an increasingly seamless interface (Biocca & Levy, 1995). As broadband networks spread, processor capacity increases, and interfaces become more refined, VR technology will become more widespread in application.

Bibliography

Astheimer, P., & Rosenblum, L. (1999, November-December). A business view of virtual reality. *IEEE Computer Graphics and Applications, 19* (6).

Biocca, F., & Delaney, B. (1995). Immersive virtual reality technology. In F. Biocca & M. Levy (Eds.). *Communication in the age of virtual reality*. Hillsdale, NJ: Lawrence Erlbaum Associates.

Biocca, F., & Levy, M., Eds. (1995). *Communication in the age of virtual reality*. Hillsdale, NJ: Lawrence Erlbaum Associates.

Brooks, F. (1999, November-December). What's real about virtual reality? *IEEE Computer Graphics and Applications, 19* (6).

Burdea, G., & Coiffet, P. (1994). *Virtual reality technology*. New York: John Wiley & Sons.

Corless, J. (1999, April 1). Virtual reality steers carrier procurement. *Jane's Navy International, 104* (3).

Cyberedge Information Services, Inc. (2000, April 30). *Second virtual reality market study released*. [Online]. Available: http://www.cyberedge.com/3a2.html.

Dixon, P. (1999, September 5). Computer conference is a window on a world of wonder. *San Diego Union-Tribune*, E4.

Earnshaw, R. A., Vince, J. A., & Jones, H. (Eds.). (1995). *Virtual reality applications*. San Diego, CA: Academic Press.

Galuszka, P. (1999, February 22). Putting real feeling into virtual reality. *Business Week*, 617.

Heudin, J. (Ed.). (1999), *Virtual worlds: Synthetic universes, digital life, and complexity*. Reading, MA: Perseus Books.

Higgins, K. (1999, October 1). Full stream ahead. *Internet World.*

Hill, J., & Jensen, J. (1998, March). Telepresence technology in medicine: Principles and applications. *Proceedings of the IEEE, 86* (3).

Hill, S. (1999, October 11). The real thing: After years of false starts and unrealized potential, virtual reality is finally ready to make its mark on the real world. *Time*, 154.

Hillis, K. (1999). *Digital sensations*. Minneapolis: University of Minnesota Press.

In the eye of the beholder. (1999, May 1). *The Economist.*

Iwata, H. (1999, November-December). The torus treadmill: Realizing locomotion in VEs. *IEEE Computer Graphics and Applications, 19* (6).

Larijani, L. C. (1994). *The virtual reality primer*. New York: McGraw-Hill.

Loeffler, T., & Anderson, T. (Eds.). (1994). *The virtual reality casebook*. New York: Van Nostrand Reinhold.

Mahoney, D. P. (1999a, July 1). A new kind of "get together." *Computer Graphics World, 22* (7).

Mahoney, D. P. (1999b, February 1). Better than real. *Computer Graphics World, 22* (2).

Making an appearance. (1999, April 9). *Building Design.*

Moukheiber, Z. (2000, March 6). Dr. Robot: Virtual reality is set to transform the field of cardiac surgery. *Forbes.*

Nordwall, B. (Ed.). (1999, June 7). Intersense, a small Burlington Mass., company. *Aviation Week and Space Technology, 150* (23).

Regan, E. C. (1995). Some human factors issues in immersive virtual reality: Fact and speculation. In R. Earnshaw, J. Vince, & H. Jones (Eds.). *Virtual reality applications*. San Diego, CA: Academic Press.

Rheingold, H. (1991). *Virtual reality*. New York: Touchstone.

Satava, R. M. (1995). Virtual reality for the physician of the 21st century. In R. Earnshaw, J. Vince, & H. Jones (Eds.). *Virtual reality applications*. San Diego, CA: Academic Press.

Satava, R. M., & Jones, S. B. (1998, March). Current and future applications of virtual reality for medicine. *IEEE Computer Graphics and Applications, 86* (3).

Slater, M., & Usoh, M. (1995). Modeling in immersive virtual environments: A case for the science of VR. In R. Earnshaw, J. Vince, & H. Jones (Eds.). *Virtual reality applications*. San Diego, CA: Academic Press.

Smith, B. A. (Ed.). (1999, July 19). Station familiarization. *Aviation Week and Space Technology, 151* (3).

Stephenson, N. (1992). *Snow Crash*. New York: Bantam Books.

Stowell, C. (1998, April 22). Improved technology sets start-up in motion. *Boston Globe*, F4.

Sun, H. (1995). Hand-guided scene modeling. In R. Earnshaw, J. Vince, & H. Jones (Eds.). *Virtual reality applications*. San Diego, CA: Academic Press.

Sutherland, I. E. (1965). The ultimate display. *Proceedings of the IFIP Congress, 2*, 506-508.

Tatham, E. W. (1999, September). Getting the best of both real and virtual worlds. *Communications of the ACM, 42* (9).

Third year of important virtual reality market study. (1999, January 11). *Business Wire.*

Vince, J. (1998). *Essential virtual reality fast*. New York: Springer-Verlag.

Web3D Consortium and Blaxxun Interactive license VRML browser code. (1999, August 1). *Business Wire.*

15

Home Video

Bruce C. Klopfenstein, Ph.D.*

Perhaps the most significant new and existing home communication technology from the 1980s to the present day is home video. For the purposes of this chapter, home video includes the videocassette recorder (VCR), personal video recorder (PVR), and DVD (digital video or versatile disc). As is the case with all media technologies, home video is progressing from the analog technologies of the past to today's digital technologies. Home video also has an established history of market success (VCR), market failure (RCA CED videodisc, Sony Betamax, and Divx), and limited adoption (laserdisc) (Klopfenstein, 1989a). The camcorder more recently became a household fixture, first in households with children (41% of 1997 households with children owned a camcorder versus 19% of households without children according to Statistical Research [1997]). Consumer level digital video cameras are now available. Despite continuing predictions for its demise, the market for both VCRs and videocassettes continues to be stronger than ever, although their days are certainly numbered.

Background

Home video recording dates back to the earliest days of magnetic recording technology (Schoenherr, 1996), while VCRs can trace their heritage to the professional Sony U-Matic, the first videotape *cassette* recorder (Klopfenstein, 1985). The first consumer Betamax was introduced in 1975 at a list price of $2,295. A recurring and continuing theme in the history of home video would soon ensue: format standards battles. The most famous is the Sony Betamax versus JVC VHS battle of the 1970s that was won rather handily by VHS. Since then, there has been a camcorder battle where the tide has turned: Sony's 8mm camcorders have been displacing VHS and VHS-C camcorders. In 1998, the new digital disc format (DVD) faced competition from incompatible recording and limited play versions of the standard read-only system, Divx. Today, Divx has nearly vanished, and multiple DVD player variations are finding great success in the market.

* Professor, Department of Telecommunications, University of Georgia (Athens, Georgia).

Many industry leaders lament the lack of standards when they occur in home video because of the belief that this slows market adoption (indeed, lack of a technical standard prevented the adoption and diffusion of AM stereo radio broadcasting) (Klopfenstein & Sedman, 1990). The standards battle between Beta and VHS led both sides to technological innovations such as longer recording times, high-fidelity sound, lower prices, and others that probably accelerated VCR adoptions (Klopfenstein, 1985). The same drive to innovation was evidenced in the early battle between the Netscape and Microsoft Explorer Web browsers. Many industry experts thought that Divx would hurt DVD; however, there is evidence to the contrary. It is reasonable to assume that DVD advocates may have increased their efforts to innovate DVD features and hold prices as low as possible to prevent Divx from upending DVD.

Copyright issues are another concern of home video. The relevance of this concern is dramatized by the plight of the digital audiotape (DAT) format. Challenges by recorded music copyright holders slowed the diffusion of this audio technology and may have effectively killed it as a consumer audio format (Cohen, 1991). More recently, a Chinese-made DVD player from Apex that could break copy protection codes was made available at a bargain price in the United States (Apex DVD deck, 2000). The Motion Picture Association of America (MPAA, http://www.mpaa.org/) and the Consumer Electronics Manufacturers Association (CEMA, http://www.ce.org) often represent competing interests in the home video industry; however, they both have concerns regarding copyright and piracy. Look to these organizations' Websites to see their latest positions on recording and anti-piracy technologies.

The VCR

Despite the onslaught of new and competing technologies related to home video such as cable and direct broadcast satellite (DBS), demand for VCRs remains high. Contrary to forecasts continually predicting the decline and fall of the VCR, total VCR sales to dealers in 1999 were 22.8 million, up 28% over 1998's figure of 21.3 million (State of the industry, 2000). This includes TV/VCR combination sets (State of the industry, 2000). The Consumer Electronics Industry Association predicted a slight increase in total sales to dealers in 2000 (CE sales projected, 2000).

While the VCR always had the potential to be disruptive to the U.S. system of commercial broadcasting (because users can easily eliminate commercials on playback), most VCR use remains playing prerecorded tapes (e.g., movies). Historical data have shown that, on any given night, only between 1% and 5% of the Nielsen television program ratings pie is made up of those taping shows. The 1.3 million viewers who taped the NBC hit program *ER* in 1995 made up about 3% of the 40 million who viewed it (Bash, 1995). These statistics are quite instructive because they tells us something about the media consumer's interest in controlling his or her media environment and gives pause to those who believe consumers have a strong desire to do so.

History of Home Video

Videotape technology was originally developed for the broadcast industry in the 1950s (Klopfenstein 1985, 1986, 1989a). By 1970, Sony had developed the first compact videotape recorder, the U-Matic 3/4-inch VCR, which used an easily inserted, book-sized cassette. Too expensive for consumers, it found a home as a video-training device in schools and businesses. Sony refined the technology and introduced its famous 1/2-inch Sony Betamax in 1975.

A rival Japanese manufacturer, Japan Victor Corporation (JVC), developed its own incompatible 1/2-inch VCR, the VHS (video home system). JVC designed VHS to work at a slower speed than Beta, made its cassette larger to hold more tape, and made the tape thinner. Although VHS produced a lower-quality picture than Beta, VHS could record for longer periods. RCA obtained a license to market the VHS and introduced its VCR in August 1977 priced $300 less than the Betamax. Within two years, VHS controlled 57% of the U.S. market. VCR sales rose from 400,000 in 1978 to four million in 1983 and nearly 12 million in 1985 (see Table 15.1), with VHS accounting for 90% of the market. The future of Betamax was sealed when several electronics manufacturers that used to produce Beta models switched to VHS in 1986, leaving Sony virtually on its own.

The growth of the VCR in the late 1980s was impressive. Indeed, it equaled and then exceeded comparable sales of color television sets in the 1960s. Although the number of total U.S. households differed in the two periods, Table 15.1 shows this intriguing growth.

Table 15.1
VCR Versus Color Television Set Sales

Total VC R Sales		Color TV Sales	
1975	30,000	1959	90,000
1976	55,000	1960	120,000
1977	160,000	1961	147,000
1978	402,000	1962	438,000
1979	475,000	1963	747,000
1980	805,000	1964	1,404,000
1981	1,361,000	1965	2,694,000
1982	2,035,000	1966	5,012,000
1983	4,091,000	1967	5,563,000
1984	7,616,000	1968	6,215,000
1985	11,853,000	1969	6,191,000
1986	12,500,000	1970	5,320,000

Note: The VCR was introduced in 1975. The color TV was introduced in 1954. The total number of households is not directly comparable.

Sources: Electronic Industries Association and U.S. Department of Commerce

In the early 1980s, the VCR was an expensive item that seemed destined for elite households (Klopfenstein, 1989b). A less expensive alternative that could exceed the video quality (but could not record) was the videodisc player (VDP). One VDP was developed by RCA, which needed a new product to follow the success of color TV in the market (Graham, 1986). Another videodisc player was developed by MCA (in partnership with Philips) as an outlet for its movie inventory. MCA's optical Laservision videodisc could hold up to 54,000 separate video images or up to one hour of moving pictures on a side. Philips's Magnavox rushed its $695 Magnavision to market in Atlanta in late 1978. Two hundred disc titles were available, mostly old movies. Technical problems cropped up with both the hardware and the discs (as many as 90% were defective), and the player price was raised to $775.

RCA introduced its $500 Selectavision VDP in March 1981. Blitzed by the more expensive but versatile VCR, RCA announced in 1984 that it was abandoning its VDP at a loss of $580 million. About 550,000 players had been sold, and far fewer laser VDPs had been sold to consumers. Laser advocates announced in late 1984 that they too would retreat from the home market. A third VDP format, Matsushita's VHD, found some success in Japan, but none in the United States.

While about 40% of U.S. households had a VCR by the end of 1986, most experts predicted only five years earlier that the VDP would be more popular than the VCR. What many of the experts did not appreciate was the usefulness of a player that could also record programs from television. While many lamented the lack of a standard, the competition between the Beta and VHS formats actually led to lower prices and additional features. The VDP with its limited software was no match for the VCR, which, ironically, also had more prerecorded software than both VDP formats (Klopfenstein, 1985).

Recent Developments

VCR Use

Current VCR user research is surprisingly difficult to locate in the public domain. One fairly recent study showed that, as of 1997, about 89% of TV homes also owned a VCR, and penetration was projected to rise to 93% by the year 2000. VCR ownership is particularly high among the most affluent and best-educated households. With the reduction of VCR prices, however, even the lowest-income homes are beginning to acquire their first machines (Everything about, 1998). Multiple VCR households have also increased dramatically in the last five years.

In 1997, average VCR use totaled about six hours per week. This was divided between 1.5 hours for recording programming and 3.5 to 4 hours for playing both home-recorded and rented or purchased tapes. One-third of program recordings were never being played back. Over half of taping (55% to 60%) was done while the TV set was off, 25% with the set tuned to the same channel, and 15% to a different channel. Most recording occurs in prime time (30%), followed by the weekday daytime hours (30% to 35%) and fringe hours (13%). About 60% of recordings are of shows aired by the big three networks or their affiliates. Serials (soap operas) are the most commonly-taped genre, accounting for half of all shows taped (Everything about, 1998).

Video Rentals

Early studies showed that the primary use of VCRs was to record programs off the air for later viewing, a practice known as time shifting, but the VCR really made its presence felt in the U.S. film industry. In 1985, 100 million people went to the theater to see a movie, and about 103 million movie cassettes were rented each month (Green, 1986). Video rentals in 1995 reached about $8 billion, with cassette sales climbing to over $7 billion (Rental stores, 1995). Home video rentals dropped in the mid-1990s, while revenues reached a high plateau (King, 1996). The enormous number of tapes now available would seem to assure the VCR some life despite the encroachment of DVD. Home video revenue for studios exceeded $9 billion in 1999, with one relatively conservative projection for 2005 reaching nearly $13 billion (Arnold, 1999).

Consumers purchased nearly 700 million videocassettes in 1995, an 18% increase from 1994 according to one market research study (King, 1996). Research data from 1993 cited by *American Demographics* showed families with children under age six spent the most on videotapes. A typical household spent more than $55 in 1993 on blank tapes to capture baby's first smiles and prerecorded tapes to entertain the child. As infancy wanes, purchases taper off. On average, a family with children aged six to 17 spent $43 on videotapes in 1993, and one with only teens and no younger children spent $41 (Mogelonsky, 1995). According to a number of research sources, video stores supply about 54% of movie studios' revenues by paying more than $60 per video. The store sees a profit after renting the tape 25 times, which usually happens, as demand remains high up to six weeks after a hit is released. About 8% of store revenue comes from late fees.

One estimate of average home video use is surprisingly low. According to data reported in "Consumer Media Use by U.S. Adults" (2000), the average American spent 57 hours with, and spent $97.51 on, home video in 1999. This statistic includes adults who do not use home video (which is probably skewed toward the oldest segments of the population).

Billboard magazine called 1999 the year of the DVD. Not only did the number of players sold in its first three years exceed those of VCRs, DVD rentals became a significant portion of the home video rental market (Fitzpatrick, 2000b). Home video sales and rentals in 1999 returned an estimated $9.5 billion to studios and suppliers, nearly three times what studios generate from their share of theatrical box office receipts. Of that total, DVD contributed $1.3 billion, with Warner Home Video dominating the rapidly-growing market (Hettrick, 2000). With escalating DVD sales and revenue-sharing increasing overall business, studios will continue to derive more than half of their revenue from home video for at least another six years (Arnold, 1999). The consensus among studios and distributors is that DVD will replace VHS within five to seven years (Nichols, 2000).

While viewing movies at home has become a way of life, the movie theater still has some advantages:

(1) Going to the movies is a social occasion.

(2) Movies appear in theaters before they appear on cassette.

(3) Theaters offer the large, wide screen.

(4) Theater sound systems dwarf what the vast majority of people have at home.

New digital home theater technology may threaten to erode the non-social reasons for going to the movie theater.

Digital VCRs

A digital VCR standards group representing about 50 companies announced an agreement on technical specifications for recording transmission signals from the U.S. high-definition television (HDTV) system. The U.S. HDTV system will operate at 19.4 megabits per second (Mb/s). JVC jumped ahead of digital VCR competitors by rolling out its own digital VHS (D-VHS) format in 1994. D-VHS players will initially work with Thomson Consumer Electronics' Digital Satellite System (DSS) set-top boxes, which are deployed as part of Hughes' DirecTV digital broadcasting satellite system. JVC demonstrated a prototype D-VHS deck operating in HD (high-definition) mode needed for terrestrial DTV (digital television) in January 1998, but offered no timetable for announcing a standard.

At the same exhibition, Hitachi touted the merits of its current D-VHS model, co-developed with Thomson, for recording HD-DSS signal (Novel VCRs, 1998). VCR manufacturers believe the VCR is on track to survive its 25th anniversary and life beyond the millennium, if only due to the lack of consensus on a rewritable DVD format as a replacement for analog tape. The prospects for advanced digital VCR formats such as D-VHS or W-VHS were negligible in 1998, but they may be introduced along with a new M-DVD magnetoresistive format (VCRs proliferate, 1998).

While we wait for pure digital VCRs to take the scene, a recent breakthrough will allow VCR technology some new life. New VCRs with 19-micron heads offer nearly the same picture quality when recording at the slow speed as at the fast speed. According to JVC, their new VCRs will have a 19-micron head gap to eliminate crosstalk interference from adjacent tracks. In addition to precision tracking, the high-performance materials used in these heads help reduce contact noise and increase electromagnetic efficiency. This development can triple tape economy, while ensuring maximum picture quality.

Smart VCRs came on the market in 1998. These VCRs automate many aspects of installation, including setting the clock and locating the available channels. Some even accept voice commands, an interface likely to begin showing up almost as an afterthought on many home appliances in the next few years. Not only are VCRs available that skip commercials in recording and/or playing back, Thomson markets one that also fast forwards past the prerecorded promotional material at the beginning of most feature films on cassette (Cole, 1998). Some Sanyo VCRs have a feature called Speed Watch that allows users to watch a tape at two times the normal tape speed while the audio remains at normal levels. It seems likely that only the huge installed base of videotapes will assure some future for the VHS VCR, but its fate is tied to the success of recordable disk technology (VCR near end, 1998).

Digital Personal Video Recorders

The latest home video technology set to revolutionize television is the personal video recorder (PVR). This digital recording technology will allow users to record programs based on their interests. TiVo (http://www.tivo.com) and Replay TV (http://www.replaytv.com) are two of the combatants in intelligent home video recording. These machines are basically stripped-down PCs with large hard drives that dial in and download program listings each week.

TiVo was founded in August 1997. The TiVo service lets consumers take control of their television viewing by allowing them to watch what they want, when they want it. As part of TiVo's easy-to-use service and unique technology, viewers can time-shift their favorite television shows and create a customized television lineup for viewing at any time. With this technology, viewers can teach TiVo what shows they like and dislike. Using these preferences, TiVo automatically records programs the viewer may want to see. TiVo also enables consumers to pause, rewind, instant replay, and play back in slow motion any live television broadcast. In addition, TiVo has developed a technology that serves as a platform for new and interactive entertainment content and services. To ensure viewer privacy, TiVo has established strict policies to help protect the personal information of consumers, while providing a valuable personalized TV experience.

TiVo offers customers a choice in service options. A lifetime subscription to the TiVo service with no monthly fees can be purchased with a single payment of $199. For customers who want to keep the upfront costs to a minimum, personal TV is available as a monthly subscription to the TiVo service for $9.95. TiVo is currently available on stand-alone personal video recorders manufactured by Philips for $299 and can be purchased at retail outlets nationwide including Best Buy, Circuit City, and Sears, or online at Amazon.com, ROXY.com, or 800.com.

ReplayTV is another digital recorder that can automatically record favorite shows and organize them according to themes. This product lets viewers pause, rewind, fast-forward, and replay live TV broadcasts. Up to 28 hours of TV signals are digitally recorded onto a hard drive. The list price in fall 1999 for set-tops ranged from $699 to $1,499. There are no monthly fees for basic service, and point-and-click programming is reported to be easy (Hale, 1999). Users can pause live broadcasts and resume viewing using the remote control. During playback, Replay's Quick Skip and TiVo's jump button allow users to skip over virtually all commercials (Kay, 2000).

At the most basic level, TiVo and Replay work the same way. The major difference is how the program guide services are offered. Replay does not charge any fee for program listings, but they plan to eventually accept advertising to defray the cost. At the time of this writing, TiVo charged a monthly, annual, or lifetime membership fee. Some question this approach unless the TiVo device is virtually given away. In an era of advertiser-supported, free Internet services, charging for a television recording service makes little sense to the vast majority of potential adopters.

While TiVo and ReplayTV are currently receiving much of the attention, they are not the only players in the game. Metabyte's (http://www.metabyte.com/) MbTV promises similar functionality to TiVo and ReplayTV, and all may run into the same privacy concerns raised by Divx. MbTV announced its collaboration with hard drive manufacturer Seagate in 1999 (Tarr, 1999). As the cost of hard disk storage drops and compression technologies improve, consumers will enjoy new features and a significant increase in the number of manufacturers and models available.

Camcorders

NPD Intellect, a market research firm, found that digital camcorders posted huge gains in 1999, ending the year with 6% of the overall camcorder market, with sales of $532 million, up from the $93.2 million in 1998. Digital cameras comprised less than 1% of camcorder sales in 1998 (Olenick, 2000b).

New digital camcorders rival professional broadcast equipment. For as low as $1,000, digital camcorders using the standard digital video format are available and allow still-frame digital photography as well as full-motion video.

Entry-level camcorders are loaded with new features. These home movie machines have now shrunk to the point that the smallest of them are hardly bigger than a deck of cards. Picture and audio quality on today's camcorders varies quite widely, although even an inexpensive unit will produce more-than-acceptable video and sound under normal shooting circumstances.

Consumer camcorders come in several incompatible formats: full-size, inexpensive VHS (although these are becoming obsolete), VHS-C (compact tapes that, with an adapter, play back on a VHS VCR), 8mm, and the new digital format. There are also high-resolution versions of the VHS and 8mm formats, called SuperVHS (or S-VHS), S-VHS-C, and Hi-8. DV gives the best picture quality, and the price is beginning to be competitive with other camcorders. Sony also introduced their Digital 8 format in 1999. The camera records a digital signal on standard 8mm videotape. In perhaps a sign of things to come, digital "tapeless" camcorders are now available for professional users.

Each format has its advantages and disadvantages. Full-size VHS camcorders are physically stable, although small camcorders can have "image stabilization" circuitry built in, and the tapes are completely compatible with any VHS VCR. VHS camcorders are much larger and bulkier than the other formats. VHS-C camcorders are smaller, but their tapes only record up to half an hour on the fast speed, and an adapter is needed to play back the tapes in your VHS VCR.

About the size of an audiocassette, at standard speed, 8mm tapes provide two hours of recording. The camcorder itself has to be used for playback, adding wear-and-tear on the machine. VHS and 8mm (high and "regular" resolution) have comparable quality to each other. Sound quality can vary widely, but a hi-fi camcorder will provide very good audio.

Death of Analog Laserdisc

The venerable laserdisc player (LDP) is a technology that, despite many impressive technical achievements, never really caught on. The technology found its place in educational media centers and was used in intensive training applications in both government and industry. Had it caught on in the consumer market, less expensive and fuller featured machines would have evolved more quickly. Given the introduction of the DVD, the days of the analog laserdisc are numbered, and the industry has announced it will no longer track aggregate laserdisc player sales. The Consumer Electronics Association estimated that only 20,000 LDPs were sold in 1998, and that number dropped to 6,000 in 1999 (CE sales projected, 2000).

DVD History

In 1995, consumer electronics manufacturers—including old VCR rivals Sony and Matsushita—bombarded the media with descriptions of their next-generation recording medium: DVD. The technology was touted as a digital replacement for VCRs, VHS tapes, laserdiscs, video game cartridges, CDs, and CD-ROMs. As with VCRs, DVD vendors proposed different standards. Sony's Multimedia Compact Disc (MMCD) was pitted against Matsushita's DVD, called Super Density DVD (SD-DVD) (D'Amico, 1995). A format war was avoided when the industry players agreed to support a format that combined a Toshiba design with a Sony/Philips (the original CD partners) encoding scheme (Braham, 1996).

The DVD is a variation on the now-ubiquitous compact disc. DVD, introduced in March 1997, can store any kind of multimedia content. This new disc is identical in shape and size to an audio CD and CD-ROM, but DVD has a great deal more storage capacity. While the 1998 vintage DVD discs could hold 133 minutes or 4.7 gigabytes (GB) of video per side, double-layered DVDs capable of holding 241 minutes or 8.5 GB of video are expected to be available soon, with increased storage at least theoretically possible after that. These higher-capacity discs are more likely to be used in computers. For example, if stereo music is the stored information, a single DVD can hold the contents of more than a dozen CDs. The movie studios consider the DVD a major opportunity because it will allow them to sell many of their existing films all over again, just as music companies have done with the audio CD.

Table 15.2 shows a comparison of the specifications for DVD and a standard CD:

Table 15.2
Specifications for DVDs and CDs

	DVD	CD
Diameter	120 mm	120 mm
Thickness	0.6 mm	1.2 mm
Track Pitch	0.74 nm	1.6 nms
Minimum Pit Length	0.40 nm	0.83 nm
Laser Wavelength	640 nm	780 nm
Data Capacity (per layer)	4.7 GB	.68 GB
Layers	1, 2, 4	1

Sources: Balkanski & Sony

A key to the early success of the DVD may be the perceived difference in the sharpness of the user's television picture. The durable CD was clearly an improvement over increasingly fragile LP records. DVD creates images with smaller and more varied pixels (720 pixels per horizontal line versus the standard 240). This advancement allows clearer shapes and forms that are more detailed than today's best VCRs can produce. If viewers can easily see the difference, this will bode well for DVD. Frank Vizard (1997) came to that conclusion in his review of DVD for *Popular Science*. Given the steady increase in television picture resolution, these differences will become more visible. An easily-overlooked problem, however, could be the durability of DVDs and how well they can stand up to the rigors of video rentals. On the other hand, DVDs may allow longer archiving than is possible with videotape.

Pearse (1999) defined the various DVD formats as of fall 1999, and Table 15.3 shows the recordable DVD formats as of 2000 (Technicolor, 2000).

Table 15.3
Recordable DVD Formats

DVD-RAM	The second version of DVD-RAM, which has been ratified by the DVD Forum, has a capacity of 4.7 GB. This format, supported by Panasonic, is at a disadvantage as double-sided discs come in sealed cartridges.
DVD-R+W	This format is backed mainly by Sony, but also has the support of Philips and Hewlett-Packard. It has not been ratified by the DVD Forum, and is effectively a direct competitor to DVD-RAM. Reflectivity differences, among other problems, mean that the disc can't be read by read by DVD-Video players. It has a capacity of 2.8 GB, with an upgrade to 4.7 GB by 2001 expected.
DVD-RW	Developed by Pioneer, this phase-change format is based on DVD-R and can be rewritten about 1,000 times. Pioneer plans to make the format compatible for playback with DVD-Video. DVD-RW offers up to six hours of recording time and has a 4.7 GB capacity.
DVD-R	A write-once standard with limited applications, DVD-R is unlikely to be economic for end users. DVD-R is useful for organizations that do a great deal of archiving or need to store huge amounts of data that won't need to be altered and is possibly useful for games developers. Currently drives are prohibitively expensive. Capacity is 4.7 GB.

Source: J. Pearse

Interactive DVD

JVC and Aiwa announced in 2000 that they would manufacture Internet-enabled DVD players in mid-2000 (Code-free DVD, 2000). Australia-based Internet service provider (ISP) Eisa.com planned to introduce a $99 combination DVD movie player and Internet set-top device called the Neo iDVD in the United States in mid-2000. The low initial cost of the Neo iDVD is subsidized with a two-year contract that would charge customers a $23.95 per month ISP fee for Web access through Eisa.com. Company proponents see the technology as a way for those who cannot afford a home computer to gain Web access. An iDVD device enabling the use of peripherals was planned for the end of 2000 (Olenick, 2000a).

The Divx Debacle

One of the more interesting phenomena since the last edition of this book was the birth and death of Divx, short for Digital Video Express. (Some manufacturers already had combination

DVD/Divx players in the works and decided to introduce them early in 2000.) Having written my doctoral dissertation on home video technology successes and failures, I was comfortable penning my doubts about Divx technology in that edition. I learned through my doctoral research, for example, that consumers rejected the laser videodisc with its unquestionably superior audio and video quality in favor of the lower picture quality that comes with the VHS VCR. There were several reasons why people rejected the seemingly attractive laserdisc alternative:

- There were far more programs available for tape than there were for disc.
- People knew they were losing an important option (recording) with the laserdisc.
- The VCR was already established, while the laserdisc was an unproven and unknown technology.

Divx discs looked just like DVDs and also offered superior video quality to videotapes. Unlike DVDs that listed for $30 to $35 and were offered for sale, Divx worked on something closer to a pay-per-view model: People could buy Divx discs for $4.50 and watch the disc any time they chose. Once they started watching, however, they only had 48 hours in which they could watch the movie as often as they wanted until digital codes made it "expire." After that, they could discard the disc or pay to "recharge" it either for two more days or for life. The Divx player dialed into a billing center over phone lines and handled all activities, sparking legitimate privacy concerns by requiring people to disclose information about their viewing habits (Fost, 1999).

Those who are unfamiliar with history are doomed to repeat it. The Divx machine offered an alternative of sorts to the video rental store, promising better audio and video quality than the common VHS tape. Consumers quickly would learn there were stiff prices to be paid in purchasing a Divx player and, in fact, were lukewarm from the start (Consumer poll, 1997). Divx proponents say the lack of software was the reason they decided to discontinue the product in mid-1999.

The price for a limited-view Divx disc, which, in essence, self destructed after a short period of time, was higher than that for a cassette rental. The local home video store had hundreds if not thousands more cassette titles available in far more locations than Divx. Although a Divx player could also play regular DVDs, even the number of DVD titles still paled dramatically in comparison to the number of VHS titles available. Divx offered fewer than 500 titles when it folded (Consumer Reports, 1999). Anyone comparing a Divx player to a new VCR would have been hard pressed to conclude that the Divx player was the way to go. Finally, the Divx player was more expensive than comparable DVD players. Theoretically, the Divx held the promise of having new releases out more quickly than DVD and possibly VHS, but Divx proponents had to make a Herculean effort to be sure this happened. Even if it did, Divx discs could not be played on regular DVD players or DVD-ROM devices.

Consumer electronic retailer Circuit City took an after-tax loss of $114 million related to the cancellation of the Divx business after the sale of approximately 100,000 players (Wiley, 1999). The fact that one consumer electronics chain was pushing Divx did nothing to ensure its success. Just as CBS dragged its feet in adopting NBC parent RCA's color television system, it's hard to imagine why Circuit City's competitors would have been anxious to sell or promote a competitor's product.

Another lesson from this failure is the potential limitations of consumer market research. Although we do not have access to proprietary market research from the companies involved, it is reasonable to assume that Divx tested this concept with consumers. RCA ignored weak consumer reaction to the testing of its VDP player in the early 1980s that suggested a lackluster reception awaited its doomed non-optical disc player (Klopfenstein, 1985). In fact, one publicly reported consumer poll showed a lack of interest in Divx (Consumer poll, 1997). Sometimes, market research has its greatest impact when it supports the existing views of decision makers.

A fascinating phenomenon was also seen in the demise of the Divx. Opponents aggressively sought to bring it down and created Websites such as Ban Divx (http://www.bandivx.com) and the Dead Divx Society (http://www.ddsociety.com). Some (but not all) videophile Websites also joined the call. It's unclear what impact these groups may have had, but such negative electronic "word of mouth" could have easily played a significant role in the process, especially with potential early adopters.

DVD and Copy Protection Controversies

Because DVD is digital, copyright owners are concerned that counterfeiters will be able to either use DVD movies as masters for VHS tapes or simply copy the data onto another disc using a computer. By encrypting a movie when it is mastered and putting a decryption circuit in consumer DVD players, digital copying can be prevented. In addition, DVD players can be designed with a copy protection feature that prevents a movie from being copied onto videotape without severe degradation in quality. This feature does not interfere with playback to a TV monitor (Balkanski, 2000).

DVDs are encrypted to prevent unauthorized copying of content using the Content Scrambling System (CSS) developed by Matsushita and Toshiba. The CSS mechanism consists of two parts: authentication and encryption. Authentication restricts user access to the encryption keys needed for decryption and to some of the actual data sectors of the disc. There are three keys on the disc. Each DVD decoder has a unique 40-bit player key that must be used to descramble the corresponding segment of the disc key to unlock the movie for playback. While it has always been possible to copy the encrypted contents of a DVD to a hard drive or other storage medium, it was impossible to play them back without the original disc in the drive for authentication (Parker, 1999). Initially, there was no commercial software available to play discs on systems other than Mac- and Windows-based PCs (Grossman, 2000).

DeCSS came to refer to the general process of defeating CSS, as well as computer source code and programs. Despite the perceived threat of piracy this poses to owners of copyrighted video, the genesis of DeCSS seems to have been innocent enough. Several groups of open-source Linux programmers, in the absence of DVD movie support for the Linux operating system, were working to create a DVD player for Linux. They reverse engineered CSS to figure out how to watch DVD movies on Linux machines. When they did, they created a computer program that would automatically disable CSS, and they called it DeCSS (Parker, 1999). Lawsuits were filed when the DeCSS program that breaks CSS was posted on the Web. U.S. and international cases are pending as of this writing (Grossman, 2000).

No matter how much effort goes into creating anti-piracy encryption schemes, it seems ways to defeat them are also found, whether deliberate or accidental. A Chinese manufactured DVD player from Apex was introduced at bargain prices to the U.S. market in 1999. The machine included many new features such as multi-standard output (plays more than NTSC discs), the ability to play MP3 compressed music recorded on blank recordable CDs, and Super VCD playback, which allows the use of karaoke discs. An amazing feature showed up in an onscreen menu called "Loopholes Menu," which allowed the user to turn off the copy protection scheme and copy DVDs to videotape and other media (Apex DVD deck, 2000). A Website told visitors about the feature, and some retailers quickly ran out of stock.

Sony introduced its PlayStation 2 game console in 2000, and with it came the ability to play DVD disks (which most initial adopters did) as well as audio CDs. PlayStation 2 users in Japan were able to override "regional coding" software, thus enabling them to read DVD software not being sold in their market. This is a violation of an agreement that says DVD players must only read software sold in the same market (PlayStation 2 DVD bug, 2000).

Current Status

Home video today has a very mature analog technology, the VHS VCR, selling nearly as many units as television sets. The growth of TV/VCR combinations may foreshadow what's ahead for all digital television receivers and storage devices. TV/VCR combos sold 4,300,000 units in the United States in 1999, with television sets selling another 23 million more (CE sales projected, 2000).

The 1999 TV/VCR subcategory led all color TV areas, with a 36.6% growth rate to 4.3 million units, and a 23.7% growth in dollar volume over the previous year. CEA anticipates sales growing 8.1% to $4.7 million in 2000. VCR decks continued their amazing resurgence, with a mix that was almost evenly split between stereo models (11.3 million units) and mono (10.9 million). Average VCR prices dropped 22.6% to $103 in 1999 (Tarr, 2000).

Early in 2000, CEA forecast modest annual growth (2.8%), projecting total video hardware sales of $17.2 billion. Category leaders in U.S. dollar volume growth were expected to include DVD players (44.4%), camcorders (13.3%), digital still cameras (11.2%), and set-top Internet devices (9.4%). DVD player sales grew 270%, while factory sales climbed 161% over 1998 numbers to $1.1 billion. This revenue growth came despite a 29.5% drop in average retail price from $390 to $275. Projections for 2000 show DVD unit volume climbing to almost 6.6 million on a 44.4% increase in factory dollars to $1.59 billion. The average player price is expected to drop another 12.4% to $241 per player (Tarr, 2000). The DVD player forecast appears to be conservative and may not include Sony PlayStation 2.

Thanks to technological advances and lower prices for digital models, in 1999, camcorders registered a 51.1% increase in factory dollars and a 24.4% rise in unit sales. Consumers paid 21.5% more in average camcorder prices over 1998, making it one of only two categories to register an increase in average pricing in 1999. The CEA forecast is for a sales increase of nearly 10%, with prices beginning to stabilize in 2000 (Tarr, 2000).

Sales of DVD players have been especially phenomenal in the United States. Sales in Japan are far less impressive due to the relative lack of disc titles there. At the end of 1999, there were almost five million DVD-Video players in the United States, and about 30 million DVD PCs (DVD frequently, 2000). Brisk DVD player sales continued into 2000. According to the May 1, 2000 issue of *Television Digest*, DVD player sales to dealers in the first 15 weeks of 2000 were 1,274,864, up nearly 170% over the same period in 1999. The first DVD players priced at under $100 after a rebate were projected to be available in mid-2000 (Code-free DVD, 2000). For the year 1999, DVD was the darling of the industry. Figures for U.S. sales to dealers in units and factory value are shown in Table 15.4 (CE sales projected, 2000):

Table 15.4
DVD Sales (U.S)

Year	Units	Value (In Millions)	Average Price
1998	1,079,000	$ 421	$273
1999	4,000,000	$1,100	$264
2000 (proj.)	6,590,000	$1,588	$242

Source: Television Digest

One market research firm (NPD) predicts that consumers will buy eight million DVD players in 2000, just the third year DVD has been on the market. It took VCRs nine years from the first year of general availability in 1975 to reach that same annual sales level. NPD says the success of DVD is traceable in part to the fact that buyers are mainstream consumers seeking higher-quality home entertainment, rather than just early adopters and hobbyists who need the latest thing for their expensive home theater systems. The research firm also notes that most video software retailers have added DVD sales and rentals, and that some software titles are now being made available on DVD before the VHS version comes out (1.1 million, 2000). The reach of DVD has quickly moved beyond early adopters and is being embraced by ordinary consumers because of rapidly-dropping prices for the players (under $200), a wider selection of movie titles (4,500) by January 2000, and a growing awareness that the sound and visual quality is far superior to videotape (Pristin, 2000).

The DVD Entertainment Group expected the industry to double its 1999 sales by shipping at least 200 million DVD movies and music videos in 2000. This represents $4 billion in revenue for the studios and music labels, which is approximately half of the total revenue generated from the VHS sell-through. With the rate at which VCR prices have been declining and DVD unit sales are growing, DVD player sales dollars are expected to surpass those of hi-fi VCRs in 2000 (Home theater sales, 2000).

Factors to Watch

The sales of DVD players have generally exceeded expectations and appear to be quickly reaching VCR levels (perhaps within five years). We are likely to see the beginning of a product substitution of DVDs for VCRs eventually, although it will remain cheaper to record on VHS tape over DVD-R in the near future. However, VCRs cannot match the random access capability of the DVD, and video movie buffs will not miss having to rewind a tape. If the CD market is any indication, software manufacturers will keep DVD prices higher than those for prerecorded videocassettes. This move may well be based on psychological rather than economic considerations.

Suppliers of DVD players will continue to add features, perfect the technology, and lower the player price. Recordable DVD machines are in development, although exact predictions of market introduction are all but impossible. Until one standard recording technology is established, DVD recorders will not be an important factor. Indeed, if incompatible standards are introduced, the life of the VHS VCR will be extended. The day is coming, however, when we will be transferring our home movies from analog tape to digital disc. If VCR use is any indication, however, it is not reasonable to expect DVD users to record programming in the near future (Joly, 1999).

By 2004, VCRs will follow the earlier pattern of multiple television sets in the home, while a significant number of homes will add a DVD player to their home entertainment systems. Home video stores are rapidly expanding their selections of discs as well. While the proliferation of VCRs into various rooms in the household bodes well for future digital television sets with built-in storage devices, it also points to the demand for prerecorded fare.

DVD is one of those unusual communication technologies where market growth actually exceeds optimistic projections. Even if the booming U.S. economy slows down, there is reason to believe that DVD growth will meet and probably exceed expectations, at least in 2000. Software availability will be key, of course, but the studios now understand what's at stake. The parallel to CDs is unmistakable: People will buy DVD versions of movies they already own on cassette. Although DVDs are less expensive to produce than VHS tapes (which must be dubbed), the price of DVDs ($25) is similar if not higher than it is for cassettes (another analogy to the audio industry). Dramatic growth can be expected in DVD rentals, as more titles (including multiple copies of individual titles) are available at home video stores.

Another interesting DVD opportunity can be seen in the ease with which movie sound tracks, currently sold separately, can be placed on the same disc (Fitzpatrick, 2000b). This is a question of economics rather than technology (i.e., which is more lucrative for the producers: selling them separately or on one disk at a higher price?). DVD audio may soon debut, but the question will be how much better the sound is rather than how much can you record on the disc.

On the PVR front, Moore's Law seems likely to continue upping the ante on low-cost digital storage. It is not clear that today's market leaders (TiVo and ReplayTV have very few customers) will be the ones who bring personal video recording into the home. It should not be surprising to see combination TV/PVRs eventually make this happen when the cost of the PVR is relatively insignificant to the price of the television set. Given the high initial prices for digital television sets, this scenario seems especially possible. For the time being, PVRs appear to be slightly ahead of their time.

You don't have to be an engineer to understand that the days of tape recording are numbered, at least in the home video arena. Random access on a disc is virtually instantaneous, while VCR tapes must still be rewound and fast-forwarded. Even as the first videocassette recorders were created for industry rather than home use, digital disc recording technologies continue to advance in professional media circles. It is only a matter of time before the relative advantages of disc recorders and disc recordings begin to push the VCR the way of the LP. Those familiar with the consumer electronics industry will further note that it bases its profitability on continuously introducing new products. The profit margin on VCRs has been thin for years. Just as the music industry has recouped billions of dollars as consumers replaced their LP collections with CDs and cassettes during the last 20 years, the home video industry understands the same can be true with DVD and other recordable disc technologies.

Another area to watch is the Web. Where will digital video on the Web be in 2002? Who will promote it? Who will view it? What will be the window of new feature film releases on the Web versus pay cable versus cassettes and DVDs? The telephone industry spent a great deal of time and research energy looking at video into the home 10 years ago, but trial after trial failed. As we move into an era in which video on demand becomes a technological possibility, the question becomes what non-technical obstacles will slow its potential adoption and diffusion.

Although it's unclear whether great strides will be made in increased broadband availability and further advances in compression technology, downloading home video content on the Web will become a niche market at the very least (Block, 2000) and relatively soon. As high-speed access becomes more common, some experts think that movies will be available for viewing in real time on the Internet (Husted, 2000).

Bibliography

1.1 Million DVD decks sold during December. (2000, February 7). *TWICE, 15* (4), 4.

Apex DVD deck defeats anticopy coding, our lab test shows. (2000, February 28). *Television Digest, 40* (9), 11-12.

Arnold, T. K. (1999, August). Video is still half of the pie for the studios. *Video Store. 21* (31), *1+*.

Balkanski, A. (2000). Digital videodisc: The coming revolution in consumer electronics. *C-Cube*. [Online]. Available: http://www.c-cube.com/files/dvd.pdf.

Bash, A. (1995, November 28). VCRs can't rescue faltering shows. *USA Today*.

Bismuth, A. (1998, March 30). PCs, players seek one DVD solution. *EETimes*. [Online]. Available: http://www.tech-web.com/search/search.html.

Block, D. G. (2000, March 18). Disc replication: DVD technology in demand. *Billboard. 112* (12), 61, 68+.

Braham, R. (1996, January). Consumer electronics. *IEEE Spectrum, 33* (1), 46-50.

CE sales projected to rise 5.8% in 2000 by CEA. (2000, January 10). *Television Digest, 40* (2), 15-17.

Code-free DVD deck gone, but issues linger. (2000, April 20). *Television Digest, 40* (15), 12-13.

Cohen, J. (1991, December). Making sense of new electronics products. *Consumer's Research, 74*, 34.

Cole, G. (1988, February 5). VCRs get smart (about time). *Electronic Telegraph (London Daily Telegraph)*. [Online]. Available: http://www.telegraph.co.uk:80/et?ac=000647321007942&rtmo+flfs-vDVs&atmo=flfsvDVs&pg=/et/98/2/5/ecvid05.html.

Consumer media use by U.S. adults rose 22 hours from 1998-1999. (2000, February 18). *Research Alert, 18* (4), 1+.

Consumer poll gives thumbs down to Divx. (1997, September 26). *TWICE*. [Online.]. Available: http://www.twice.com/domains/cahners/twice/archives/webpage_1084.htm.

Consumer Reports. (1999, July 27). DVDs may be on a roll, but don't dump that VCR just yet. *The Washington Post*, C4.

D'Amico, M. (1995, June 5). Digital videodiscs. *Digital Media, 5*, 13.

Director, K. M. (1998, March). Zoom in, zoom out. *Videomaker.* [Online]. Available: http://www.videomaker.com/scripts/article.cfm?id=3600.

DVD frequently asked questions. (2000, February 13). *Video Discovery*. [Online.]. Available: http://www.videodiscovery.com/vdyweb/dvd/dvdfaq.html.

Everything about television is more. (1998, March 6). *Research Alert, 16* (5), 1+.

Fitzpatrick, E. (2000a, January 8). 1999: The year of DVD. *Billboard, 112* (2), 92-95.

Fitzpatrick, E. (2000b, January 29). Steeplechase, Warner Video plan VHS/DVD collector's sets. *Billboard, 112* (5), 77.

Fost, D. (1999, June 18). Divx's death pleases opponents. *San Francisco Chronicle.*

Gilroy, A. (2000, March 13). DVD in the fast lane of mobile navigation. *TWICE, 15* (7), 27.

Graham, M. B. (1986). *RCA and the videodisc: The business of research.* New York: Cambridge University Press.

Grossman, W. (2000, May). DVDs: Cease and DeCSS? *Scientific American.* [Online]. Available: http://www.sciam.com/2000/0500issue/0500cyber.html.

Hale, K. (1999, September 6). Personal TV: Replay TV. *Broadcasting & Cable, 129* (37), 26.

Hettrick, S. (2000, April 14). BV vid victorious. *Variety*. [Online]. Available: http://www.variety.com/article.asp?articleID=1117780612.

Home Recording Rights Coalition. (Undated). *Selected chronology of the home taping controversy*. [Online] Available: http://www.hrrc.org/history.html.

Home theater sales post record year. (2000, March 13). *TWICE. 15* (7), 5, 36.

Husted, B. (2000, April 2). Movies may come to a computer screen near you. *Atlanta Journal and Constitution*, 5.

Joly, C. (1999, June 7). Changing DVD trends. *Electronic News, 45* (23), 32.

Kay, R. (2000, January 24). Easy backup, but difficult digital video. *Computerworld, 34* (4), 61.

King, S. (1996, April 19). Home video rentals drop but revenues are still up. *Los Angeles Times*, 26.

Klopfenstein, B. C. (1985). *Forecasting the market for home video players: A retrospective analysis*. Unpublished doctoral dissertation. Columbus: Ohio State University.

Klopfenstein, B. C. (1986). Forecasting the market for home video players: A retrospective analysis. In T. A. Shimp, S. Sharma, et al. (Eds.). *American Market Association marketing educator's conference proceedings*. Chicago: American Marketing Association.

Klopfenstein, B. C. (1989a). The diffusion of the VCR in the United States. In M. Levy (Ed.). *The VCR age*. Newbury Park, CA: Sage Publications.

Klopfenstein, B. C. (1989b). Forecasting consumer adoption of information technology and services—Lessons from home video forecasting. *Journal of the American Society for Information Science, 40* (1), 17-26.

Klopfenstein, B. C., & Sedman, D. (1990). Technical standards and the marketplace: The case of AM stereo. *Journal of Broadcasting & Electronic Media, 34* (2), 171-194.

Nichols, P. M. (2000, March 10). Home video. *New York Times*, E33.

Mogelonsky, M. (1995, December). Video verite. *American Demographics, 17* (12), 10.

Novel VCRs and camcorders make CES debut. (1998, January 19). *Consumer Electronics, 16* (36).

Olenick, D. (2000a, April 3). Eisa DVD player/Web device planned for June shipment. *TWICE, 15* (9), 15.

Olenick, D. (2000b, March 13). NPD: Digital camera sales in '99 topped $1 billion for first time. *TWICE, 15* (7), 6, 17.

Parker, D. J. (1999, November 4). Cease and DeCSS: DVD's encryption code cracked. *E Media Live*. [Online]. Available: http://www.emediapro.net/news99/news111.html.

Pearse, J. (1999, November 29). *DVD formats at a glance*. [Online]. Available: http://www.zdnet.co.uk/news/1999/47/ns-11817.html.

PlayStation2 DVD bug may prompt disk recall. (2000, March 20). *Wall Street Journal*, A27.

Priston, T. (2000, January 7). DVD takes a big step in competition with VHS. *New York Times Online*. [Online]. Available: http://www.nytimes.com/library/tech/00/01/biztech/articles/07video.html.

Rental stores keep packing in the crowds. (1995, October 23). *USA Today.*

Schoenherr, S. (1996). *Recording technology history: A chronology with pictures and links* [Online.]. Available: http://ac.acusd.edu/History/recording/notes.html.

Skriba, L. (2000, March). DVD home recording: Superhighway or dead end? *E Media, 13* (3), 20-22.

Snow, S. (1996, March 7). Morning report; TV & video. *Los Angeles Times*, p F-2.

Sony (2000). *DVD technical information.* [Online]. Available: http://www.sel.sony.com/SEL/consumer/dvd/about_faq.html.

State of the industry. (2000, January 24). *Television Digest, 40* (4), 11.
Statistical Research, Inc. (1997). *1997 TV ownership survey.* [Online]. Available: http://www.sriresearch.com/pr970616.htm.
Tarr, G. (1999, July 26). MbTV, Seagate team on HDRs. *TWICE, 14* (17), 17.
Tarr, G. (2000, January 24). Digital revolution propels '99 video sales. *TWICE, 15* (3), 3.
Technicolor. (2000, January). *DVD for "not-so" dummies: Your DVD technical reference guide.* Camarillo, CA: Technicolor. [Online]. Available: http://www.technicolor.com/services/DVD2000v1.pdf.
VCR near end of long and rewinding road. (1998, January 10). *Atlanta Journal-Constitution.* [Online]. Available: http://wwwaccessatlanta.com/business/news/1998/01/10/ces2.html.
VCRs proliferate, as do their brands. (1998, January 5). *Video Week, 16* (33).
Vizard, F. (1997, August). DVD delivers. *Popular Science, 251* (2), 8-72.
Wiley, L. (1999, August). Divx: Gone, but not forgotten. *E Media. 12* (8), 24.

16

Digital Audio

Ted Carlin, Ph.D.*

The audio industry's tempo slowed during the 1990s, but this sector is poised to experience renewed growth. The foundation for technological advances and marketing innovation laid in 1998 and 1999 should support new demand for digital audio products throughout the next several years. The industry did modestly well in 1998 and 1999, as combined sales of home and portable audio products rose an estimated 3.1% to $6.1 billion (CEMA, 2000). This increase was driven by the increasing affordability and performance of products, as well as by growing demand for home theater systems, which simulate the sonic and visual impact of the cinema experience.

Advances in home theater technology are not the only developments pointing a renewed, vibrant audio industry. Technological developments in 1998 and 1999 gave consumers compelling reasons to buy products to rekindle their passion for listening to music—the passion that launched the hi-fi industry in the 1950s. During 1999, for example, key manufacturers worked assiduously to develop DVD-based multichannel replacements for the two-channel compact disc (CD). Combination audio and video players incorporating one or both of the two standards under development, DVD-Audio and Super Audio CD, arrived in U.S. stores. In another positive development, two startup companies, CD Radio and XM Satellite Radio, continue to work toward their goal of launching satellite digital audio radio services in late 2000 or in early 2001. Each company will offer subscription-based 100-channel services that consist of talk programming and near-CD-quality stereo music channels beamed to car and home receivers (Anderson, 2000).

Three other developments, all related to digital recording, came together in 1998 and 1999 to stimulate demand: the relaunch of the recordable digital minidisc (MD) format, the introduction of mainstream-priced CD-recorders (CD-Rs), and the sale of the first Rio stereo MP3 player that uses solid-state flash memory to store music that can be downloaded from the Internet. The audio industry's strongest growth decades—the 1970s and 1980s—were energized in part by consumer passion for recording *custom* music compilations onto a convenient medium—analog tape. That

* Assistant Professor of Radio/Television, Department of Communication and Journalism, Shippensburg University (Shippensburg, Pennsylvania).

format may be replaced at some point by recordable digital discs and flash-memory media that deliver additional convenience and better sound quality.

Audio suppliers, however, do not appear to be banking solely on technological advances to further their fortunes. Also in 1998 and 1999, there were continued changes in tradition-defying product designs that simplified purchase and hookup, fit unobtrusively with home décor, or filled a need that reflected new music-listening habits at home. Whereas the core audio customer was once a serious music listener who assembled a complex system of standard-sized components in a room set aside for serious music listening, audio consumers today are listening to music in more than one room in the house—often as background to other activities. In addition, over 200 electronic games exist in surround and high-quality sound (CEMA, 2000).

As a result, traditional suppliers of home audio components have developed multimedia PC speakers, tabletop radio/CD players, under-cabinet CD/radios for the kitchen, and a broader selection of stylish "microsize" stereo systems for use in offices and homes. Microsize systems are a key driver of audio sales since the stereo can be placed anywhere from a dorm room to a bedroom, office, or den. Suppliers also have improved the price/performance ratio of custom-installed whole-house audio systems that distribute music throughout the house to in-wall speakers powered from a single rack of audio components (CEMA, 2000).

Convenience, portability, and sound quality have, therefore, re-energized the audio industry for the 21st century. Competition, regulation, innovation, and marketing will continue to shape and define this exciting area of communication technology.

Background

Analog vs. Digital

Analog means "similar" or "a copy." An analog audio signal is an electronic copy of an original audio signal as found in nature. For instance, an analog audio signal follows the same pattern as the vibration in the air pressure caused by the original sound, or is a copy of the vibration of your eardrum as the air pressure hits it. Microphones turn audible sounds into electronic analogs of those sounds through mechanical reproduction and transduction. For example, an original sound wave travels through a dynamic microphone's port and causes an internal wire to vibrate. This vibration is transduced (changed) into electronic pulses that are sent through the audio system to a recorder, loudspeaker, etc., where they are stored or reproduced as analog sound waves.

As this analog copy is created, some unwanted system distortion (noise) is also being recorded or broadcast. This is due primarily to the amount of electrical impedance present in the system components and cables. In an analog audio system, this distortion can never be totally separated from the original signal; in the digital domain, it can be eliminated (Watkinson, 1988). Further, subsequent copies of the original analog sound suffer further signal degradation called generational loss as signal strength lessens and noise increases for each successive copy.

In a digital audio system, the original sound is encoded in binary form as a series of 0 and 1 "words" called bits. The process of encoding different portions of the original sound wave by digital words of a given number of bits is called "pulse code modulation" (PCM). This means that the original sound wave (the modulating signal, i.e., the music) is represented by a set of discrete values. In the case of music CDs using 16-bit words, there are 2^{16} word possibilities (65,536). The PCM tracks in CDs are represented by 2^{16} values, and hence, are digital. First, 16 bits are read for one channel, and then 16 bits are read for the other channel. The other bits are used for data management. The order of the bits in terms of whether each bit is on (1) or off (0) is a code for one tiny spot on the musical sound wave (Watkinson, 1988). For example, a word might be represented by the sequence 1001101000101001. In a way, it is like Morse code, where each unique series of dots and dashes is a code for a letter of the alphabet (see Figure 16.1).

Figure 16.1
Analog Versus Digital Recording

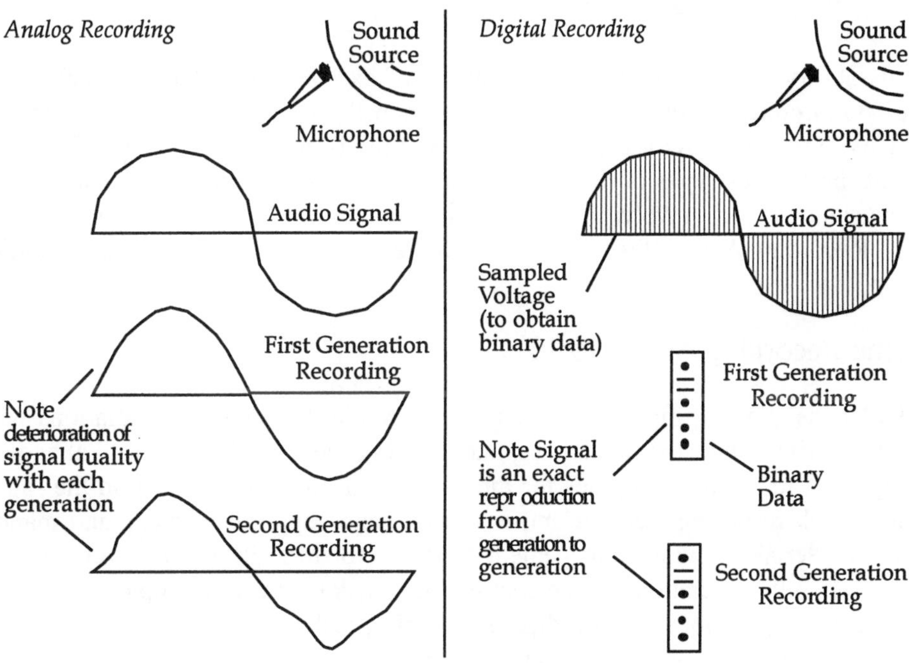

Source: Focal Press

Digital audio systems do not create bit word copies for the entire original sound wave. Instead, various samples of the sound wave are taken at given intervals using a specified sampling rate. Three basic sampling rates have been established for digital audio: 32 KHz for broadcast digital audio, 44.1 KHz for CDs, and 48 KHz for professional digital audiotapes (DATs) and videotape recorders (VTRs).

So, as a CD spins along, 44.1 thousand samples, each represented by a series of 16 "1" or "0" bits, are collected each for the left and right stereo channels every second. These samples are then sent to a digital-to-analog converter (DAC). The job of the DAC is to connect the samples together end-to-end in order to produce a smooth representation of the original musical sound wave, sort of like connecting the dots to make a picture. However, instead of connecting the dots (samples) with straight lines, the DAC uses sophisticated algorithms (mathematical equations) to fit the samples into a curved sound wave that is more representative of the way music truly sounds. Early digital music sounded rather harsh, in part because the DAC had a hard time making accurate guesses about the curve of the sound wave in between the samples (Luther, 1997).

This digitalization process, then, creates an important advantage for digital audio versus analog audio:

> A digital recording is no more than a series of numbers, and hence can be copied through an indefinite number of generations without degradation. This implies that the life of a digital recording can truly be indefinite, because even if the medium (CD, DAT, etc.) begins to decay physically, the sample values can be copied to a new medium with no loss of information (Watkinson, 1988, p. 4).

With this ability to make an indefinite number of exact copies of an original sound wave through digital reproduction comes the incumbent responsibility to prevent unauthorized copies of copyrighted audio productions to safeguard earnings of performers and producers. Before taking a closer look at the various types of digital audio systems in use, a brief examination of the two important legislative efforts involving this issue of digital audio reproduction is warranted: the Audio Home Recording Act of 1992 and the Digital Performance Right in Sound Recordings Act of 1995.

Audio Home Recording Act of 1992

This 1992 legislation exempts consumers from lawsuits for copyright violations when they record music for private, noncommercial use, and it eases access to advanced digital audio recording technologies. The law also provides for the payment of modest royalties on the sale of both hardware and blank recording media to music creators and copyright owners, and mandates the inclusion of the Serial Copying Management Systems (SCMS) in all consumer digital audio recorders to limit multi-generational audio copying (i.e., making copies of copies). This legislation was also written to apply to all future digital recording technologies, so Congress will not be forced to revisit the issue as each new products become available (HRRC, 2000).

The Digital Performance Right in Sound Recordings Act of 1995

For more than 20 years, the Recording Industry Association of America (RIAA) has been fighting to give copyright owners of sound recordings the right to authorize digital transmissions of their work. Before the passage of the Digital Performance Right in Sound Recordings Act of 1995, sound recordings were the only U.S. copyrighted work denied the right to authorize public performance.

This bill allows copyright owners of sound recordings the right to authorize certain digital transmissions of their works, including interactive digital audio transmissions, and to be compensated for others. This right covers, for example, interactive services, digital cable audio services, satellite music services, provision of commercial online music, and future forms of electronic delivery. Most non-interactive transmissions are subject to statutory licensing at rates to be negotiated or, if necessary, arbitrated.

Exempt from this bill are traditional radio and television broadcasts and subscription transmissions to businesses. The bill also confirms that existing mechanical rights apply to digital transmissions that result in a specifically identifiable reproduction by or for the transmission recipient, much as they apply to record sales.

The advent of digital audio transmissions via the Internet in the late 1990s led to the development of two initiatives to deal with unauthorized digital reproductions of copyrighted material: the Digital Millennium Copyright Act of 1988 and the Secure Digital Music Initiative (SDMI).

Digital Millennium Copyright Act of 1988

On October 28, 1998, the Digital Millennium Copyright Act (DMCA) became law. The main goal of the DMCA was to make the necessary changes in U.S. copyright law to allow the United States to join two new World Intellectual Property Organization (WIPO) treaties that update international copyright standards for the Internet era. Among its provisions, the DMCA specifically addresses the licensing for Webcasting.

The DMCA amends copyright law to provide for the efficient licensing of sound recordings for Webcasters and other digital audio services. In this regard, the DMCA does the following:

The DMCA creates a new statutory license for public performances by certain non-interactive, non-subscription digital audio services (in addition to the subscription services already covered). This statutory license applies only to those services with the primary purpose of providing audio or other entertainment programming and not the sale or promotion of products. The royalty rate for the statutory license will be set through negotiation or, if necessary, arbitration. Key terms of the license include:

(1) A Webcaster may not play in any three-hour period more than three songs from a particular album, including no more than two consecutively, or four songs by a particular artist or from a boxed set, including no more than three consecutively.

(2) Advance song or artist playlists generally may not be published.

(3) Parameters for archived, looped, and repeated programming are set by the license.

(4) When performing a sound recording, a Webcaster must identify the sound recording, the album, and the featured artist.

(5) In addition to the above, Webcasters must meet other conditions such as accommodating technical measures, taking steps not to induce copying, and not transmitting bootlegs.

The DMCA expands the current exemption for "ephemeral recordings" (copies of recordings to facilitate performances) to cover digital audio services that pay royalties for public performances under a statutory license. In addition, a statutory license is created to allow a Webcaster to make more than one ephemeral copy (Band, 2000).

Secure Digital Music Initiative (SDMI)

The Secure Digital Music Initiative is a forum of more than 160 companies and organizations representing a broad spectrum of information technology and consumer electronics businesses, Internet service providers, security technology companies, and members of the worldwide recording industry working to develop voluntary, open standards for digital music. A DMAT® mark is the trademark and logo specifying that products bearing this mark (including hardware, software, and content) meet the SDMI guidelines. A list of SDMI members is available on the SDMI Website (SDMI, 2000).

The voluntary guidelines offered by SDMI will speed the development of a new market for digital music, offering consumers new ways to enjoy music. The ability to use this open platform will encourage new business models such as "try before you buy," the option of purchasing an artist subscription, distribution that enables consumers to recommend music to friends, and other plans. DMAT systems (software and hardware) will allow consumers to use the music they already have on their computers, or other currently available digital music, in the same ways they do today. In addition, DMAT products will allow consumers to access digital content that may not otherwise be available on the Internet.

These guidelines will permit artists to distribute their music in both unprotected and protected formats—the choice is up to them. This means that those artists who choose to allow their music to be freely distributed without security will be able to do so, with the knowledge that devices carrying the DMAT mark will be able to play that music. At the same time, those artists who do wish to protect their own creative work can do so. Artists will benefit by the ability to reach and expand their fan base through new technologies, while respecting and protecting their copyrights to the extent they choose (SDMI, 2000).

Therefore, devices marked with the DMAT symbol may:

- Play currently-available digital music.
- Play music released in the future that does not carry the DMAT mark.
- Play music released in the future that carries the DMAT mark.

With the assurance of DMAT protection, more and more artists and labels should be willing to distribute music over the Internet. DMAT devices will be able to play this and other music. However, portable devices that do not carry the DMAT mark may not be able to play this music.

The SDMI guidelines permit consumers to copy their CDs or personal use (on their PCs, on their portable devices, on their portable media, etc.). In fact, the guidelines enable consumers to do so as many times as they wish—as long as they have the original disk.

Digital Audio Systems

Compact Disc (CD)

In the audio industry, nothing has revolutionized the way we listen to recorded music like the compact disc. Commercial availability of compact discs and players is less than two decades old, and, over this time, CDs have slowly replaced the long playing (LP) record to the point that music stores (except for the specialty market) carry only CDs and cassette tapes. LPs and cassettes represent the last vestiges of analog music reproduction, where the musical sound wave is recorded continuously.

Before starting the CD business, many engineers engaged in the development of the CD solely for its improvement in sound quality. After the introduction of the CD player, however, the consumer became aware of the quick random-access characteristic of the optical disc system. In addition, the size of the 12-cm disc was easy to handle compared with the LP. The longer lifetime for both the media and the player strongly supported the acceptance of the CD format. The next target of development was the rewritable CD. Sony and Philips jointly developed this system and made it a technical reality in 1989. Two different recordable CD systems were established. One is the write-once CD named CD-R and the other is the rewritable CD named CD-RW (Disctronics, 2000a).

With CDs, music is recorded in digital fashion. Instead of a continuous recording of the waveform, the music is sampled 44.1 thousand times per second (44.1 KHz) for each of the two channels (left and right stereo). The information is stored on a spiral track, beginning near the center and winding its way to the outside edge. LPs are recorded in a spiral groove beginning at the outside edge and winding their way to the center. The difference between a CD and an LP is what is in the spiral. The spiral track of a stereo LP is a series of peaks and valleys engraved at 45-degree angles to the surface, one for the left channel and one for the right. In a CD, the spiral contains a series of small bumps (usually called "pits") and flat areas (called "land") in between and beside the bumps. A small laser beam is aimed onto the spiral as the disc spins, and light is reflected from an area about three times as wide as one of the bumps. The reflected light is directed onto a sensor that reads the light as a series of "on" (when only reflected from land) or "off" (when reflected from a bump and surrounding land) "bits" of digital data. An on bit is transmitted or stored as a one (1), and an off bit is transmitted or stored as a zero (0) (Disctronics, 2000a).

Digital Audiotape

Digital audiotape (DAT) is a recording medium that spans the technology gulf. On the one hand, it uses tape as the recording medium; on the other, it stores the signal as digital data in the form of numbers to represent the audio signals. A DAT cassette is about half the size of a standard analog cassette, measuring 73 × 54 × 10.5 mm. Its tape is 3.81 millimeters (1/8") wide. DATs can record up to two hours of audio on a tape about 60 meters long, but they come in various lengths. Blank tapes use a metal-powder oxide, while prerecorded tapes use barium-ferrite oxide.

Computer-grade DATs are more reliable than audio-grade DATs. Although originally designed for computer backup systems, computer-grade DATs make an excellent audio recording tape. They use a different formulation, run cleaner, are easier on the heads, and are certified error free. Current DAT recorders support all three digital sampling rates, and do not use compression to deliver excellent sound reproduction. However, as a tape format, DAT does not offer the random access capability of a disc-based medium such as the CD or MD (TargetTech, 2000).

Minidisc (MD)

Minidiscs were announced in 1991 by Sony as a disc-based digital medium for recording and distributing consumer audio that is "near CD" in quality. In 1993, Sony announced MD Data, a version of the minidisc for storing computer data. There are two physically distinct types of discs: premastered MDs similar to CDs in operation and manufacture, and recordable MDs that can be recorded on repeatedly and employ magneto-optical technology. The disc itself is enclosed in a small (7cm × 7cm) convenient cartridge (Yoshida, 2000).

Sales of cassette tapes have been decreasing since 1989, and Sony felt that the compact cassette system was approaching the end of its format life. Even if recordable CDs were to be accepted by the consumer, it would still be difficult to break into the portable market. Here, the portable compact cassette dominated because of its strong resistance to vibration and its compactness. A clear challenge for a new disc system *was* to overcome these weaknesses. Sony was able to achieve this by introducing the minidisc (Sony Electronics, 2000).

Magneto-optical disc recording technology had been used for computer data storage systems for several years. Based on this technology, Sony developed a direct overwriting technology with a similar recording density as the compact disc. Additionally, they employed a shock-resistant memory control for portable use and applied a high-quality digital audio compression system called ATRAC (Adaptive TRansform Acoustic Coding) that enabled the use of a 64-mm disc. With a diameter of 64 mm, which is smaller than a CD, a minidisc can hold only one-fifth of the data. Therefore, data compression of 5:1 is necessary in order to offer a comparable 74 minutes of playback time. Continuous technological improvement of semiconductors also helped to realize this technology (Sony Electronics, 2000).

The advantages of minidisc over a tape format like DAT include its editing capabilities and quick random access. Unlike any tape-based format, you can instantly jump to a specific track on an MD and tracks can be deleted, split, and reordered. Minidisc also supports disc and track titles and has a wider availability of portable and car players.

Prices have dropped significantly since Sony's reintroduction of the MD in 1998, with an MD recorder audio component and portable player combo widely available at stores such as Best Buy, Circuit City, and The Good Guys for under $500. These stores also started advertising MD units in their weekly advertising circulars—a good sign that sales are going well. Finally, MD blanks have dropped dramatically in price, from $8 apiece by mail order at the beginning of 1997, to less than $5 apiece at retail stores. Nothing is certain, but the future of the MD is more promising than ever (Yoshida, 2000).

Recent Developments

Home Theater

Since 1998, key innovations in digital audio processing and digital audio distribution have helped to create a vibrant consumer marketplace for audio products. The primary home entertainment experience today is delivered by a quality home theater system. Home theater has grown increasingly affordable over the years, and a respectable system with all the necessary audio and video gear costs as little as $800. At a minimum, a home theater system is a TV set with a diagonal screen size of at least 25 inches, a video source such as a hi-fi/stereo VCR or DVD player, a surround-sound-equipped stereo receiver or compact shelf system, and four or more speakers.

According to a 1998 Consumer Electronics Manufacturers Association (CEMA) survey, 20% of American households own a home theater system meeting this definition. On average, their systems cost $3,000. Home theater's impact is such that 80% of the audio receivers sold at the manufacturer level in 1998 and 1999 were equipped with some form of surround-sound decoding. More than 100 current and syndicated TV series are broadcast in surround sound, as are dozens of special programs and sporting events. In addition, when broadcast in stereo to a home theater's TV set, any of the almost 9,000 movies recorded in surround sound can be viewed in a theater-like setting. At the very least, surround sound puts the viewer in the center of the action, enveloping the viewer with the ambient background sounds (i.e., a driving rainstorm, a thunderous explosion, or a fast-paced chase scene). Surround sound also enhances dialog intelligibility and realism by channeling dialogue to a TV-top center-channel speaker, making the voices of on-screen actors come from the same direction as their images (CEMA, 2000).

Three companies have taken the lead in developing digital audio processing technologies, such as surround sound and 3-D audio, which are being used to enhance the media experience of the consumer: Q-Sound, Dolby, and DTS.

Q-Sound Digital Audio. Providing "audio processing technologies to meet your needs," Q-Sound has developed a number of digital audio processing enhancements for music, video and arcade games, the Internet, computers, and even hearing aids (QSound Labs, 2000). Companies such as Aiwa, Sharp, RealNetworks, Boston Acoustics, and Sega Dreamcast are among its clients. Q-Sound's key innovations are QSurround, Q3D, and QXpander.

QSurround reproduces multichannel audio formats on almost any playback system. By creating virtual speakers, QSurround reproduces spatially correct, multichannel output on regular two-channel equipment and enhances sound reproduction on five- and six-channel systems. Designed for professional audio recording and video game applications, Q3D places multiple individual sounds in specific locations outside the bounds of conventional stereo reproduction to provide a three-dimensional listening environment. QXpander is a stereo to 3-D enhancement process available for both headphones and speakers. A robust algorithm allows it to process any stereo signal, making QXpander suitable for a broad range of consumer electronic applications (QSound Labs, 2000).

Dolby Digital 5.1 Audio. Dolby revolutionized tape recording in the late 1960s and early 1970s with Dolby A (for professional applications) and Dolby B (for consumer applications) noise reduction. Later in the 1970s, Dolby updated film sound with the Dolby Stereo analog sound system. Dolby Stereo brought four-channel sound to the movie theater, with three channels of sound in the front (left and right for music and effects and the center for dialog) and a surround channel for effects and atmosphere. Then, in the 1980s, both tape recording and film sound saw significant improvements through the use of Dolby SR (Spectral Recording). In the late 1980s and early 1990s, Dolby Surround and Dolby Pro Logic home theater systems (basically using the Dolby Stereo technology in the home environment for videotapes and laserdiscs) entered the marketplace. These systems allowed home viewers to create the same four-channel theater-type setup in the home (Dolby Labs, 2000).

Today's Dolby Digital 5.1 audio system takes the next step, and provides six channels of digital surround sound. Dolby Digital 5.1 is an advanced form of digital audio coding (AC-3 perceptual coding) that makes it possible to store and transmit high-quality digital sound far more efficiently than was previously possible. First used in movie theaters in 1992, it exploits the characteristics of human hearing. (Dolby noise reduction works by lowering the noise when no audio signal is present, while allowing strong audio signals to cover or mask the noise at other times.)

Dolby Digital 5.1 delivers surround sound with five discrete full-range channels—left, center, right, left surround, and right surround—plus a sixth channel for those powerful low-frequency effects (LFEs) that are felt more than heard in movie theaters. As it needs only about one-tenth the bandwidth of the others, the LFE channel is referred to as a ".1" channel (and more commonly as the "subwoofer" channel) (Dolby Labs, 2000).

Dolby Digital 5.1 audio is available via laserdiscs, DVD videodiscs, DVD-ROM discs for computers, digital cable systems, direct broadcast satellite (DBS) systems, and digital broadcast TV (DTV). One reason for the abundance of Dolby digital programming is that Dolby Digital is the audio standard for the movie industry and for these new digital media applications. Also, for DVD-Video, Dolby Digital is called a "mandatory" audio coding format, meaning that a Dolby Digital soundtrack can be the only one on a disc. Discussed below, DTS, by comparison, is an "optional" coding format, meaning that the disc must have a mandatory-format soundtrack as well.

DTS Digital 5.1 Audio. Another new technology for surround-sound entertainment, DTS Digital Surround is an encode/decode system that delivers six channels (5.1) of 20-bit digital audio. In the encoding process, the DTS algorithm encrypts six channels of 20-bit digital audio information in the space previously allotted for only two channels of 16-bit linear PCM. Then, during playback, the DTS decoder reconstructs the original six channels of 20-bit digital audio. Each of these six channels is audibly superior to the 16-bit linear PCM audio found on CDs, with the ability to provide effective channel separation and a wide dynamic range of sound (DTS, 2000).

CD Technology

High Density Compatible Digital (HDCD), developed by Pacific Microsonics, is a recording process that enhances the quality of audio from compact discs, giving an end result more acceptable to audiophiles than standard CDs. HDCD discs use the least significant bit of the 16 bits per channel to encode additional information to enhance the audio signal in a way that does not affect the playback of HDCD discs on normal CD audio players. The result is a 20-bit per channel encoding system. HDCD is claimed to provide more dynamic range and a very natural sound. Many

HDCD titles are available, particularly in the United States. By the end of 1999, over 4,000 titles were available. Discs can be recognized by the presence of the HDCD logo (Disctronics, 2000a).

CD-ROM XA (for eXtended Architecture) discs were designed to allow audio and other data to be interleaved and read simultaneously. This avoids the need to load images first and then play CD audio tracks. The CD-ROM XA specification also defines certain image and audio formats to use:

- The graphics formats include 256 color modes, which are compatible with PC formats and CD-i.
- The audio used is ADPCM (Adaptive Delta Pulse Code Modulation), which is also defined for CD-i.

This CD-ROM XA format has not been successful in itself, but there are three important formats based on it: Photo CD, Video CD, and CD EXTRA (Disctronics, 2000a).

Super Audio CD. When Phillips and Sony developed the CD in the 1980s, the PCM format was the best technology available for recording. Nearly two decades later, huge strides in professional recording capabilities have outgrown the limitations inherent in the PCM's 16-bit quantization and 44.1 KHz sampling rate. Philips and Sony have been working on an alternative specification called Super Audio CD that uses a different audio coding method, Direct Stream Digital (DSD), and a hybrid disc format. Like PCM digital audio, DSD is inherently resistant to the distortion, noise, and other limitations of analog recording media and transmission channels. DSD samples music at 64 times the rate of a compact disc (64 × 44.1 KHz) for a 2.8224 MHz sampling rate. As a result, music companies can use DSD for both archiving and mastering (Super Audio CD, 2000).

The result is that a single Super Audio CD can contain three versions of the music stored on two separate layers. First is CD-compatible stereo, stored on the reflective layer (top). High-resolution stereo is stored on the semi-transmissive layer (bottom-center) as is six-channel sound, where available (bottom-outside). The result is a disc that sounds noticeably better than conventional CDs when played in a conventional CD player. The Super Audio CD disc makes better use of the full 16 bits of resolution that the CD format can deliver (Super Audio CD, 2000).

In terms of copyright protection, Philips and Sony have developed a multi-faceted technology called Digital Watermarking. Using a technology called Pit Signal Processing (PSP), the system can actually put a faint image or "watermark" on the signal side of the disc. This image, which can take the form of text or graphics, is extremely hard for pirates to duplicate clearly, no matter what duplication strategy is used. Visibly corrupted watermarks then become a sure sign of piracy. They alert consumers and retailers that something is wrong, and they help prosecutors trace illegal copies back to the source. The Super Audio CD Digital Watermarking system also embraces disc bar codes, plus invisible, non-removable information embedded on the disc (Super Audio CD, 2000).

In essence, Sony and Philips designed DSD to capture the complete information of today's best analog systems. "The best 30ips half-inch analog recorders can capture frequencies past 50 KHz. DSD can represent this with a frequency response up to 100 KHz. To cover the dynamic range of a good analog mixing console, the residual noise power was held at -120 dB through the audio band. This combination of frequency response and dynamic range is unmatched by any other recording system, digital or analog" (Super Audio CD, 2000).

A few smaller music labels have begun to release Super Audio CDs. Independent British classical recording label, Hyperion Records, launched its first titles in January 2000. The release of its first Super Audio CDs is a natural progression for Hyperion Records since it has been using the DSD recording system for the past 18 months and has completed 20 DSD recording projects. Following the successful debut of Super Audio CD releases earlier this year, independent Dutch record label, Challenge Records, has announced plans to release 15 Super Audio CDs in 2000 (Super Audio CD, 2000).

In November 1999, Sony Music Entertainment was the first of the big record companies to release Super Audio CD software. At the time of the European launch, the company released 15 titles spanning a wide range of music from Leonard Bernstein to Miles Davis and Mariah Carey. In December 1999, the company announced 20 additional releases before Christmas. These included: George Szell (Dvorak: Slavonic Dances, Conducts Strauss and Conducts Wagner), Glenn Gould (Goldberg Variations 1981), Michael Jackson (Thriller), Celine Dion (Greatest Hits), Billy Holliday (Lady in Satin), Thelonious Monk (Straight, No Chaser), and Yo-Yo Ma (Solo) (Super Audio CD, 2000).

DVD-Audio

DVD-Audio is the latest member of the DVD family of prerecorded optical disc formats designed for higher-quality audio than current CDs. The initial version was released in April 1999, and discs and players began appearing in the second half of 1999. DVD-Audio discs can be comprised of one or two layers. The capacity of a dual-layer DVD-Audio is up to at least two hours for full surround-sound audio and four hours for stereo audio. Single layer capacity is about half these times (Disctronics, 2000b).

The DVD-Audio specification makes use of a scalable, linear PCM multichannel and stereo encoding format, down-mixing control, and optional audio formats. DVD-Audio offers a range of features including even higher quality surround sound, longer playing times, and additional features not available on CDs. DVD-Audiodiscs are capable of carrying video, like DVD-Video titles, as well as high-quality audio files with limited interactivity. Additional content includes still pictures, text information, menus, and navigation (Videodiscovery, 2000).

Text is used for the contents, artists' names, Internet URLs (uniform resource locators), lyrics etc. Static text information can be used for the overall content, while dynamic text is suitable for lyrics that change during the audio presentation. Video clips follow the DVD-Video specification, but certain functions (including multi-story, parental management, region control, and user operation control) are not supported. The audio part of the video may be presented without the video. Panasonic, Technics, and Yamaha have released universal DVD-Audio/DVD-Video players costing $700 to $1,200 (Videodiscovery, 2000).

MP3

Since long before MP3 came onto the scene, computer users have been recording, downloading, and playing high-quality sound files using a format called WAV. The trouble with WAV files, however, is their enormous size. A two-minute song recorded in CD-quality sound would eat up about 20 MB of a hard drive in the WAV format. That means a 10-song CD would take up more than 200 MB of disk space!

The file-size problem for music downloads has changed thanks to the efforts of the Moving Picture Experts Group, a consortium that develops open standards for digital audio and video compression. Its most popular standard, MPEG, produces high-quality audio (and full-motion video) files in far smaller packages than those produced by WAV. MPEG filters out superfluous information from the original audio source, resulting in smaller audio files with no perceptible loss in quality. WAV, on the other hand, spends just as much data on superfluous noise as it does on the far more critical dynamic sounds, resulting in huge files (MPEG, 1999).

Since the development of MPEG, engineers have been refining the standard to squeeze high-quality audio into even smaller packages. MP3—short for MPEG 1 Audio Layer 3—is the latest of three progressively more advanced coding schemes, and it adds a number of advanced features to the original MPEG process. Among other features, Layer 3 uses entropy encoding to reduce, to a minimum, the number of redundant sounds in an audio signal. Thanks to these features, the MP3 standard will take music from a CD and shrink it by a factor of 12, with no perceptible loss of quality (MPEG, 1999).

All the copyright laws that apply to vinyl records, tapes, and CDs also apply to MP3. Just because a person is downloading an MP3 of a song on a computer rather than copying it from someone else's CD does not mean he or she is not breaking the law. The Secure Digital Music Initiative (SDMI) is the recording industry's main effort to prevent unauthorized duplication of digital audio using MP3 technology.

Authorized and unauthorized MP3 files are literally all over the Internet. Finding a particular song or an album by a specific artist in the jumble of MP3s, however, can be quite difficult. A general search engine such as Yahoo! or Snap can be used to locate MP3s, but the results are often less than desirable. There are plenty of Web pages that happen to have the artist's name and MP3 on them, without any MP3 files to speak of.

A better option is to use a dedicated MP3 search engine. Sites such as 2look4, FileQuest.com, and Audiofind scour sites all over the Internet for MP3 files that match desired search terms. A stand-alone MP3 search engine can be downloaded, such as MP3 Fiend, which will query multiple MP3 search sites simultaneously for the desired songs. Both Web-based and desktop MP3 search engines will locate both legal and pirated MP3 files (CNET, 2000).

Another option is an MP3 file directory, such as MP3.com, EMusic.com, or listen.com. These file libraries, which typically categorize songs by genre, are the best way to find legal MP3s. Most of the songs listed on these directories are by new artists hoping to get their music heard through the free MP3 format, by big-name musicians who have licensed their songs for MP3 distribution, or the songs are older ones that are no longer under copyright protection (CNET, 2000).

A firestorm of controversy involving the downloading of MP3s has recently centered on two issues: (1) the unauthorized downloading of copyrighted music and (2) the decision by several colleges and universities to prohibit access to MP3 hosting sites, such as Napster and Gnutella, on campus networks. The heavy metal band Metallica filed a lawsuit against Napster in U.S. District

Court, Central District of California, alleging that the company encourages piracy by enabling and allowing its users to trade copyrighted songs (Jones, 2000). Other artists, such as Dr. Dre, have asked Napster to remove listings of their songs from host servers (Foust, 2000).

In terms of college access, student use has been tying up network bandwidth, making other non-MP3 Web activities impossible—or extremely slow—for other campus network users. Some schools found that Napster use was accounting for between 40% and 60% of total campus network usage. Citing the violation of campus network acceptable use policies, these schools have enacted firewalls to prohibit student access (Oakes, 2000).

To play MP3s, a computer-based or portable MP3 player is needed. Dozens of computer-based MP3 players are available for download. Winamp, the most popular, sports a simple, compact interface that contains such items as a digital readout for track information, an appealing but unobtrusive sound level display, and intuitive controls. Other popular computer-based players include Sonique, AudioCatalyst, MusicMatch Jukebox, and RealJukebox (CNET, 2000).

A computer-based MP3 player can be customized by using skins, which are tiny files that let the user change the appearance of the MP3 player's interface. By far, the majority of skins are developed for Winamp, although other players are increasingly getting more skins. Skins do much more than just change the color of a player's user interface: They can actually change the entire look and feel. Skins are available, for example, that will make an MP3 player look like a slick car stereo, a pencil-and-paper sketch, or a space-age masterpiece. Some skins are even based on TV shows, sports teams, and movie stars (CNET, 2000).

The Diamond Rio portable player receives the distinction as the first portable MP3 player (S3, 2000). Using flash memory cards about the size of a stick of gum, users can build a library of MP3 titles on these storage disks and interchange them with ease. Available memory on these flash cards is currently 4 MB, 8 MB, 16 MB, and 32 MB, with Sony already developing a 256 MB card. The Rio has spawned several models within the Diamond product line over the last couple of years, as well as competing portable units from companies such as Creative Labs (Nomad), e-go, I-Jam, Jukebox (PJB-100), and RCA (Lycra). Several Websites, such as MP3.com, compile reviews and links to the rapidly-growing list of portable MP3 players.

One intriguing use of flash technology is being implemented by Sony, which has launched several consumer electronics products based on its Memory Stick flash technology, including digital still cameras, photo printers, and its Vaio line of PCs. In October 1999, Sony announced the Memory Stick Walkman, a flash device for playback of MP3 music files. By mid-2000, Sony had shipped nearly two million Memory Sticks, a number that is expected to reach over three million by late 2000 (Sony, 1999).

Another intriguing part of the MP3 world is CD rippers. These are programs that extract—or rip—music tracks from a CD and save them onto a computer's hard drive as WAV files. This is legal as long as the MP3s created are used solely for personal use, and the user owns the CDs. There is a large number of CD rippers available to download. Once the CD tracks have been ripped to the hard drive, the next step is to convert them into the MP3 format. An MP3 encoder is used to turn the WAV files into MP3s. Many CD rippers have MP3 encoders built in (such as MusicMatch or Jukebox), or they can be downloaded as a separate encoder utility, such as MP3Enc (CNET, 2000).

Factors to Watch

CEMA hosted the first Audio Industry Summit in 1998 to develop strategies for the digital audio industry's future. Key industry executives met with top marketing and retail experts during a three-day brainstorming session, and they concluded that the industry must change to reflect demographic and lifestyle changes. The Summit's report found that people are no longer listening to music in the same way they have been for the past 30 years. "With the introduction of the LP in 1948, listening to music [at home] became more available and more affordable. But sitting and intently listening to music...is not part of today's shared entertainment experience," the report revealed. The report also found that "there is no longer a desire for simply a single-sensory experience, such as listening to music." The growing popularity of satellite TV, high-definition television (HDTV), multimedia PCs, and electronic games provides "plenty of chances for the innovative marketer to sell the impact of the audio experience" of these home entertainment products, according to the report (CEMA, 2000).

Summit attendees outlined marketing changes that the industry must make, such as promoting the experience rather than price. Based on the Summit's conclusions, CEMA sponsored the first in a series of marketing seminars intended to help suppliers focus on establishing consumer needs and solving their problems rather than promoting new "bells and whistles." In 1999, CEMA launched a major direct mail and public-relations campaign to promote awareness that audio is critical to the home entertainment experience under the theme, "Audio is the Soul of the System" (CEMA, 2000).

By the end of 2000, the two most important factors in digital audio will be (1) which digital audio format(s) consumers will adopt and (2) the protection of copyrighted material. With a host of digital audio choices making it to the marketplace, consumers will be evaluating products on ease-of-use, audio quality, storage capacity, and price. The successful worldwide adoption of the CD, with 500 million existing CD players and 10 billion existing CDs, will be a challenge for any new digital audio format to top. Will consumers replace a perfectly good CD audio system with a new technology? Will they supplement their CD system with additional new technologies that fill specific needs such as Internet audio or home theater systems? Are there just too many new technologies for consumers to evaluate?

For example, two companies, Liquid Audio and AT&T's A2b, are already offering MP3-like Internet audio formats. They are *not* compatible with MP3 players as they use different digital compression techniques (i.e., MPEG-2 AAC Low Complexity Profile Audio Coding for A2b), forcing users to try yet another format. However, they are promising quicker downloads that use a smaller amount of disk space, while maintaining sound quality that is virtually indistinguishable from the CD (i.e., a three-minute song downloaded from A2b would take up only 2.25 MB of disk space) (A2b, 2000).

The issue of copyright protection has been a part of the digital audio world since its inception. Each new digital audio technology has been forced to tackle this issue at some point. In 2000, with SDMI support from CEMA and RIAA, it appears that most, if not all, new digital audio products are making copyright protection a priority. Hoping to attract content providers (i.e., record labels) to develop materials for their products and avoid alienation or litigation, new digital audio technologies are being forced to state their copyright protection measures from the outset.

For example, the introduction of MP3s, which occurred before the establishment of SDMI, did not follow this model and the result has been a lawsuit between RIAA and MP3.com, an MP3 music provider. The lawsuit involves RIAA's claim that MP3.com copied 45,000 copyrighted CDs onto computer servers. Whether the copies on the servers are in MP3 or any other format is irrelevant—the lawsuit concerns unauthorized copying. The lawsuit also attacks two new creative applications of Internet music created by MP3.com that may become commonplace in years to come. The new services are called Instant Listening and Beam-it, and are part of My.MP3.com. They are, according to MP3.com, designed to allow users to listen to their CDs anywhere they have a Net connection.

- With Instant Listening, when a user buys a CD from an online retailer partnered with MP3.com, the user can choose to have the album immediately put into his or her MP3.com "locker" for immediate listening.
- With Beam-it, when a user puts a copy of a CD into his or her CD-ROM drive, MP3.com will put that album into that user's MP3.com locker on their Website.

According to RIAA, however, users are not actually copying their CDs into their MP3.com lockers. Instead, MP3.com is giving those users access to a digital music library of over 45,000 albums that MP3.com had previously created. According to RIAA, the only issue in the lawsuit is the propriety of MP3.com launching a commercial business with music it does not own and has not licensed (RIAA, 2000).

After almost a decade of few digital audio developments, expect to see continued innovation, marketing, and debate in the 2000s. Which technologies and companies survive will largely depend on the evolving choices made by consumers, and the continued growth of the Internet and digital technology.

Bibliography

A2b. (2000, April). *A2b music technology*. [Online]. Available: http://www.a2bmusic.com/faq.asp#two.

Anderson, M. (2000, April). *From CB to CD radio*. [Online]. Available: http://abcnews.go.com/sections/tech/NextFiles/nextfiles990628.html.

Band, J. (2000, April). *The Digital Millennium Copyright Act*. [Online]. Available: http://www.hrrc.org/JB-Memo.html?.

Consumer Electronics Manufacturers Association. (2000, April). *Market overview*. [Online]. Available: http://www.ce.org/index.cfm?area=market_overview.

CNET. (2000, April). *The MPEG Resource Center*. [Online]. Available: http://home.cnet.com/internet/0-4004.html?st.cn.sr1.ssr.tc_mp3&tag=st.cn.sr.bb.3.

Disctronics. (2000a, April). *Digital audio on CD*. [Online]. Available: http://www.disctronics.co.uk/cdref/cdframe.htm.

Disctronics. (2000b, April). *DVD-Audio*. [Online]. Available: http://www.disctronics.co.uk/dvd/dvdaudio/dvdaud-spec.htm.

Dolby Labs. (2000, April). *Dolby Digital*. [Online]. Available: http://www.dolby.com/digital/diggenl.html.

DTS. (2000, April). *DTS technology*. [Online]. Available: http://www.dtsonline.com/consumer/index.html.

Foust, S. (2000, April 18). *Dr. Dre added to list of complete morons*. [Online]. Available: http://www.dmusic.com/news/news.php?id=2500.

Goudsmit, J. (1997, November 23). *DCC FAQ*. [Online]. Available: http://www.xs4all.nl/~jacg/dcc-faq.html.
Home Recording Rights Coalition. (2000, April). *HRRC'S Summary of the Audio Home Recording Act*. [Online]. Available: http://www.hrrc.org/ahrasum.html.
Jones, C. (2000, April 13). *Metallica rips Napster*. [Online]. Available: http://www.wired.com/news/politics/0,1283,35670,00.html.
Luther, A. (1997). *Principles of digital audio and video*. Boston: Artech House.
Moving Pictures Expert Group. (1999, December). *MPEG audio FAQ*. [Online]. Available: http://tnt.uni-hanover.de/project/mpeg/audio/faq/#a.
Oakes, C. (2000, February 10). *Time for a Napster rest*? [Online]. Available: http://www.wired.com/news/technology/0,1282,34201,00.html.
QSound Labs. (2000, April). *Technology*. [Online]. Available: http://www.qsound.com/tech/.
Recording Industry Assocation of America. (2000, April). *FAQ about RIAA's lawsuit against MP3.com*. [Online]. Available: http://www.riaa.com/tech/tech_pr.htm.
S3. (2000, April). *Company profile*. [Online]. Available: http://www.diamondmm.com/anouncement/profile.html.
Secure Digital Music Initiative. (2000, April 10). *Frequently asked questions about DMAT® and the Secure Digital Music Initiative*. [Online]. Available: http://www.sdmi.org.
Sony Electronics. (1999, February 19). *Expanded AV/IT applications create a world of new possibilities for "Memory Stick" IC recording media*. [Online]. Available: http://www.sony.com/SCA/ press/feb_19_99.html.
Sony Electronics. (2000, April). Minidisc. [Online]. Available: http://www.sel.sony.com/SEL/consumer/md/.
Super Audio CD. (2000, April). *What is Super Audio CD?* [Online]. Available: http://www.superaudio-cd.com.
TargetTech. (2000, April). *DAT (digital audio tape)*. [Online]. Available: http://www.whatis.com/dat.htm.
Videodiscovery. (2000, April). *DVD FAQ*. [Online]. Available: http://www.videodiscovery.com/vdyweb/dvd/dvd-faq.html#1.12.
Watkinson, J. (1988). *The art of digital audio*. London: Focal Press.
Yoshida, T. (2000, April). *What are minidiscs?* [Online]. Available: http://www.minidisc.org/ieee_paper.html.

IV

TELEPHONY & SATELLITE TECHNOLOGIES

Local and long distance telephone revenues in the United States exceed those of all advertising media combined. Clearly, point-to-point transmission of voice, data, and video represents the single largest sector of the communications industry. The sheer size of this market has two effects: companies in other areas of the media want a piece of the market, and telephone companies want to grow by entering other media.

Until 1996, federal regulations kept the telephone, cable television, and broadcast television industries separate from each other. In an effort to encourage competition and technical innovation, the Telecommunications Act of 1996 changed the playing field, allowing levels of cross-ownership across these media that were unthinkable a few years ago.

These changes in the regulation and organizational structure of communication media have been forced by rapid advances in digital technology that are erasing the distinctions in the transmission process for video, audio, text, and data. Because all of these types of signals are transmitted using the same binary code, any transmission medium can be used for almost any kind of signal (provided the needed bandwidth is available). Furthermore, the advance of digital compression technologies reduces the bandwidth needed to transmit a variety of signals, further blurring the lines dividing communications media.

The immediate result of these technological innovations is more competition. The most dramatic site of new competition will be telephony, where consumers will soon have a choice of service providers, and service will consist of more than the transmission of voices. The chapters in this section explore different technologies used for point-to-point communication, thus far dominated by the telephone industry.

The first chapter in this section discusses the basics of today's telephone network in the United States, and the following chapter explores the most important technological innovations in the switched networks used for the transmission of data. In addition to explaining how these technologies work and how much information they can transmit, this chapter discusses a variety of organizational, economic, and regulatory factors that will influence when and how each becomes part of the telephone network.

The application of these network technologies for home use is then explored in Chapter 19. This chapter introduces residential gateways, which many predict will become a key enabling factor to give you control over the flow of telephone, Internet, and entertainment services to and throughout the home.

Satellites are a key component of almost every communication system. Chapter 20 explains the range of applications of satellite technology, including the history of the technology and the range of equipment needed (on the ground and in space) for satellite communication. Chapter 21 then explores one of the most important applications of early satellite technology—distance learning—which has since evolved to encompass virtually every communication medium.

The rapidly evolving (and lucrative) cellular telephone industry is then reviewed in Chapter 22, along with explanations of the differences between traditional cellular telephony and newer incarnations such as PCS. Chapter 23 explores the range of personal communication devices that are serving important niches in our daily lives.

The final chapter in this section offers glimpses of telephone technologies that add video to telephone service. The videoconference chapter discusses videoconferencing and videophone systems that are primarily designed to facilitate face-to-face communication over distances, discussing the rapid evolution of group-based videoconferencing systems and the continued failure of one-to-one videophones.

In studying these chapters, you should pay attention to the compatibility of each technology with current telephone technologies. Technologies such as cellular telephone are fully compatible with the existing telephone network, so that a user can adopt the technology without worrying about how many other people are using the same technology. Other technologies, including the videophone and ISDN (Integrated Services Digital Network) are not as compatible. Consumers considering purchase of a videophone or ISDN service have to consider how many of the people with whom they communicate regularly have the same technology available. (Consider: If someone gave you a videophone today, whom would you call?)

Markus (1987) refers to this problem as an issue of "critical mass." She indicates that adoption of interactive media, such as the telephone, fax, and videophone, is dependent upon the extent of adoption by others. As a result, interactive communication technologies that are not fully compatible with existing technologies are much more difficult to diffuse than other technologies. Markus indicates that early adoption is very slow, but once the number of adopters reaches a "critical mass" point, usage takes off, leading quickly to use by nearly every potential adopter. If a critical mass is not achieved, adoption of the technology will start to decline, and the technology will eventually die out.

One of the most important concepts to consider in reading this chapter (and the other chapters that include satellite technology) is the concept of "reinvention." This is the process by which users of a product or service develop a new application that was not originally intended by the creator of the product or service. Satellite technology is being reinvented almost daily as enterprising individuals devise new uses for these relay stations in the sky.

The final consideration in reading these chapters is the organizational infrastructure. Because of the potential risks and rewards, even the largest companies entering the market for new telephone services are hedging their bets with strategic partnerships and experimentation with multiple, competing technologies. In this manner the investment needed (and thus the risk) is spread over a number of technologies and partners, with the knowledge that just one successful effort could pay back all the time and money invested.

Bibliography

Markus, M. L. (1987). Toward a "critical mass" theory of interactive media: Universal access, interdependence, and diffusion. *Communication Research, 14* (5), 491-511.

17

Local & Long Distance Telephony

David Atkin, Ph.D.*

As we enter a new millennium, the venerable telephone medium stands poised at the brink of a technology revolution. Bates, Jones, and Washington (in press) note that the telephone, focused on delivery of voice services for its first century, is being reinvented to deliver a myriad of services through wired and wireless channels. The telephone network is undergoing a transition from analog transmission over copper wire to digital fiber optic transmission, and now lies at the heart of the emerging global information infrastructure.

Those authors recount the expansive Department of Commerce definition of the telecommunications industry, which ranges from local exchange and long distance/international services to cellular telephony and paging. (The latter elements are explored in other chapters of this book.)

More expansively, scholars note the telephone may help facilitate the rise of that information superhighway, helping cultivate user skills in interactivity and scalability that are crucial in the operation of emerging information technologies (Neuendorf, Atkin, & Jeffres, 1998; in press). The technology convergence generating this inertia flows from the highly deregulatory Telecommunications Act (P.L. 104-104, 1996), notable for its removal of entry barriers between local, long distance, and cable service providers. Yet, insofar as cross-sector competition in telecommunications is only beginning to take root, commentators and regulators (Chen & Wilke, 1999) suggest that it remains a "long distance" off (Cauley, 1997).

That such media convergence is so difficult to achieve underscores its rapid departure from the era of plain old telephone service (POTS), already one of the most ubiquitous and lucrative communications technologies in the world. In the United States, the $230.5 billion in gross revenues from local and long distance companies—30% of the world market—exceeds the combined revenues of all advertising media, even surpassing the GNP (gross national product) of most nations

* Professor & Assistant Chair, Department of Communication, Cleveland State University (Cleveland, Ohio).

(Atkin, 1999). The profit potential of these two markets is the primary force behind the revolution in telephony, as other media companies seek to enter the lucrative telephone market, and telephone companies seek to enter other media. Using history as a guide, this chapter outlines the influence of changing regulations on the conduct of telephone companies, including implications for cross media competition.

Background

The Bell system dominated telephony in the century after A. G. Bell won his patent for the telephone in 1876. Under the early leadership of Theodore Vail, Bell, Inc. staved of Western Union's bid to become a phone company. After leaving and then rejoining the company, Vail guided Bell through a series of acquisitions and mergers that became major avenues of growth for the firm, which came to be known as American Telephone & Telegraph (AT&T).

After the original phone patent lapsed in 1893, over 6,000 independent phone companies entered the fray, providing phone service and selling equipment (Atkin, 1996; Weinhaus & Oettinger, 1988). However, competition was not always a virtue. In Hawaii during the 1890s, competition between Mutual and Bell led to confusing allegiances of customers who could not be connected to clients of the competing phone company. Similar concerns over gaps in standardization and interconnection prompted government oversight in 1910 (Mann-Elkins Act, 1910).

Meanwhile, acquisition of independents intensified after 1910, forming the building blocks for the regional Bell operating companies (RBOCs). Concerned over Bell's acquisition of an independent (Northwestern Long Distance Company), the Justice Department threatened its first antitrust suit against AT&T in 1913 (*U.S. v. Western Electric, Defendant's Statement*). As a preemptive strike against antitrust remedies, AT&T Vice President Nathan Kingsbury sent a letter to the Attorney General. Known as the "Kingsbury Commitment," this letter outlined an AT&T promise to dispose of its stock in Western Union, cease acquisition of independents, and interconnect them to the Bell network (*Kingsbury Commitment*, 1913). Content with those concessions, the Justice Department ended its antitrust plans. Before long, AT&T was allowed to resume its acquisition of independent companies, pending Interstate Commerce Commission approval.

By the 1920s, Congress was actually in favor of a single monopoly phone system (Willis-Graham Act, 1921). In the meantime, Bell worked to accommodate remaining independents, allowing interconnection with 4.5 million independent telephones in 1922 (Weinhaus & Oettinger, 1988). This industry rapprochement enabled Bell to focus its energies on new ventures, such as "toll broadcasting" on radio stations (e.g., WEAF; see Brooks, 1976; Briggs, 1977).

Yet, fearing telco domination of the nascent broadcast industry, Congress formalized a ban on telco/broadcast cross-ownership in the Radio Act of 1927 (and the succeeding Communication Act of 1934). The 1934 Act also granted AT&T immunity from antitrust actions, in return for a promise to provide universal phone service; 31% of U.S. homes had a phone at that time (Dizard, 1989).

After investigating complaints concerning AT&T's market dominance in the 1930s, the FCC endorsed the industry's structure and characterized it as a "natural monopoly" (FCC, 1939). By the late 1940s, however, the Justice Department began to feel uneasy about the sheer magnitude of

AT&T's empire. It initiated antitrust proceedings against the phone giant in 1948, which culminated in a 1956 Consent Decree. Under the decree, the government agreed to drop its lawsuit in return for an AT&T pledge to stay out of the nascent computing industry.

The next year saw the first serious challenge to AT&T's notorious exclusive dealing practices, euphemistically known as the "foreign attachment" restriction. Under the guise of protecting its network from problems of incompatibility and unreliability, AT&T alienated several non-monopoly companies seeking to attach consumer-owned equipment to the network. Hush-A-Phone, Inc. petitioned the FCC to vacate this policy, which had been applied against the company's mouthpiece "hushing" device (Hush-A-Phone, 1955). On appeal, the court held that AT&T's ban of a device that emulates the natural cupping motion of one's hand "is neither justifiable nor reasonable" (*Hush-A-Phone Corp. v. U.S.*, 1956, p. 266).

This pattern of deregulation accelerated in 1968, when the FCC ruled against AT&T's ban of an acoustic coupler connecting radio telephones to AT&T's telephones. When allowing connection of this "Carterfone," the FCC and courts signaled that equipment not made by AT&T's manufacturing subsidiary, Western Electric, could be used in its network (*Carterfone*, 1968). During the following year, MCI was given permission to operate a long distance line, despite AT&T's objections (*MCI, Inc*., 1969).

Having sustained these legal setbacks, various segments of the Bell network began to feel the pressure of competition. This exogenous shock was compounded by inflation during the 1960s, which further reduced AT&T's profitability. The company responded by postponing maintenance and reducing labor costs during the 1970s in order to assure profitability. In the meantime, demand for telephone service in the U.S. skyrocketed, prompting long waits for equipment and charges of sloppy service (Dizard, 1989).

Dissatisfied with this industry conduct, the Justice Department in 1974 initiated proceedings to dismember AT&T, reminiscent of its 1948 action. In particular, the complaint against AT&T alleged that they had:

(1) Denied interconnection of non-Bell equipment to the AT&T network.

(2) Denied interconnection of specialized common carriers with the Bell network.

(3) Foreclosed the equipment market with a bias toward the Western Electric subsidiary.

(4) Engaged in predatory pricing, particularly in the intercity service area (see *U.S. v. AT&T*, 1982; Gallagher, 1992).

Only this time, with the aid of several interested non-monopoly firms, the Justice Department was in a much stronger position.

Thus, as demand for telecommunications services grew, the Bell system's regressive monopoly structure proved an impediment to growth and innovation (Jussawalla, 1993). After initially enjoying government protection during the first part of this century, the Bell monopoly gradually fell into disfavor with regulators for inefficient and anticompetitive market conduct.

In defense of AT&T, it is fair to say they achieved the burdensome goal of providing universal service, with the highest reliability in the world. The company employed a million workers in 1982, claiming that it lost about seven dollars a month on its average telephone customer (Dizard, 1994). While that plea may be debatable, such costs necessitated the practice of cross-subsidization via pricier long distance service. Known as the behemoth that worked, Bell was the largest company in the world, subsuming 2% of U.S. gross national product. The company carried over a billion calls per day at the time of the Modified Final Judgment (*U.S. v. Western Electric*, 1980).

Whether or not AT&T was well compensated for its trouble, by 1982, even they viewed their regulated monopoly as an impediment to progress. AT&T's willingness to consent to divestiture was, arguably, as much a function of self-interest as exogenous government pressure. In fact, Dizard (1989) notes that they had the financial and political clout to litigate the 1982 decree for another 30 years, had they been so inclined. Yet AT&T recognized, instead, that it was in their own interest to discontinue the "voice-only" monopoly in which they had become encased. So they negotiated a divestiture settlement, or consent decree, known as the Modified Final Judgment (MFJ) in 1982 (*U.S. v. AT&T*, D.D.C. 1982).

Effective in 1984, AT&T's local telephone service was spun off into seven new companies, known as the regional Bell operating companies, sometimes referred to as the "Baby Bells." The divestiture allowed Bell a convenient vehicle to exchange pedestrian local telephony for entry into the lucrative computer market. Perhaps more important, it enabled them to cut most of their labor overhead without fear of union unrest or "bad press."

The question remains, though, whether AT&T's conduct was an anomaly of that particular monopoly, or characteristic of vertically integrated capital-intensive phone utilities in general. In gaining a better understanding of that question, it's useful to examine the post-divestiture conduct of the telephone industry and the implications of their entry into allied fields.

Recent Developments

After a series of FCC and court rulings relaxed restrictions on telco entry into cable (Brown, 1993; *Chesapeake*, 1992), the RBOCs requested that the MFJ be vacated in 1994 (Atkin, 1996). As 1995 came to a close, the Supreme Court was hearing arguments to lift the ban on telco-provided video on appeal in *Chesapeake*. The case was preempted when Congress and the President removed the ban as part of the Telecommunications Act of 1996. Since the law also removes the ban on telco purchases of cable companies in communities of over 35,000, we're likely to see new alliances between these industries. Although U S WEST was one of the most aggressive entrants into cable, they recently moved to split their phone and cable businesses into two companies. AT&T's bold move to deliver local service and become the nation's largest cable operator, bolstered by its acquisition of Tele-Communications, Inc. (TCI) and MediaOne, was dealt a setback when a deal to reach Time Warner's 12 million subscribers was tabled. It seems, then, that the strategy of simultaneously pursuing an in-region and out-of-region service in the two industries hasn't succeeded (Atkin, 1999). Telcos can also provide video programming in their own service areas, but the new law prohibits joint telco/cable ventures in their home markets.

While this pending competition should lead to lower long-distance rates, local rates may not decline until AT&T, MCI, et al. receive favorable terms for reselling Bell connections as their own services. New York City offers the first laboratory for local loop competition, as NYNEX competes against TCI and Cox's MFS Communications. Those cable-based competitors now control over 50% of Manhattan's "special access" market—which involves the routing of long-distance calls to and from local lines—and will soon face further competition from MCI (Landler, 1995). The Telecommunications Act of 1996 also maintains universal service mandates, but allows states and the FCC to decide how they should be funded.

With the new law resonating like the starting gun at a race, several companies issued threats about invading each other's markets. AT&T, for instance, issued a declaration of war on local phone companies everywhere, pledging to enter the market in all 50 states and win over at least one-third of the $100 billion sector in the next few years. The company plans to offer service via alternative access providers, while using its cable lines as a platform for speedy Internet access.

Despite this competitive rhetoric, it seems that telcos are much more interested in pursuing markets in telephony than in cable. Before being allowed to enter long distance markets, RBOCs must prove they've opened their local phone networks to new rivals, following a 14-point checklist contained in the 1996 Act. The FCC set rules for implementing RBOC entry into long distance in August 1996, promulgating 742 pages of guidelines. The rules were immediately challenged by several Baby Bells. The rules were consolidated in the Eighth Circuit Court of Appeals in St. Louis, which granted a stay of the FCC's order. In late December 1997, U.S. District Court Judge Joe Kendall issued a ruling that undermined those FCC rules, claiming they trampled on state's rights (Mehta, 1998). In December 1999, Bell Atlantic became the first Baby Bell to win regulatory approval for its application to provide long distance service. Providing service to New York State, this arrangement represents the first time since the breakup of AT&T that consumers can receive both local and long distance service from a former Bell operating company.

All of this activity prompted Judge Greene to wonder whether the new law can deter phone giants from essentially reconstructing the Bell monopoly (Keller, 1996). Counter to that concern, AT&T actually initiated a split into three different companies in 1995. They include:

(1) A communication services firm, which retains the name AT&T.

(2) A network services firm named Lucent Technologies.

(3) A computer unit named Global Information Solutions.

Yet, if telco entry into other industries is any indicator, this move toward computing and video won't prove as revolutionary as some might fear. As Bates, et al. (in press) note, new services can be grouped into several categories:

- Adding new types of content/uses that can utilize existing telephone networks.
- Value-added services complementing the basic service.
- New mechanisms for delivering existing and emerging services.
- Those new uses that were not feasible under the technical limitations of the old networks, but which may be under new broadband networks.

Years after winning the right to provide information services, the Baby Bells had little to show for their efforts, as most were in the process of "regrouping" (Singer, 1993; Chen & Wilke, 1999). Ameritech recently offered ISDN (Integrated Services Digital Network) services in Ohio, including high-speed data and video applications, along with enhanced facsimile service—complete with mailboxes for storing and sorting fax messages. Pacific Bell has offered voice mail subscribers customized reports delivered on a daily basis.

NYNEX and Dow Jones entered a joint venture to deliver text information over phone lines in New York (Potter, 1992). Dow Jones has also worked with BellSouth to serve cellular phone customers in Los Angeles. Beyond that, the RBOCs have engaged in a few modest projects involving information delivery. The scope of these activities pales in comparison to those involving video delivery, although cross media competition has yet to reach levels envisaged under the Act.

In addition to numerous video on demand and pay-per-view ventures, phone companies such as AT&T and MCI have engaged in DBS (direct broadcast satellite) experiments (the latter in conjunction with Rupert Murdoch). MCI also recently entered an alliance with Microsoft for the delivery of online products and services. AT&T's plans to develop high-speed Internet services were dealt a setback when an Oregon judge ordered the company to open its high-speed network to rival services such as America Online (Stern, 1999).

Ancillary Services. Local and long distance service continues to dominate telco activities, generating $111.3 billion and $69.2 billion, respectively, in 1999. There has also been prolific growth in wireless ($40 billion revenue) and Internet/broadband ($10 billion revenue) services, along with a host of others, including call waiting, call forwarding, call blocking, computer data links, and audiotext (or "dial-it") service.

Perhaps the most explosive growth has been in an area involving mass-audience applications of the telephone—800 and 900 services (Atkin & LaRose, 1994). Virtually nonexistent prior to divestiture, these services generated 10 billion calls in 1991, as over 1.3 million 800 numbers earned an estimated $7 billion in revenues for the industry in 1993 (Atkin, 1995). It's striking to note that this ancillary revenue source for telephony was nearly equal to that of the entire radio industry that year. One study of audiotext (LaRose & Atkin, 1992) reported that, of 44 Dial-It services available in one local exchange, one-fifth were sexually-oriented; the remainder involved other types of entertainment (e.g., soap opera updates), while a few served information and self-help needs.

Aside from dial-a-porn, audiotext presents hundreds of information options, including daily TV listings, national and international news, celebrity information, recordings of dead (and undead) celebrities, sports scores, horoscopes, and updates of popular television soap operas (Ameritech Publishing, 1995; Atkin, 1993; 1995). Audiotext today is roughly a $1 billion per year industry (Neuendorf, et al., 1998). Even so, the Psychic Friends Hotline, which targeted low-income minority households with extensive promotions on cable, went out of business in 1998; they apparently didn't see the end coming!

As telcos implement DSL (digital subscriber line) technology and/or move to install fiber optics to the "last mile" of line extending to the home, providers will be able to provide a broad range of information and entertainment channels. Many are also developing personal communication service (PCS) systems, which effectively turn pagers into mini-wireless telephones.

In the meantime, the proliferation of new telephone services is having another effect: telephone companies are running out of available phone numbers. The traditional solution of splitting an area code in two and allocating a new area code is proving more difficult to implement as the geographical areas for these area codes becomes smaller and smaller. One solution being tested in many areas is an "overlay" of the new area code in the same area as the old one, requiring that 10 digits be dialed for every local call.

A related issue is "local number portability." As competition emerges in the local telephone market, the FCC has been concerned that businesses and residences be able to keep their phone numbers when switching to a new carrier. To address the issue, the FCC has mandated local number portability, adding slightly to the cost of local telephone service.

The ability to transmit digital signals such as DSL on the same phone lines that carry ordinary POTS signals created another controversy, as competitive phone companies (CLECs, competitive local exchange carriers) argued that existing phone companies should share the copper wires from their switching offices to homes or businesses. The incumbent phone companies (ILECs, incumbent local exchange carriers) argued against sharing, a position that could have delayed the introduction of digital services because of the lack of availability of enough existing telephone lines to allow additional service to everyone who wanted it. In early 2000, the FCC solved the problem by mandating line sharing among local phone companies. As a result, U.S. residents can now get multiple services (POTS and DSL) provided by different companies on the same phone line.

Current Status

The telephone network is often referred to as a "star" network because each individual telephone is connected to a central office. The heart of the network (see Figure 17.1) is the central office that contains the switching equipment to allow any telephone to be connected to any other telephone. Central offices are, in turn, connected to each other by two networks. The local phone companies interconnect all central offices within a service area (known as a LATA), and long distance companies provide a network of interconnections for central offices located in different LATAs.

The telephone companies are now in a formidable position to dominate video and information services, as they control more than half of U.S. telecommunications assets (Dizard, 1994). The industry handled 620 billion phone calls in 1994 and has a long history of providing mass entertainment and information services (Atkin, 1995). The value of AT&T and its former segments has tripled since divestiture, as the telecommunications sector will soon subsume $1 trillion worldwide, and one-sixth of the U.S. economy (King, 1994). Taken together, telephony and information services account for roughly one out of every 10 dollars spent in the United States, or over $2,000 per household annually.

While the Baby Bells are anxious to compete in long distance, the long distance companies express concern that local companies still command 98% of local revenues over their regions and retain control over 98% of the 155 million phone lines in the United States. Thus, these companies possess "bottleneck control of the local loop, and they'll be able to squash competition" (Berniker, 1994, p. 38). AT&T's plans for cable telephony, and 92 million wireless access customers have the potential to break this bottleneck control. Given that MCI pays $0.46 out of every dollar it earns to the RBOCs in access charges, they have an interest in entering the $25 billion access charge business. This helped motivate GTE to enter into a $70 billion merger with Bell Atlantic, which was being reviewed by the government at the time of this writing (Mehta, 2000).

Figure 17.1
Traditional Telephone Local Loop Network Star Architecture

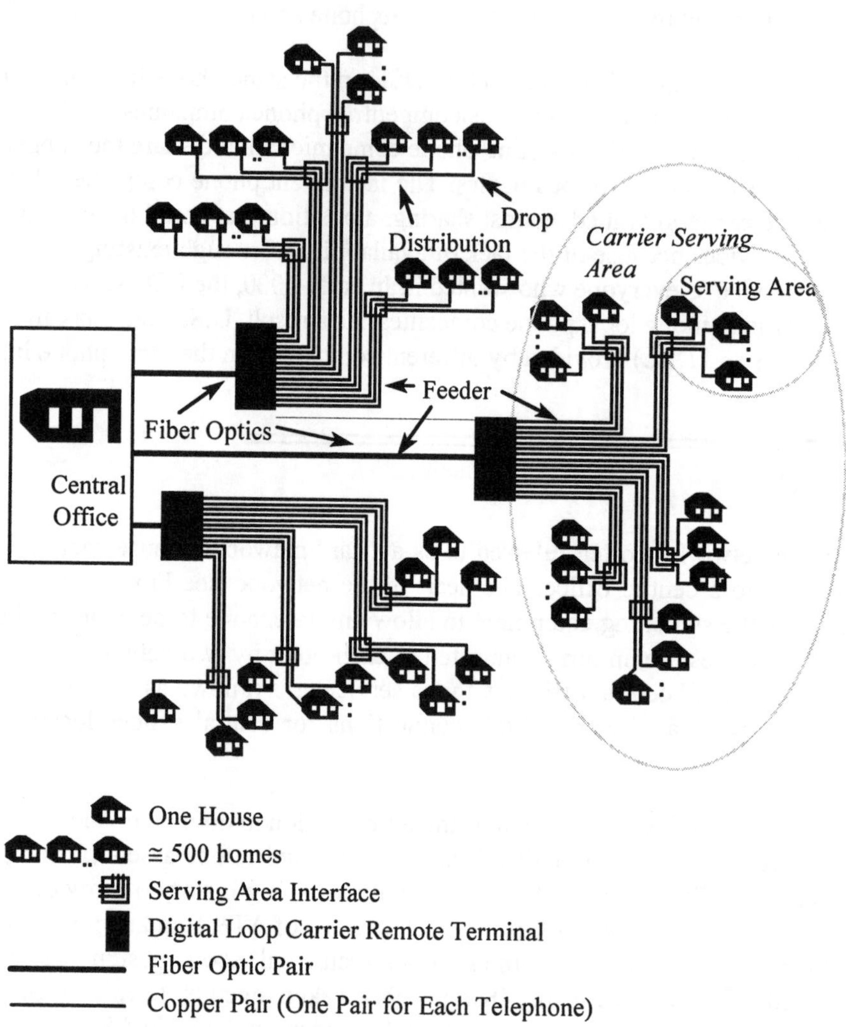

Source: Technology Futures, Inc.

At the time of divestiture, the Baby Bells represented roughly 75% of AT&T's assets. Between 1940 and 1980, the number of independent companies decreased by 77%, owing chiefly to consolidation among themselves, although some were acquired by AT&T (Weinhaus & Oettinger, 1988). By 1982, there were 1,459 independent telephone companies, producing 15.9% of the industry's revenues. Although merger activity has ebbed and flowed since that time, the United States had over 1,200 telephony providers in mid-2000.

Although the Telecommunications Act of 1996 returns telephone regulation to the federal government, the Baby Bells were previously governed by the judiciary through Judge Harold Greene's review of the MFJ. The MFJ imposed several line-of-business restrictions on the RBOCs, including (1) the manufacture of telecommunications products, (2) provision of cable or other information services, and (3) provision of long distance services (*U.S. v. A.T.& T.*, 1982). Selected waivers to the MFJ were granted in the years leading up to the Act, including decisions allowing regional phone companies to transmit information services owned by others (*U.S. v. Western Electric Co.*, 1987) as well as their own such services (*U.S. v. Western Electric Co.*, 1991).

In 1988, the FCC reified telco/cable cross-ownership restrictions out of fear that telcos would (1) engage in predatory pricing characteristic of capital intensive industries, and (2) use their natural monopoly over utility poles and conduit space to hinder competition with independent cable operators. Under these rules, former RBOCs were banned from providing cable service outside of their local access transport area (47 C.F.R. S. 63.54-.58, 1988).

Factors to Watch

As the above review suggests, the long distance segment of the U.S. telephone market has been open to competition for years, while "the industry's other basic service market—local exchange service—seems to be shifting from a natural regulated monopoly to an open competitive marketplace" (Bates et al., in press). The latter was spurred, in part, by landmark legislation (104 Pub. L. 104, 1996), which was passed amidst heady optimism concerning the benefits of deregulation for competition, consumer prices, the construction of an information superhighway, and the resulting job creation in the coming information age. This legislation was designed to encourage unprecedented media cross-ownership and competition by enabling cable companies to enter local telephone markets and local and long distance companies to enter each other's markets.

Shortly after its passage, however, the Act encountered heavy criticism in the popular press (Schiller, 1998). Prominent Congressional supporters now seek hearings to investigate problems with industry conduct, such as service and pricing, that the measure was designed to remedy (Atkin, 1999).

Although it's too early to render a definitive judgment, preliminary returns suggest that the Act has not succeeded in that regard. Gene Kimmelman, co-director of Consumers Union, concludes, "It's an abysmal failure so far. The much-ballyhooed opening of markets to competition was a vast exaggeration" (Schiller, 1998, p. A1).

So even without faulty regulation, consumers are thus far paying more for telecommunications services in the era of deregulation ushered in by the Act. In the absence of concerted competition from telcos, cable companies raised rates nearly three times the general inflation rate during the Act's first year, while rates for intrastate toll calls rose a hefty 6.1%. Although long distance phone rates increased only an average of 3.7% in 1996, that proportion was closer to 12% for the two-thirds of Americans not on any discount calling plan (Mills & Farhi, 1997).

While former FCC Chair Reed Hundt expressed dismay that competition was delayed by legal wrangling and industry "détente," he suggested that higher rates may be a necessary first step toward competition (Mills & Farhi, 1997). One competitive bright spot involves a group of smaller companies, such as Teligent, which are building their own advanced, high-speed communications facilities to give them direct access to their customers, bypassing remnants of the old Bell system.

The ongoing litigation over other pending RBOC long-distance service applications, mentioned earlier, will have strong ramifications for the structure of the phone industry. It remains to be seen whether long distance companies can compete with the RBOCs, given the latter's advantages in capitalization and switching equipment control. While politicians might debate the need to maintain regulatory oversight until competition takes root, few would dispute that competition will eventually be in the offing for telephony. The decade following divestiture saw competitive market forces dramatically reshape the telecommunications landscape. Consumer long distance rates dropped from $0.40 a minute in 1985 to just $0.14 a minute by 1993 (Wynne, 1994). New long distance carriers helped spawn thousands of new jobs, with millions more likely to accompany a doubling of domestic spending on telecommunications projected by the decade ending in 2004, subsuming nearly 20% of gross domestic product.

The recent merger of SBC and Ameritech, in conjunction with those involving NYNEX and Bell Atlantic as well as Pacific Telesis and SBC, raise the issue of whether the remaining independent Baby Bells (U S WEST and BellSouth) can go it alone. Although their territories range in profitability from the fast-growing South to the depopulated West, analysts expect that there remains plenty of opportunity for all RBOCs in the local phone industry. In fact, RBOCs continue to enjoy profit margins nearly double the 20% or so that their counterparts register (Cauley, 1997).

GTE, a non-Bell entity, remained free to pursue mergers with long distance carriers. Another pending merger, involving MCI WorldCom's $115 buyout of Sprint, represents the largest takeover in history (Chen & Wilke, 1999). For the present, AT&T continues to dominate long distance, commanding 62% of the market, followed by MCI (12.3%) and Sprint (5%) (Chen & Wilke, 1999). There are 100 smaller long-distance players (e.g., Metromedia). Table 17.1 reviews the top 10 worldwide merger and acquisition deals, the five biggest of which rank among the 10 largest business deals in history. This new consolidation, combined with obvious telco cross-subsidization concerns, would present a basis for continued vigilance by regulators.

Telco entry into the video marketplace could create the much-anticipated "revolution" in the broadcast and cable industries. This move could also open up a window of opportunity for various independent programmers that may supply specialty entertainment and information services, much like specialization in the magazine industry.

In sum, with the myriad avenues for telco entry into other media, the greatest challenge involves preventing the undue consolidation of our communications capabilities into ever fewer hands. If we are to maximize innovation, quality, and diversity in tomorrow's information grid, regulators must help channel telephony's vast capital and human resources into allied fields in a way that augments, rather than depletes, the existing cast of players.

Table 17.1
Top Telecommunications Mergers through 1999

Target	Acquirer	Value (in Billions)	Announced
Sprint	MCI WorldCom	$127.3	10/5/99
Ameritech	SBC Communications	$72.4	5/11/98
GTE	Bell Atlantic	$71.3	7/28/98
AirTouch	Vodafone Group	$65.9	1/18/99
U S WEST	Qwest Communications	$48.5	6/14/99
MCI Communications	WorldCom	$43.4	10/1/97
Orang	Mannesmann	$35.3	10/20/99
Telecom Italia	Olivetti	$34.8	2/20/99
NYNEX	Bell Atlantic	$30.8	4/22/96
Pacific Telesis	SBC Communications	$22.4	4/1/96

Source: FCC

Bibliography

Ameritech Publishing, Inc. (1995). *Ameritech pages plus Cleveland area white/yellow pages, 1995*. Chicago, IL: Ameritech Publishing.

Ameritech's telco-cable exchange. (1993, June 14). *Broadcasting & Cable*, 70-73.

Atkin, D. (1993). Indecency regulation in the wake of Sable: Implications for telecommunications media. *1992 Free Speech Yearbook, 31*, 101-113.

Atkin, D. (1995). Audio information services and the electronic media environment. *The Information Society, 11*, 75-83.

Atkin, D. (1996). Governmental ambivalence toward telephone regulation. *Communications Law Journal, 1*, 1-11.

Atkin, D. (1999). Video dialtone reconsidered: Prospects for competition in the wake of the Telecommunications Act of 1996. *Communication Law and Policy Journal, 4*, 35-58.

Atkin, D., & LaRose, R. (1994). Profiling call-in poll users. *Journal of Broadcasting & Electronic Media, 38* (2), 211-233.

Bates, B., Jones, K. A., & Washington, K. D. (in press). Not your plain old telephone: New services and new impacts. In C. Lin & D. Atkin (Eds.). *Communication technology and society: Audience adoption and uses of communication technologies*. Cresskill, NJ: Hampton.

Berniker, M. (1994, December 19). Telcos push for long-distance entry. *Broadcasting & Cable*, 38.

Briggs, A. (1977). The pleasure telephone: A chapter in the prehistory of the media. In Ithiel de Sola Pool (Ed.), *The social impact of the telephone*. Cambridge, MA: MIT Press.

Brooks, J. (1976). *Telephone: The first hundred years*. New York: Harper & Row

Brown, R. (1993, February 8). U S WEST answers video dialtone call. *Broadcasting & Cable*, 14.

Carterfone. In the matter of use of the Carterfone device in message toll telephone service. (1968). FCC Docket Nos. 16942, 17073. *Decision and order*, 13 FCC 2d 240.

Cauley, L. (1997, December 10). Genuine competition in local phone service is a long distance off. *Wall Street Journal*, A1, 10.

Chen, K. J., & Wilke, J. R. (1999, October 6). Kennard signals telecom deal will get tough FCC scrutiny. *Wall Street Journal*, A10.

Chesapeake and Potomac Telephone Co. v. U.S. (1992). Civ. 92-17512-A (E.D.-Va).

Dizard, W. (1989). *The coming information age*. New York: Longman.

Dizard, W. (1994). *Old media, new media.* New York: Longman.

Federal Communications Commission. (1939). *Investigation of telephone industry*. Report of the FCC on the investigation of telephone industry in the United States, H.R. Doc. No. 340, 76th Cong., 1st Sess. 602.

Felsenthal, E. (1998, January 13). Court to speed review of ruling in local phone competition case. *Wall Street Journal*, B17.

Gallagher, D. (1992). Was AT&T guilty? *Telecommunications Policy*, *16*, 317-326.

Hush-A-Phone Corp. v. AT&T, et al. (1955). FCC Docket No. 9189. *Decision and Order*. 20 FCC 391.

Hush-A-Phone Corp. v. United States (1956). 238 F. 2d 266 (D.C. Cir.). *Decision and Order on Remand.* (1957). 22 FCC 112.

Jessell, H. A. (1992, July 20). FCC calls for telco TV. *Broadcasting & Cable*, 3, 8.

Jussawalla, M. (1993). *Global telecommunications policies: The challenge of change*. Westport, CT: Greenwood Press.

Keller, J. (1996, February 12). AT&T and MCI explore local alliances. *Wall Street Journal*, A3.

King, J. F. (1994, February 20). Running its own race: Ameritech in the slow lane of the information superhighway. *Cleveland Plain Dealer*, F1.

Kingsbury Commitment. (1913, December 19). Letter from N. C. Kingsbury, AT&T to J. C. McReynolds, Attorney General, Justice Department.

Landler, M. (1995, April 3). The man who would (try to) save New York for NYNEX. *New York Times*, C1, C6.

LaRose, R., & Atkin, D. (1992). Audiotext and the re-invention of the telephone as a mass medium. *Journalism Quarterly, 69*, 413-421.

Mann-Elkin's Act. (1910). Mann-Elkins Act, Pub. L. No. 218, 36 Stat. 539.

Mehta, S. (1998, January 5). Baby Bells cautious on quick entry to long-distance market after ruling. *Wall Street Journal*, B8.

Mehta, S. (2000, January 20). Bell Atlantic and GTE file with FCC to split off GTE's Internet backbone. *Wall Street Journal*, A4.

Microwave Communications. Microwave Communications, Inc. (MCI). (1969). FCC Docket No. 16509. *Decision*, 18 FCC 2d 953.

Mills, M., & Farhi, P. (1997, January 27). A year later, still lots of silence: Dial up the Telecommunications Act and get bigger bill and not much else. *Washington Post National Weekly Edition*, 18.

1956 Consent Decree. *U.S. v. Western Electric Co. and AT&T.* (1956). Civil action no. 17-49, 13 RR 2143; 161 USPQ (BNA) 705; 1956 trade case. (CCH) Section 68246, at p. 71134 (D.C. N.J.).

Naik, G. (1996, February 9). Landmark telecom bill becomes law. *Wall Street Journal*, B3.

Neuendorf, K., Atkin, D., & Jeffres, L. (1998). Understanding adopters of audio information services. *Journal of Broadcasting & Electronic Media, 42*, 80-95.

Neuendorf, K., Atkin, D., & Jeffres, L. (in press). Adoption of audio information services in the United States. In C. Lin & D. Atkin (Eds.). *Communication technology and society: Audience adoption and uses of communication technologies*. Cresskill, NJ: Hampton.

Potter, W. (1992, May). Bells start to adopt new services. *Presstime*, 42.

Schiller, Z. (1998, February 9). Local phone competition is still just a promise. *Cleveland Plain Dealer*, A1, A6.

Singer, J. (1993). *Fight for the future: Congress, the Brooks bill and the Baby Bells.* Paper presented to the annual meeting of the Association for Education in Journalism and Mass Communication, Kansas City.

Stern, C. (1999, June 21). Cabler's angst rises as courts fret 'net. *Variety*, 23-24.

Telecommunications Act of 1996, 104 Pub. L. 104, 110 Stat. 56, 111. (1996). (Codified as amended in 47 C.F.R. S. 73.3555).

Telephone Company/Cable Television Cross Ownership Rules (1992), 47 C.F.R. 63.54 - 63.58. *Second Report and Order, Recommendation to Congress* and *Second Further Notice of Proposed Rulemaking*, 7 FCC Rcd. 5781.

U.S. v. Western Electric, Defendant's Statement, U.S. v. Western Electric Co. (1980). Civil Action No. 74-1698, pp. 169-170.

U.S. v. Western Electric Co., 767 F. Supp. 308. (D.D.C. 1991).

U.S. v. Western Electric Co., 673 F. Supp. 525 (D.C.C. 1987).

U.S. v. A.T.& T., 552 F. Supp. 131, 195 (D.D.C. 1982), aff'd sub nom. *Maryland v. U.S.*, 460 U.S. 1001 [1983).

Weinhaus, C. L., & Oettinger, A. G. (1988). *Behind the telephone debates*. Norwood, NJ: Ablex.

Willis-Graham Act. (1921). Pub. L. No. 15, 42 Stat. 27.

Wynne, T. (1994, November 16). An earshocking proposition. *Cleveland Plain Dealer*, 11-B.

Your new computer: The telephone. (1991, June 3). *Business Week*, 126-131.

18

Broadband Networks

Lon Berquist*

The highly-touted information superhighway is finding its way into homes across the United States as high-speed connections offered by cable operators and telephone companies continue their rapid diffusion. Coupled with promising wireless broadband technologies, the vision of a "network society" advanced by futurists appears to be close at hand. The need for greater network capacity is becoming increasingly evident as the number of users connected to the Internet grows and the popularity of bandwidth-intensive multimedia applications soars. As we begin the new millennium, there are 200 million Internet users worldwide, and three million Websites consisting of 800 million Web pages (Lippis, 1999).

Background

A network can simply be defined as a "hardware and software data communications system." Put more precisely, "networks exist to facilitate the exchange of communication and the delivery of goods or services, and they evolved from the need to speed up and enhance this exchange" (Aden & Stanard, 1998, p. 42). The evolution from analog to digital technologies has led to a convergence of all transmitted content (video, voice, audio, graphics, text, and data) to digitized bits. This development has necessitated a significant change from the traditional circuit-switched network used by phone companies to the packet-switched network that is far more efficient in sending data (see Figure 18.1). A circuit-switched network requires a dedicated phone line to each sender/receiver during calls, while a packet-switched network transmits data in separate small data packets that travel over various routes through whatever line is available. These packets of bits are then reassembled into the original message at its destination (Copeland, 2000).

* Telecommunications and Information Policy Institute, University of Texas at Austin (Austin, Texas).

Figure 18.1
Packet & Circuit Switching

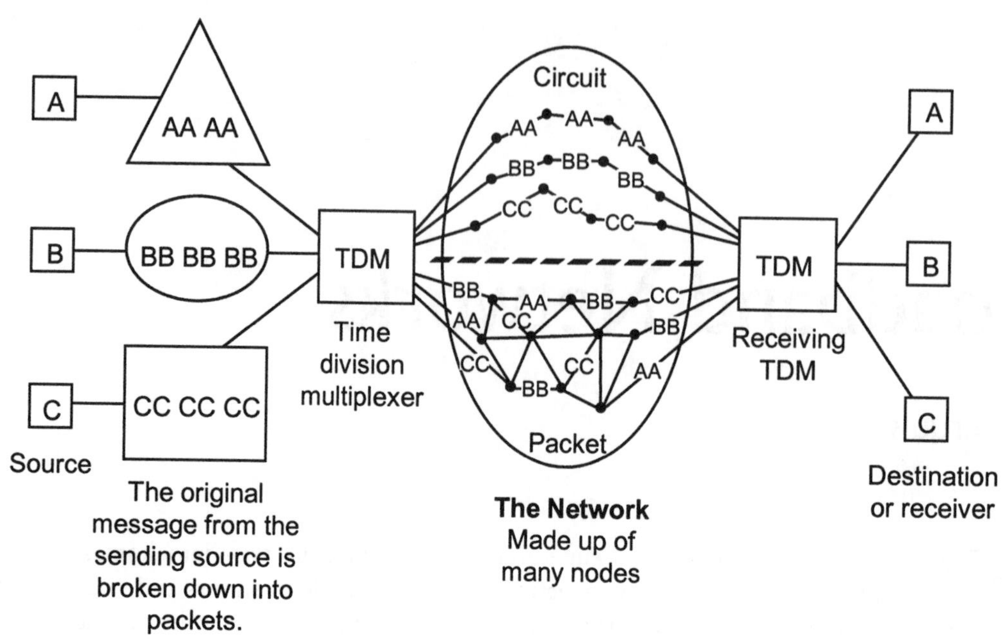

Circuit Switching—In a circuit switched network, the source establishes a path to the sender and then transmits all the packets along the same path. Circuit switching is used to carry voice and video to ensure the order is maintained when received by the destination site.

Packet Switching—In a packet switched network, individual packets must pass through the nodes on the network to be passed to the receiver. Throughout its route to the receiver, packets traverse any path and do not have to travel together. Packet number three of message one can be received before packet number one. If this does occur, the receiver will wait until all the packets arrive and then assemble them into the correct order.

Source: J. Hassay (1996)

The concept of broadband and bandwidth are terms left over from the legacy of analog networks in which the "space" available, or bandwidth (typically in megahertz or MHz), on the network determined the network's capacity. More recently, because of the proliferation of digital networks, network capacity is described in terms of transmission speed or the number of bits transmitted per second (b/s). In the United States, the Telecommunications Act of 1996 officially defines "advanced telecommunications capability" as a "high-speed, switched, broadband telecommunications capability that enables the users to originate and receive high-quality voice, data, graphics, and video telecommunications using any technology" (Werbach, 1997, p. 79). Subsequently, the Federal Communications Commission (FCC) more precisely defined broadband as, "the capability of supporting, in both the provider-to-consumer (downstream) and the consumer-to-provider (upstream) directions, a speed (in technical terms, 'bandwidth') in excess of 200 kilobits per second (Kb/s) in the last mile" (Lathen, 1999, p. 14).

The standard twisted-pair phone line offers analog transmission speeds of 64 Kb/s, with digital Integrated Services Digital Network (ISDN) phone lines reaching 128 Kb/s—both of which fail to meet the FCC's definition of broadband. In order to provide higher transmission speeds over the limited capacity of phone lines, special hardware and network upgrades are required. Coaxial cable utilized for cable TV provides up to 160 megabits per second (Mb/s); however, a traditional cable television system only transmits downstream from the headend to the customer. In order to become a two-way network, cable systems must add two-way switching capability.

To enhance their bandwidth capacity, telephone networks and cable systems are upgrading much of their infrastructure with fiber optic cable. Fiber offers tremendous advantages over copper cable, since it is virtually immune to electrical interference, can transmit over long distances, and provides almost unlimited transmission capacity. Because of dense wavelength division multiplexing (DWDM), even greater bandwidths are obtainable with a single fiber optic strand capable of supporting over 1,000 wavelengths at transmission speeds of up to 160 gigabits per second (Gb/s) (Montgomery, 1999).

In addition to improving bandwidth, technological advancements have allowed for faster and more efficient switching. Over the last decade, switching throughput has increased from 150,000 packets per second to over 6.4 terabits per second (Breidenbach, 1998). Those that marvel at the rapid change in computing power often cite Moore's Law, which recognizes that computer chips double their processing speed every 18 months while halving in price. With the increase in bandwidth and the improvements in switching capabilities, networking speeds are now increasing at two to three times the rate of Moore's Law (Lippis, 1999).

Recent Developments

The growth of the Internet and World Wide Web, combined with the proliferation of intensive Web-based applications, has created a demand for residential networks with bandwidth requirements previously available only on business networks. These emerging broadband networks utilize the existing infrastructure of the public switched telephone network (PSTN), and the cable television, terrestrial microwave, and satellite delivery systems.

DSL

Digital subscriber line (DSL) technology is rapidly diffusing throughout the United States. DSL is a digital technology provided by local phone companies that uses the existing two-way telephone network and squeezes as much bandwidth as possible out of the few thousand feet between the telephone network's central office (CO) and the subscriber's home.

There are many varieties of DSL technologies (see Table 18.1), and their transmission capacity is determined in large part by their distance from the central office. DSL modems provide a dedicated, always-on link to the Internet and offer the sort of bandwidth desired by teleworkers, small office/home office (SOHO) professionals, and heavy Web users (Hamblen, 2000).

Table 18.1
DSL Technologies

		Maximum Data Rate		Maximum Distance
Technology	**Name**	**Downstream**	**Upstream**	**CO to User**
G.dmt	Full-rate DSL	1.5 to 8 Mb/s	1.5 Mb/s	18,000 feet
G.Lite	DSL Lite	1.5 Mb/s	384 Kb/s	22,000 to 25,000 feet
HDSL	High Data Rate DSL	1.5 Mb/s	1.5 Mb/s	12,000 feet
RADSL	Rate Adaptive DSL	8 Mb/s	1 Mb/s	18,000 feet
SDSL	Symmetric DSL	768 Kb/s	768 Kb/s	10,000 feet
VDSL	Very High Data Rate DSL	51.8 Mb/s	2.3 Mb/s	4,000 feet

Source: D. A. Lathen

Some forms of DSL utilize a splitter near the customer premises in order to separate the DSL data and the phone signal. This device allows a single phone line to continue to provide voice service at the same time it transmits high-speed data (Rodbell, 1999). In 1999, the International Telecommunications Union (ITU) approved a consumer oriented DSL technology, G.Lite, that does not rely on a splitter and offers speeds up to 1.5 Mb/s (Menezes, 1999). Because G.Lite does not utilize a splitter, consumers can easily install their own DSL modems without requiring a service technician, which should help create a retail DSL modem market.

Ironically, the consumer-friendly nature of G.Lite may lead to its undoing. Shortly after G.Lite was developed, advocates of higher-speed DSL (dubbed G.dmt after its use of discrete, multitone transmission) realized that they could make the higher-speed (8 Mb/s) G.dmt service available in a splitterless form as well, allowing customers to install the service themselves (Romano, 2000). As of mid-2000, monthly fees for DSL service can be as little as $29, with an installation fee ranging from free to $100 or more.

Cable Modems

There are two types of cable modems available depending on the technical sophistication of the cable system. For those systems that have upgraded to a two-way hybrid fiber/coax (HFC) network, a cable modem can receive download speeds in the 3 Mb/s to 10 Mb/s range, with upload speeds of 128 Kb/s to 10 Mb/s. For older, one-way cable systems, a cable modem can download 2 Mb/s, but must use a dial-up phone modem for upload of data (Freed & Derfler, 1999). The cable modem is connected to a PC with an Ethernet connection, requiring that an Ethernet card be installed in the PC. The cable industry has established a standard for all cable modems, Data Over Cable Service Internet Specification (DOCSIS), which allows retail competition among modem vendors.

A major drawback of cable modems is that the available bandwidth is shared by the 500 or so subscribers on a particular network node. More alarming is the fact that subscribers setting up file sharing on their computers can have their files viewed by others subscribers on the node. Because DSL provides a dedicated line from the central office to the subscriber, it does not suffer from this security problem. Costs for cable modems are around $350, with installation fees of around $175. Monthly fees for cable Internet service are $39 to $49 as of mid-2000, not including cable TV subscription fees.

Terrestrial Wireless

In 1998, the FCC adopted rules to allow multipoint multichannel distribution service (MMDS) licensees to provide digital two-way communications services (FCC, 1999). MMDS, known as "wireless cable," is in the 2.5 GHz to 2.7 GHz spectrum and provides limited two-way distribution at speeds of 10 Mb/s (Pappalardo, 1999).

Local multipoint distribution service (LMDS) is another fixed wireless broadband technology touted by the FCC as a wireless broadband solution. LMDS is found at the extremely high end of the spectrum in the 27.5 GHz to 31 GHz range, and it has the potential to transmit at speeds of 1 Gb/s. Like MMDS, LMDS data service is being targeted at small and medium-sized businesses (Weinschenk, 2000). A major advantage for fixed wireless is that it can be deployed quickly and inexpensively. Problems for LMDS include signal interference due to rain fade and the need for line-of-sight for successful transmission.

Satellite Broadband

Hughes' DirecPC provides satellite Internet service with downstream data rates to homes at 400 Kb/s, with business customers receiving rates of up to 3 Mb/s. Because of prohibitive costs and still-developing technology, upstream signals must be transmitted over a customer's existing phone line (or other wireline service such as DSL). As of mid-2000, DirecPC is available via the 21-inch DirecTV DBS dish (Roberts-Witt, 1999a) and costs $29.99 per month for residential service and $129.99 per month for business customers. Subscribers must purchase a DirecPC antenna (or DirecDuo antenna which includes DirecTV video service) and a DirecPC satellite modem.

FTTH

Perhaps the most anticipated broadband network for residences is fiber-to-the home (FTTH). In Palo Alto (California), the Fiber Optic Network links directly to the home and provides data rates up to 45 Mb/s for $45 per month in addition to ISP (Internet service provider) charges, but includes a hefty installation fee of $1,200. Customers desiring 100 Mb/s can connect for $100 per month with a $2,400 installation price (Metcalfe, 2000).

Current Status

There is growing competition between phone companies' DSL service and cable modem service to capture the most customers wanting high-speed Internet access. As of mid-2000, the earlier introduction of cable modems has put this technology ahead in the battle, with close to 2.5 million subscribers in North America (see Table 18.2). Not surprisingly, cable TV giant Time Warner with its heavily marketed RoadRunner Internet service leads among cable multiple system operators (MSOs) with 400,000 cable modem subscribers. AT&T Broadband, partnering with cable ISP Excite @Home, follows with 294,000 cable subscribers. MediaOne, set at this writing to merge with AT&T pending FCC approval, is third with 278,00 cable Internet customers.

Table 18.2
North America Cable Modem Market as of March 2000

Operator	Cable Modems
Time Warner	400,000
AT&T	294,000
MediaOne	278,000
Cox	259,770
Shaw	235,000
Rogers	215,200
Comcast	195,200
Charter	122,900
Cablevision	75,000
Videotron	66,000
Cogeco	57,500
Adelphia	53,000
RCN	25,000
Other	174,000
TOTAL	*2,450,570*

Source: Kinetic Strategies, Inc.

Among DSL service providers, Southwestern Bell and its west coast company, Pacific Bell, had more than 200,000 DSL customers in early 2000. U S WEST followed with 167,000 DSL subscribers, and the total number of DSL users in North America was just under one million (see Table 18.3) (Burstein, 2000).

Table 18.3
U.S. DSL Deployment as of April 1, 2000

Operator	DSL Subscribers
SBC/PacBell	201,000
U S WEST	167,000
Covad	93,000
GTE	88,000
Bell Atlantic	62,000
BellSouth	49,000
Northpoint	41,300
Broadwing	24,000
Rhythms	20,000
Other	105,000
TOTAL	*850,300*

Source: DSL Prime

The proliferation of cable modems for high-speed Internet access has led to concerns among cities that regulate local cable franchises. Cities, as part of their push for competition among ISPs have insisted that cable operators provide "open access" by allowing competing ISPs to gain access to the cable operator's cable system. Unlike phone companies that are required to open their networks under federal law, cable systems have no such requirement. Local municipalities insist they have the right under their franchising authority to require cable systems to provide local ISPs with access to their cable modem service. The battle has moved to the courts, where it will continue to be fought.

In DSL markets, a number of telephone companies have developed trials for the delivery of MPEG-2 digital video over DSL lines (Dawson, 1999). Ironically, ADSL (asymmetric DSL) was originally unsuccessfully introduced as a technology for phone companies to offer video programming over phone lines in the 1980s.

In wireless technology, MCI WorldCom and Sprint (who MCI WorldCom hopes to acquire pending FCC approval) have spent $3 billion on MMDS licenses to serve roughly 50 million people, many in rural areas without wireline high-speed connectivity (Goodman, 2000).

Soon, satellite broadband will be let loose from its wireline upstream signal by providing two-way satellite access. Lockheed Martin's Astrolink, Loral Orion's CyberStar, Hughes' SpaceWay, and start-up KaStar will all be offering small home dishes called Satellite Interactive Terminals

(SITs) (Brown, 2000) that provide a data uplink. Costs of $4,000 per dish may limit its competitiveness with cable modem and DSL service, however (Roberts-Witt, 1999b). In addition, Microsoft has invested in Israeli Gilat Satellite Networks to develop a two-way broadband satellite system, "Gilat-To-Home," with hopes of launch before the end of 2000.

A private firm, OnFiber, Inc., will expand the fiber-to-the-home trial in Palo Alto to include a number of cities in the San Francisco Bay Area with hopes of connecting 100,000 homes locally before moving to other areas of the United States (Pease, 2000).

Factors to Watch

The network of the future will be able to deliver hundreds of video and audio channels, at the same time it delivers millions of phone calls, millions of e-mail messages with attachments, chat messages, e-business transactions, and e-government transactions. However, despite the growing capacity of networks, bandwidth-heavy applications and increasing numbers of users online are creating havoc. Even though they have some of the most elaborate networks in the United States, more than 100 universities throughout the country had to block or limit access to Napster.com, which allows Internet users to download thousands of popular songs, because it was overwhelming campus networks (Schevitz, 2000). Bandwidth supply must keep pace with bandwidth demand.

In the United States, concern is growing over the emerging "digital divide," which some suggest is leading to a society of technology haves and technology have-nots. Internationally, there is alarm about the bandwidth disparity between technology-rich nations and developing countries. New York City, for example, has more Internet hosts than the entire African continent (ITU, 1999).

Not satisfied with the pace of broadband development by private U.S. telecommunications firms, a number of public and private utilities, particularly in rural areas, have entered the broadband network market by building their own fiber networks or HFC cable systems to ensure their citizens are not left behind in the quickly moving digital economy. Tacoma Power's Click! Network and Glasgow, Kentucky's Electric Plant are among those who have leveraged their existing infrastructure to provide high-speed connectivity to their citizens (Schwartz, 2000).

To address these growing concerns, the FCC has established a Federal-State Joint Conference on Advanced Services whose purpose is to assess the status of deployment of advanced telecommunications services with the goal of bringing broadband capability to all Americans.

Bibliography

Aden, D., & Stanard, J. (1998). A crash course in networks. *Government Technology, 11* (3), 42-43.
Breidenbach, S. (1998). Assessing switching. *Network World, 15* (18), 75-85.
Brown, P. J. (2000). Internet over satellite soars. *Broadcasting & Cable, 130* (12), 46-48.
Burstein, D. (2000). *The numbers: First quarter U.S. subscribers.* [Online]. Available: www.dslprime.com.
Copeland, L. (2000). Packet-switched vs. circuit-switched networks. *Computerworld, 34* (12), 74.

Dawson, R. (1999). More telcos are doing video over DSL. *Multichannel News, 20* (50), 154-155, 164.

FCC. (1999). *In the matter of annual assessment of the status of competition in markets for the delivery of video programming*. CS Docket No. 99-230.

Freed, L., & Derfler, Jr., F. J. (1999, March 31). Cable modems. *PC Magazine*.

Goodman, P. S. (2000, March 28). MCI WorldCom plans wireless test. *Washington Post*, E01.

Hamblen, M. (2000). Digital subscriber line. *Computerworld, 34* (6), 69.

Hassay, J. (1996). Broadband network technologies. In A. E. Grant, *Communication technology update*, 5th edition. Boston: Focal Press.

International Telecommunications Union. (1999). *Challenges to the network: Internet for development, 1999*. [Online]. Available: http://www.itu.int/ti/publications/INET_99/index.htm.

Lathen, D. A. (1999, October). *Broadband today*. Report No. CS 99-14. Washington, DC: Federal Communications Commission.

Lippis, N. (1999). The Internet century. *Tele.com, 4* (24), 67-74.

Menezes, B. (1999, June 28). ITU approves consumer DSL standard. *Multichannel News*, 39.

Metcalfe, B. (2000). Faster than DSL or CTM, fiber optics to the home: Build it and they will come. *Infoworld, 22* (5), 116.

Montgomery, J. (1999). Annual technology forecast: Fifty years of fiber optics. *Lightwave, 16* (13), 49-54.

Oxman, J. (1999, July). *The FCC and unregulation of the Internet*. OPP Working Paper No. 31. Washington, DC: Federal Communications Commission.

Pappalardo, D. (1999). Fixed wireless users seeing more choices. *Network World, 16* (50), 38.

Pease, R. (2000). Fiber startup pushes high-speed Internet to California homes. *Lightwave, 17* (5), 1, 32-34.

Roberts-Witt, S. L. (1999a). Highway in the sky. *Internet World, 5* (34), 73-77

Roberts-Witt, S. L. (1999b). A network's anatomy: Hughes Network Systems. *Internet World, 5* (34), 76.

Rodbell, M. (1999). The DSL solution. *Communication Systems Design, 36* (12), 72-76.

Romano, P (2000). Full-rate, splitterless ADSL: A new spin on an old technology. *The TeleChoice Report on xDSL, 4* (8), 1-4.

Schevitz, T. (2000, March 3). Download discord: Universities are frustrated as students overwhelm Internet lines to access digital music files. *San Francisco Chronicle*, A1.

Schwartz, E. (2000). Utilities build fiber links. *Infoworld, 22* (5), 30.

Weinschenk, C. (2000, April 3). Prime candidate. *tele.com*, 48.

Werbach, K. (1997, March). *The digital tornado: The Internet and telecommunications policy*. OPP working paper 29. Washington, DC: Federal Communications Commission.

Websites

Cable Modems

- http://www.Cable-Modem.net/
- http://www.catv.org/modem/
- http://www.cablemodeminfo.com/
- http://www.cabledatacomnews.com/

DSL

- http://www.2wire.com
- http://www.adsl.com/
- http://www.everythingdsl.com/
- http://www.DSLreports.com/
- http://www.xdslresource.com/
- http://www.xdsl.com/

Wireless

- http://www.wcai.com/

19

Residential Gateways & Home Networks

Jennifer H. Meadows, Ph.D.*

Just a couple of years ago, few people thought that the typical American home would include a data network. Networking was for institutions and businesses, not residences. But technology has changed, and networking has come home.

It used to be that there would be one personal computer in a home, perhaps with one Internet connection. With the rapid growth of personal computer ownership came the rapid growth of multiple computer households. Over 60% of new computers bought in 1999 were by households that already had at least one computer—and that number is growing (About HomePNA, 1999; PC penetration, 1999). With multiple computers in a household comes the problem of sharing not only the Internet connection, but also the various computer peripherals in the home such as printers and scanners. A home network solves that problem by allowing multiple computers to share one Internet connection as well as those peripherals. This sharing, though, is only the beginning of what a home network can do (Marriott, 1999).

In addition to multiple computers in the home, more and more people have access to broadband connections that can carry a great deal of information at a fast rate (PC penetration, 1999). Examples of broadband connections include digital subscriber line (DSL) and cable modem service. (See Chapter 18 for more information about broadband networks.) These connections not only carry data, but also telephony, video, and audio.

The possibility of using a single network for all of these signals has led to the creation of residential gateways. These are devices that connect the home network to the outside broadband network and allow different streams of information to be routed intelligently throughout the home.

* Assistant Professor, Department of Communication Design, California State University, Chico (Chico, California).

Here are some possible applications of a home network and residential gateway:

- A movie is sent to the downstairs television set, Internet radio is sent to a receiver upstairs, and another household member researches a school paper on the Web—all at the same time.
- Telephone calls to your teenager are routed automatically to her bedroom, while business calls go to the home office.
- The dishwasher breaks and uses the home network to contact the repair facility before you know something is wrong.
- On your way home from a trip, you hear of a deep freeze coming to your area. You contact your home network from your wireless Web phone and turn the heat on so the house is comfortable by the time you get there, and the pipes don't freeze.

This chapter will briefly review the development of home networks and residential gateways, discuss the types and uses of these technologies, and examine the current status and future developments of these exciting and life-changing technologies.

Background

As mentioned earlier, networking was once thought to be only the domain of the office or institutions, not the home. Those few homes that did have networks would have installed a traditional Ethernet network that required an expensive type of telephone wiring called Category 5 (CAT 5). Alternatively, there would be wires running all over the house. Such a network needed a server, hub, and router. Clearly, these early home networks were not for the average home computer user because extensive computer and networking expertise and finances would be required.

Several factors changed the environment to allow home networks and residential gateways to take off:

The Internet. Internet use and growth has been exponential. More and more people want access, and competition for Internet access within the home is quickly becoming a common occurrence.

Computer and Peripheral Sales. The number of multiple computer homes is growing rapidly, facilitating a need for shared peripherals and Internet access. Who wants to buy several printers and have multiple ISP (Internet service provider) accounts?

Broadband. The availability of broadband connections such as DSL and cable modems to the home is growing. These connections carry much more information than a traditional phone line and allow for faster and enhanced Web browsing.

New Consumer Electronic Devices. There have been a plethora of new consumer electronic devices that work in concert with the Internet. These include portable Web pads, Internet radio receivers, MP3 players, and even Internet enhanced washing machines (Whirlpool takes, 1999).

Technologies such as direct broadcast satellites and digital video recorders can also be used to great benefit on home networks (see Chapters 5 and 15).

The five major uses for home networks include resource sharing, communication, home control, scheduling, and entertainment (Enikia, 1999). Sharing one broadband connection and computer peripherals within the home is an example of resource sharing. Communication is enhanced when one computer user can send files to another computer in the house for review, or perhaps send reminders for chores. A family can keep a master household schedule that can be updated by each member from his or her computer. Figure 19.1 shows an example of this kind of family information center. Being able to route digital video and Internet radio to different players within the home are examples of entertainment. The residential gateway allows the home network to have these multiple functions. The broadband pipe connects the home network local area network (LAN) to the outside wide area network (WAN), via your broadband connection.

Figure 19.1
HomePortal Screen

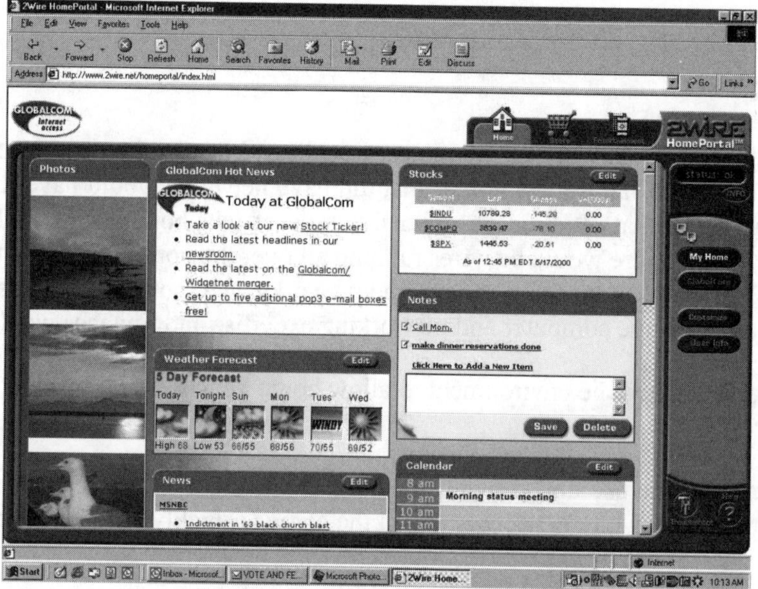

2Wire's HomePortal Residential gateway includes a household information interface accessible from any computer in the home.

Source: 2Wire, Inc.

The next section will discuss the recent growth in home networking and residential gateways including the types available and how they work.

Recent Developments

Recent changes in technology have made home networking and residential gateways affordable and easy-to-use.

Home Networking

There are four basic types of home networks:

- Traditional.
- Powerline.
- Phoneline.
- Wireless.

When discussing each type of home network, it is important to consider the transmission rate or speed of the network. High-speed Internet connections and digital audio and video require a faster network. Regular file sharing and low-bandwidth applications, such as home control, may require a speed of only 1 Mb/s (Megabit per second) or less. The MPEG-2 digital video and audio from direct broadcast satellite (DBS) services requires a speed of 3 Mb/s, and DVD requires between 3 Mb/s and 8 Mb/s. Many of the current home networking technologies can operate at 10 Mb/s, which is important considering future generations of digital media.

Traditional networks use Ethernet, which has a data transmission rate from 10 Mb/s to 100 Mb/s. This is the kind of networking commonly found in offices and universities. As discussed earlier, traditional Ethernet has not been popular for home networking because it is expensive and difficult to use. To direct the data, the network must have a server, hub, and router. Each device on the network must be connected, and many computers and devices require add-on devices to enable them to work on the Ethernet. Thus, despite the speed of this kind of network, its expense and complicated nature make it somewhat unpopular for the home networking market.

Homes under construction are more likely to have network wiring built into the home. For the pre-existing home, "no new wires" networking solutions have been developed that use existing phonelines or powerlines, or are wireless (Scherf, 1999).

The developers of powerline networking realized that all of the devices a consumer might want to be networked are already plugged into the home's power lines. Why not network over these power lines? Intellogis, which changed its name in mid-2000 to Inari (www.intelogis.com or www.inari.com) and Enkia (www.enkia.com) are two companies working on powerline home networking. Household electrical wires are used for data transmission, with data rates up to 10 Mb/s as of early 2000. With these products, a gateway is plugged into an outlet and a modem. Network interface adapters are plugged into electronic devices via their USB or parallel ports. Powerline networking is limited at this time by the hostile nature of a powerline. Data rates are affected by the change and flux created by power surges, lightning, and brown outs (Intelogis, 2000; Enikia, 2000).

Just as most homes have powerlines throughout, phone lines are also ideal for home networking. This technology uses the existing random tree wiring typically found in homes and runs over regular old telephone wire—there is no need for Cat 5 wiring. The technology uses frequency division multiplexing (FDM) to allow data to travel through the phone line without interfering with regular telephone or DSL service. There is no interference because each service is assigned a different frequency. The Home Phoneline Networking Alliance (HomePNA) has presented two open standards for phoneline networking. HomePNA 1.0 provided data transmission rates up to 1 Mb/s, but it is quickly being replaced by HomePNA (HPNA) 2.0, which allows data transmission rates up to 10 Mb/s and is backward compatible with HPNA 1.0 (HPNA embarks, 2000).

One of the most talked about types of home network is wireless. There are three major types of wireless home networking technologies as of the beginning of 2000: Wi-Fi (otherwise known as IEEE802.11 High-Rate DS), HomeRF, and Bluetooth (Hafner, 1999).

These wireless technologies are based on the same premise. Radio signals from the Instrumentation, Science, and Medical (ISM) bands of spectrum are used to transmit and receive data. The ISM bands, around 2.4 GHz, are not licensed by the FCC and are used mostly for microwave ovens and cordless telephones.

Wireless networks are configured with a receiver that is connected to the wired network or gateway at a fixed location. Transmitters are either within or attached to electronic devices. Much like cellular telephones, wireless networks use microcells to extend the connectivity range by overlapping to allow the user to roam without losing the connection (WECA, 2000).

Wi-Fi is the latest label attached to IEEE 802.11 High Rate Direct sequence, the specification for high-rate wireless Ethernet. Wi-Fi can transfer data up to 11 Mb/s and is supported by the Wireless Ethernet Compatibility Alliance, a group of companies, including Apple, Lucent, Dell, and Nokia, interested in wireless networking (Spangler, 1999). The traditional wireless Ethernet specification, IEEE802.11, is primarily cost prohibitive to home users: access points cost $1,000 to $2,000 per point and $1,000 to $3,000 per adapter (Spangler, 1999; WECA, 2000).

HomeRF, like Wi-Fi, was developed primarily for the home networking market. HomeRF is an open specification using Shared Wireless Access Protocol (SWAP). SWAP 2.0 can transmit up to 10 Mb/s of data a maximum distance of 100 feet. Developers claim that SWAP handles voice better than wireless Ethernet because it was designed for voice and data, while 802.11 was designed for data only (About the HomeRF, 2000).

While HomeRF and Wi-Fi can transmit data up to 11 Mb/s for up to 150 feet, Bluetooth was developed for short-range communication at a data rate of 1 Mb/s. Bluetooth technology is built into devices, and these enhanced devices can communicate with each other, creating an ad-hoc network. The technology works with and enhances other networking technologies. For example, students with a Bluetooth-enabled laptop computer and cellular phone in their backpack can surf the Web while sitting in the park because the computer links to the phone, which can be dialed up to an ISP (Bluetooth FAQ, 2000). (For more on Bluetooth, see Chapters 22 and 23).

Table 19.1 shows a comparison between each of the home networking technologies discussed in this section.

Table 19.1

A Comparison of Advantages and Disadvantages of Networking Technologies

Networking Technology	Advantages	Disadvantages
Conventional Ethernet IEEE 802.3	- Fastest data transmission rate up to 100 Mb/s - Reliable and standards-based - Flexible	- Costly - Requires special wiring - Difficult to install - Requires hub, router, and server for intelligent networking - Devices must be connected to the network using dedicated wires
Home PNA	- Fast data transmission rate up to 10 Mb/s - Reliable and standards-based - Flexible - Uses existing home phone wiring - Easy to install - Low cost - No hub or router needed	- Devices must be wired to the network
Wireless: Ethernet IEEE 802.11	- Fast data transmission rate up to 11 Mb/s - Reliable and standards-based - Flexible - No wiring required - Mobility - Heavily supported by computer industry	- Can be costly - Can have structural setbacks (some walls block wireless signals) - Range problems - Base station required
Wireless: Home RF	- Fast data transmission rate up to 10 Mb/s - Reliable and standards-based - Flexible - No wiring required - No cables or wires - Easy to install - Mobile - Low cost	- Range problems - Can have structural setbacks - Not as widely adopted as 802.11 - Base station required
Wireless: Bluetooth	- Reliable - Flexible - No wiring required - Mobile - Low cost - Embraced by computer industry for handheld devices	- Limited range - Data transmission rate currently only 1 Mb/s - Can have structural setbacks - Devices must have Bluetooth chip
Powerline	- Fast data transmission rate - Flexible - Uses existing powerline home wiring - Easy to install	- Can have transmission blocks and interference - "Hostile" network environment - Tied to outlets - No standard established

Source: 2Wire, Inc.

Residential Gateways

The residential gateway is what makes the home network infinitely more useful. This is the device that allows users on a home network to share access to their broadband connection. As broadband connections become more common, the one "pipe" coming into the home will most probably carry numerous services such as Internet, phone, and entertainment. A residential gateway seamlessly connects the home network to a broadband network so all network devices in the home can be used at the same time.

The current definition of a residential gateway has its beginnings in a white paper developed by the RG Group, a consortium of companies and research entities interested in the residential gateway concept. The RG Group determined that the residential gateway is "a single, intelligent, standardized, and flexible network interface unit that receives communication signals from various external networks and delivers the signals to specific consumer devices through in-home networks" (Li, 2000).

Since the RG Group's white paper, there has been significant development of the residential gateway concept, and three groups are presently working on standards which they see as enabling increased adoption and development of this technology: The Telecommunications Industry Association's TR-41.5 Committee, the International Organization for Standardization and the International Electrotechnical Commission (ISO/IEC), and the Open Services Gateway Initiative (OSGi) (Li, 2000).

The reason behind all these groups working on specifications is linked to the function of the residential gateway. The definition taken by the groups discussed above envisions a residential gateway as a single centralized device, thus the need for standardization. Another concept of the residential gateway envisions multiple gateways in and out of the home attached to specific devices such as a set-top box and a modem.

OSGi's (2000) specification notes that a service gateway should:

- Be open and independent of platform.
- Have a standard.
- Have advanced privacy and security features.
- Support all types of home networking.
- Be reliable.

Residential gateways can be categorized as complete, home network only, and simple.

A *complete residential gateway* operates independent of a personal computer and contains a modem and networking software. This gateway can intelligently route incoming signals from the broadband connections to specific devices on the home network. Set-top box and broadband centric are two categories of complete residential gateways. A broadband-centric residential gateway is an independent digital modem such as a DSL modem with IP (Internet protocol) management and integrated HomePNA ports. Set-top box residential gateways use integrated IP management and routing with the processing power of the box (The emerging, 1999).

Home network only residential gateways interface with existing DSL or cable modems in the home. These can still route incoming signals to specific devices on the home network.

Simple residential gateways provide limited routing and connectivity between properly configured devices. Also known as "dumb" residential gateways, they have limited processing power and applications.

Clearly the complete residential gateway or the whole house gateway is the technology receiving the most attention as of mid-2000. These gateways are the ones supported by OSGi and the RG Group. 2Wire, a company that develops residential gateways, has defined a set of minimum requirements for residential gateways. The residential gateway should:

- Be user installable—It should be an easy procedure that anyone can do.
- Be remotely configurable—With technical support provided by the manufacturer, the home user should be able to configure and update the RG through the Internet.
- Support all types of home networking.
- Have firewall and virtual private network capability—There should be privacy and security using a firewall to hide computers within the home. A VPN (virtual private network) secures the private network.
- Support both versions of DSL. (2Wire develops RGs for use with a DSL connection.)
- Be controllable through a browser—All controls should operate through a browser so the user who needs to change something can just use the home PC.
- Support multiple services such as Internet, voice, data, and entertainment (2Wire, 1999).

The Residential Gateway and Home Network

Working in concert, the residential gateway and home network allow multiple users to access the broadband connection at the same time. Household members will not have to compete for access to the Internet, printers, scanner, or even the telephone (Marriott, 1999). The home network allows for shared access to printers and peripherals. Household members can keep a common schedule, e-mail reminders to each other, share files, etc. The residential gateway allows multiple computers to access the Internet at the same time, and each computer can be given a "virtual" IP address. The residential gateway can route different signals to the appropriate devices in the home. For example, entertainment signals in the form of radio, video, and games can be routed to the stereo, television, or digital video recorder (such as a ReplayTV unit).

Home networks and residential gateways are the key to what industry pundits are calling the "smart home." Although the refrigerator that tells us we are out of milk may seem a bit over the top, utility management, security, and enhanced telephone services are just a few of the applications of this technology. For a nice summary and view of a smart home, check out Cisco's Smart Home video at http://www.cisco.com/warp/public/79/consumer/.

Current Status

The market for home networking and residential gateways is being primarily driven by low-cost computers, increases in the number of multiple PC homes, increases in the number of online households, and increases in homes with broadband connections. Parks Associates expects the percentage of homes with multiple PCs to almost double from 18% in 1998 to 31% in 2004. The group also predicts that Internet/online service penetration will increase to 49% in 2004 (PC penetration, 1999). Access to broadband connections is also expected to grow exponentially. Cahner In-Stat predicts there will be over 48 million U.S. households with broadband service by the end of 2004 (Kirstein, 2000).

As of mid-2000, there is anywhere from several hundred thousand to over half a million homes with home networking, depending on whose statistics you look at. The home networking and residential gateway markets are expected to have tremendous growth between 1999 and 2005. Cahners In-Stat reports that home networking sales should total more than $281 million by the end of 2000, and Forrester predicts home networking sales to reach $1 billion by 2002 (Berst, 2000). As for residential gateways, Cahners In-Stat predicts the market will grow to $3.8 billion by 2004, up from just $34 million in 1999 (Strother, 2000). Combining the two, the worldwide home networking and residential gateway market is expected to explode to $5.7 billion by 2004 from $600 million in 2000 (Ewalt, 2000).

Many of the companies that have come to be associated with networking and personal computers have become players in the lucrative home networking and residential gateway market. These technologies, however, will not reach a critical mass until consumers realize that these products are useful to them and are easy-to-use and install.

Companies such as Microsoft, Intel, and Apple have developed home networking products that plug into computers and use one of the networking technologies discussed earlier in the chapter. For example, the Apple Airport for the iMac uses 803.11 HR or Wi-Fi. This system includes a base station that is connected to the wired network such as DSL or cable modem. An iMac with an AirPort card can then link to the Base Station from up to 150 feet. Up to 10 iMacs can be connected to one Base Station and to each other (www.apple.com). Wi-Fi is affordable for home users and small businesses. For example, as of mid-2000, the AirPort Base Station cost $199 and the card was $99. Lucent makes a WaveLan network card for $200 to connect other PCs and Macs to an AirPort network (Lewis, 1999).

Cisco, 2Wire, Ericsson, and IBM are a few of the companies who have developed or are developing residential gateways (Shimogori, 2000; What is a residential gateway, 2000). For example, 2Wire sells the HomePortal 1000, a complete residential gateway, which contains a DSL modem, home networking hub, router, Web server, firewall security, and a simple browser interface (2Wire, 2000). This residential gateway uses HPNA 2.0 to create a phoneline network. Electronic devices can be plugged into any telephone jack if they have the necessary dongle, HPNA add-on, or have HPNA built in. Other home networking technology can also be used. The HomePortal has both Ethernet and USB connections. As with other DSL modems, the DSL provider is expected to absorb part of the cost of the HomePortal 1000, so the price to consumers is kept well below $400.

HomePortal 100 (without the DSL modem or the subsidy) costs about the same (see Figure 19.2) (2Wire, 2000).

Figure 19.2
HomePortal Residential Gateway

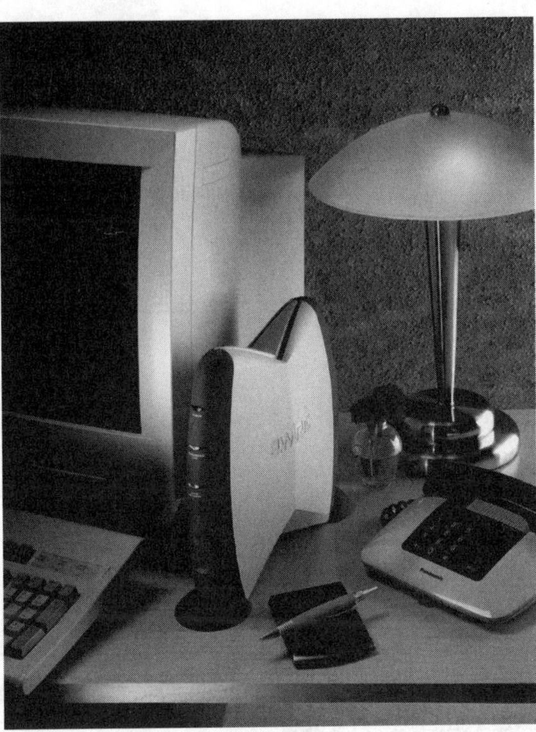

Source: 2Wire, Inc.

Sony is teaming up with General Instrument Corporation to put Sony's home entertainment network technology into GI's advanced digital set-top box, the DCT-5000. Users will be able to watch television, use the Internet, and talk on the telephone simultaneously (Greene, 2000).

SmartAMERICA introduced the Intelligent Home Gateway to markets in Australia, New Zealand, and Southeast Asia in late 1999. The product has a focus on utility management such as gas, power, and water. Developers note the product's advantages are "cost-effective delivery of telecommunications service; accurate automatic remote meter reading; and automatic shifting of non-essential electricity use to off-peak tariffs, producing savings for power company and consumer and reducing the generation of greenhouse gases" (SmartAMERICA parent, 1999).

Ericsson's e-box has a 486 processor, an Oracle database, an Ethernet port and two PCMCIA slots, and is Java-compliant. Several small-scale trials were underway as of mid-2000 in Scandinavia. Using OSGi's standard, the e-box is aimed at utility companies. Ericsson sees the utility company installing the e-box for free on its customer's homes, thus enabling utility management.

These dumb gateways could in the future have "software 'pushed' to them when needed from a central service, managed by an Oracle database" (Sullivan, 1999).

A small residential gateway trial began in July 1999 in a group of Bell Canada homes in Ottawa, Ontario. Each home had a residential service gateway that used a DSL Nortel 1M Modem and support for HomePNA phoneline networking and CBUS to allow home appliance control (Levin, 1999).

A recent spin-off from IBM called Home Director has developed a gateway product for new home development. The product is essentially a central server for the home. Telephone, Internet, and video ports are part of The Home Director Connection Center, which costs between $600 and $3,000. For $10,000 to $15,000, customers can have light, thermostat, and security control added to the package. These costs are usually included in the price of the new home. Home Director sees this as "future-proofing" the home (Wong, 1999).

Factors to Watch

The future looks bright for home networking and residential gateways. Some of the things to watch for include:

Broadband Access

The number of homes with a broadband connection is expected to grow rapidly. This is an important factor because broadband access is what makes home networking and residential gateways attractive. These technologies allow homes to use the broadband network efficiently and get the most out of it. The two major players in broadband access are cable and DSL. Which one will emerge as the dominant player remains to be seen.

DSL proponents say prices are lowering while availability is growing. The major problem with DSL is the home has to be within a certain distance of the telephone company switching facility (also known as the central office or CO). Any farther away, and the copper wires will attenuate too much of the DSL signal. Cable modem service, on the other hand, is available to any home passed by the coaxial cable. One problem with cable and cable modems is that all users within a node of up to 500 homes share the same cable. If many people are on at the same time in the neighborhood, the speed of the network can be slowed significantly. Node sizes may be reduced in the future to minimize this problem. Currently, some service providers are considering nodes of approximately 50 homes. In addition, a cable modem currently requires that all upstream traffic share a comparatively small amount of bandwidth; even within the home, there can be problems if multiple users want to upload a document at the same time.

As of mid-2000, cable is leading the competition for customers, but that may soon change. Some industry watchers predict cable modems, while others predict DSL. As of mid-2000, predictions are difficult to make. A smaller player in the fray is satellite (Jerome, 2000). (For more on broadband networks, see Chapter 18.)

Increased Rollouts of Home Networking Products and Residential Gateways

As home networks become more popular, look for new home networking products using traditional phoneline, powerline, and wireless networking technologies. Residential gateway products should also become more popular. The question now is whether users want to build their own using networking technologies traditionally used in the office, or have one of the easy-to-use, "all-in-one" gateways discussed earlier. The new consumer home products are much more user friendly, a good idea, since most Americans still have trouble setting the clock on their VCRs.

The capabilities of residential gateways will certainly expand if the product achieves widespread use. For example, the same HPNA technology that transports large amounts of data over ordinary telephone lines can also transmit additional voice lines throughout a home without the need for additional wiring. This technology can be connected through a residential gateway that has PBX-type functionality (telephone switching) to a broadband connection such as a DSL line that can also support numerous virtual phone lines. Such a set-up creates home telephone systems as sophisticated as in any office—without adding any additional wiring to the home.

One key to the success of any home networking technology is the number of companies that build connections for that type of network into their products. It is not practical to have Internet radios, Web pads, and smart appliances that must each have their own dedicated modem to connect to the Internet. Widespread adoption of any home network technology by manufacturers of peripherals will be a good predictor of success of the technology.

Broadband networks and home networks do present some security threats. An increasing concern over cyber crime makes protecting the home network all the more important in many users' minds. The residential gateway is one means of easy-to-use security. Many have firewalls that hide the home network from potential intruders. Look for these security features to become more advanced.

More Networked Devices in Homes

The term "smart home" is not likely to fade away in the near future (for better or worse). Entertainment devices, appliances, Web pads, phones, and security systems are just a few of the networked devices available as of mid-2000. Look for more devices to come out at lower prices and with greater distribution. The key here is that before users are going to want networked devices, they are going to need a home network. This fact points to the importance of standards. You wouldn't buy a PAL television set for the United States, would you? The wireless and phoneline networking industries both have organizations working on standards since the 1990s, while powerline networking companies did not begin working on a common standard until mid-2000 (Enikia champions, 2000). This fact alone could slow the adoption of powerline networking technology.

More New Home Construction with Advanced Network Wiring

The home building industry has begun to integrate networking technology into new homes. Broadband connection providers and networking technology companies are partnering with builders to put "structured wiring" into new homes. For example, Bell Atlantic Communications and Construction Services, Inc. (BACCSI) has a deal with builders to build "Bell Atlantic-ready" homes. As of mid-2000, there have been over 15,000 homes built under this program (Nelson, 2000). For more information on these homes, see BACCSI's consumer services Web page at www.baccsi.com/cat_admulti.htm.

The Home Director spin-off from IBM's home networking division is marketing Home Director wiring packages that can be incorporated into new and pre-existing homes (www.homedirector.net). These groups are also targeting SOHOs (small offices/home offices) and multiple dwelling units. Look for increases in "structured wiring" in new residential buildings.

Faster and More Robust Networking Technologies

While the current home networking technologies have speeds up to 11 Mb/s (not counting traditional Ethernet), look for faster and stronger networking technologies in the future. These will be cheaper and easier to use than anything available now and will be fast enough to accommodate whatever networked devices are yet to be developed.

While it is easy to be optimistic about the future of home networking and residential gateways, several factors must be considered when pondering their future success. First, the standards issue must be addressed. The industry may just let the market play out to see who the winner will be. Second, the industry must remember to keep the user in mind. It's wonderful to read about smart washing machines and refrigerators but, if the electronic devices are not easy-to-use and useful in the home, then what's the point? Finally, before any of this can happen, broadband connections have to be readily available at reasonable prices, and the services provided over these networks must be something people want and cannot get elsewhere for less or with more ease.

Bibliography

2Wire, Inc. (1999). *The emerging residential gateway market.* [Online]. Available: http://www.2wire.com/company/articles/article_resgatemarket.html.

2Wire. (2000). [Online]. Available: http://www.2wire.com.

About HPNA. (1999). *HomePNA FAQ.* [Online]. Available: http://www.homepna.org/faq.htm.

About the HomeRF WG. (1999). *HomeRF.* [Online]. Available: http://www.homerf.org/aboutHRF/.

Berst, J. (2000, February 7). Why Sun and Microsoft won't let you have a wired home. *ZDNet AnchorDesk.* [Online]. Available: http://www.zdent.com/anchordesk/story/story_4376.html.

Bluetooth FAQ. (1999). *Bluetooth SIG.* [Online]. Available: http://www.bluetooth.com/bluetoothguide/faq/1.asp.

Cisco Internet Home Gateways. (2000). *Cisco Systems.* [Online]. Available: http://www.cisco.com/warp/public/779/consumer/products/prodlit.html.

Enikia champions Homeplug Powerline Alliance. (2000, April 11). *Press release.* [Online]. Available: http://www.enkia.com/headlines/homeplug.html.

Enikia, Inc. (1999). *High-speed home networking over AC power lines!* Piscataway, NJ: Enikia, Inc.

Ewalt, M. (2000a, March 14). Home connectivity no longer a niche—Market to reach $5.7 billion by 2004. Press release. *Cahners In-Stat*. [Online]. Available: http://www.instat.com/pr/2000/rc001hn_pr.htm.

Ewalt, M. (2000b, April 4). Home networking market finishes 1999 with a bang! Press release. *Cahners In-Stat*. [Online]. Available: http://www.instat.com/pr.

Greene, M. (2000). *New set-top boxes move networking to higher level*. [Online]. Available: http://www.parks-associates.com/media/kurtcb&f.htm.

Hafner, K. (1999, September 30). Promising an end to cable spaghetti. *New York Times on the Web*. [Online]. Available: http://www.nytimes.com.

Home Phoneline Networking Alliance (HomePNA) embarks upon second generation networking standard effort. (1999). *HomePNA press release*. [Online]. Available: http://www.homepna.org/news/pressr9.htm.

Howard, B. (1999a, November 15). Standards time. *PC Magazine*. [Online]. Available: http://www.zdent.com/pcmag/stories/opinions/0,7802,2385242,00.html.

Howard, B. (1999b, December 13). Looking ahead in 2000. *PC Magazine*. [Online]. Available: http://www.zdnet.com/pcmag/stories/0,7802,200667,00.html.

HPNA—About HomePNA. (1999). *HomePNA*. [Online]. Available: http://www.homepna.org/docs/wp1.htm.

Intelogis shifts business focus: Changes name to Inari. (2000). *Intellogis press release*. [Online]. Available: http://www.intelogis.com/news/010400f.html.

Jerome, M. (2000). How Bluetooth works. *ZDNet PC Computing*. [Online]. Available: http://www.zdnet.com:80/pccomp/stories/all/0,6605,2453050,00.html.

Kirstein, M. (2000, March 15). Tap into the broadband era. *Press release*. Cahners In-Stat.

Levin, C. (1999, July 29). Cyberservices for the home. *PC Magazine*. [Online]. Available: http://www.zdnet.com/pcmag/stories/trends/0,7607,2304367,00.html.

Lewis, P. H. (1999, November 25). Not born to be wired. *New York Times on the Web*. [Online]. Available: http://www.nytimes.com.

Li, H. (1997). Residential gateway—The future is now? *Parks Associates*. [Online]. Available: http://www.parks-associates.com/media/hjhometechnologies.htm.

Li, H. (1998). Evolution of the residential-gateway concept and standards. *Parks Associates*. [Online]. Available: http://www.parksasociates.com/media/jhcable.htm.

Marriott, M. (1999, February 25). Computers that talk, even when families don't. *New York Times on the Web*. [Online]. Available: http://www.nytimes.com.

Nelson, K. (2000). Consumer readiness hangs in the balance. *Parks Associates* [Online]. Available: http://www.parks-associates,com/media/sysconews.htm.

Open Services Gateway Initiative. (2000). *White paper*. [Online]. Available: http://www.osgi.org/about/white-paper.html.

Open Services Gateway Initiative technology demonstrations debut at CES 2000. (2000, January 6). *Press release*. [Online]. Available: http://www.osgi.org/news/press/pressrel010600.htm.

Parks Associates releases its Residential Gateway Report. (1999). *TechMall*. [Online]. Available: http://www.tech-mall.com/techdocs/TS990804-2.html.

PC penetration, Internet access drive in-home networks. (1999). *Parks Associates*. [Online]. Available: http://www.parksassociates.com/media/secsales.htm.

Product solutions. (2000). *Cisco Systems*. [Online]. Available: http://www.cisco.com/warp/public/779/consumer/products/.

Residential gateway market set to explode. (1999). *Cahners In-Stat Group*. [Online]. Available: http://www.instat.com/pr/1999/In9909hn_pr.htm.

Scherf, K. (1999). Advanced wiring delivers future technology today. *Parks Associates*. [Online]. Available: http://www.parksassociates.com/media/kurtec.htm.

Shimogori, M. (2000, February 17). Sony, Intel eye home network alliance. *ZDNet AnchorDesk*. [Online]. Available: http://www.zdnet.com/anchordesk/bcenter/bcenter_352.html.

SmartAMERICA parent/development partner IHG debuts first commercially available intelligent home gateway. (1999, November 18). *Home toys news release*. [Online]. Available: http://www.hometoys.com/releases/dec99/smartamerica02.htm.

Spangler, T. (1999, September 16). Wireless group touts "Wi-Fi" networks. *Inter@ctive Week*. [Online]. Available: http://www.zdnet.cm/intweek.

Strother, N. (2000, February 23). Home networking goes mainstream. *ZDNet AnchorDesk*. [Online]. Available: http://www.zdent.cim/anchordesk/story/story_4496.html.

Sullivan, B. (1999, May 4). The home networking e-boxing match. *ZDNet Tech News*. [Online]. Available: http://www.zdnet.com/zdnn/stories/news/0,4586,2252765,00.html.

Thornton, C. (2000, April). Wild wired home. *PC World.com*. [Online]. Available: http://www.pcworld.com:80/shared/printable_articles/0,1440,15392,00.html.

WECA. (2000). *Wireless Ethernet Compatibility Alliance*. [Online]. Available: http://www.weca.net.

Weil, N. (1999, November 18). Picking the right home net. *PC World*. [Online]. Available; http://www1.pcworld.com/pcwtoday/article/0,1510,13773,00.html.

What is a residential gateway? (2000). *PointClark Networks*. [Online]. Available: http://www.pointclark.net/description.html.

Whirlpool takes active role in Open Services Gateway Initiative. (1999, December 20). *Press release*. [Online]. Available: http://www.osgi.org/news/press/pressrel122099.htm.

Wong, W. (1999, April 20). IBM sings its own home networking tune. *CNET News.com*. [Online]. Available: http://www.cnet.com/category/0-1004-200-341462.html.

Websites

www.2wire.com
www.cisco.com
www.bluetooth.com
www.enikia.com
www.homedirector.com
www.homerf.org
www.hpna.org
www.intellogis.com
www.inari.com
www.osgi.org
www.pointclark.net

20

Satellite Communications

Carolyn A. Lin, Ph.D.*

Artificial satellites have been serving the world's population in nearly every aspect of the cultural, economic, political, scientific, and military life during the past three decades. Since the successful launch of the very first communication satellite—Intelsat I—in 1965, the satellite was envisioned as the ultimate vehicle for linking the world together into a global village. As we begin the 21st century, that vision is being augmented, as companion communication technologies in Internet, digital voice, data, and video transmission—as well as networking systems—are evolving rapidly.

In a multimedia, multichannel communication environment, where wired and wireless communication technologies compete against each other for the finite pool of communication consumers, satellite technology as a broadband and global coverage communication delivery system has been able to maintain its favorable position in the marketplace. This is because satellite technology is uniquely suited for delivering a wide variety of communication signals to accomplish a great number of different communication tasks. There are three basic categories of non-military satellite services: fixed satellite services, mobile satellite systems, and scientific research satellites (commercial and noncommercial).

- *Fixed satellite services* handle hundreds of millions of voice, data, and video transmission tasks across all continents between fixed points on the earth's surface.
- *Mobile satellite systems* help connect remote regions, vehicles, ships, and aircraft to other parts of the world and/or other mobile or stationary communication units, in addition to serving as navigation systems.
- *Scientific research satellites* provide meteorological information, land survey data (e.g., remote sensing), and other different scientific research applications such as earth science and atmospheric research.

* Professor & Coordinator of Multimedia Advertising Certificate Program, Department of Communication, Cleveland State University (Cleveland, Ohio).

The satellite's functional versatility is imbedded within its technical components and operational characteristics. Looking at the "anatomy" of a satellite, one discovers two modules (Miller, Vucetic & Berry, 1993). First is the *spacecraft bus* or service module, consisting of five subsystems:

(1) A structural subsystem provides the mechanical base structure, shields the satellite from extreme temperature changes and micro-meteorite damage, and controls the satellite's spin function.

(2) The telemetry subsystem monitors the on-board equipment operations, transmits equipment operation data to the earth control station, and receives the earth control station's commands to perform equipment operation adjustments.

(3) The power subsystem comprises solar panels and back-up batteries that generate power when the satellite passes into the earth's shadow.

(4) The thermal control subsystem helps protect electronic equipment from extreme temperatures due to the intense sunlight or the lack of sun exposure on different sides of the satellite's body.

(5) The altitude and orbit control subsystem is comprised of small rocket thrusters that keep the satellite in the correct orbital position and antennas pointing in the right directions.

The second major component is the *communications payload*, which is made up of transponders. A transponder is capable of:

- Receiving uplinked radio signals from earth satellite transmission stations (antennas).
- Amplifying radio signals received.
- Sorting the input signals and directing the output signals through input/output signal multiplexers to the proper downlink antennas for retransmission to earth satellite receiving stations (antennas).

The satellite's operational characteristics are literally "out of this world" and reach deep into outer space. Satellites are launched into orbit via a space shuttle (a reusable launch vehicle) or a rocket launcher (a non-reusable launch vehicle). There are two basic types of orbits: geostationary and non-geostationary orbits. The geostationary (or geosynchronous) orbit refers to a circular or elliptical orbit incline approximately 22,300 miles (36,000 km) above the earth's equator (Jansky & Jeruchim, 1983). Satellites that are launched into this type of orbit at that altitude typically travel around the earth at the same rate that the earth rotates on its axis. Hence, the satellite appears to be "stationary" in its orbital position and in the "line-of-sight" of an earth station, with varying orbital shapes, within a 24-hour period. Therefore, geostationary orbits are most useful for those communication needs that demand no interruption around the clock, such as telephone calls and television signals (see Figure 20.1).

Non-geostationary (or non-geosynchronous) orbits are typically located either above or below the typical altitude of a geostationary orbit. Satellites that are launched into a higher orbit travel at slower speeds than the earth's rotation rate; thus, they can move past the earth's horizon to appear for a limited time in the line-of-sight of an earth station. Those satellites that are launched into lower orbits travel at higher speeds than the earth's rotation rate, so they race past the earth to appear more than once every 24 hours in the line-of-sight of an earth station. Non-geostationary orbits serve those communication needs that do not require 24-hour input and output, such as scientific land survey data.

Figure 20.1
Satellite Orbits

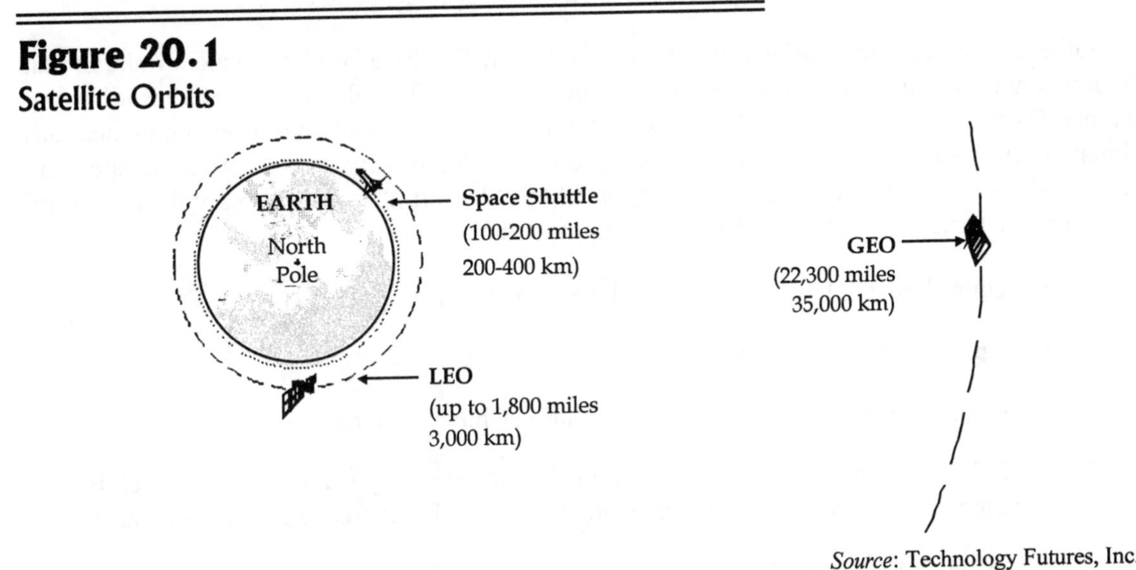

Source: Technology Futures, Inc.

Background

During the early years of satellite technology development, the most notable event was the identification of the geostationary orbit by the English engineer Arthur C. Clarke in 1945 and the successful launch of the first artificial satellite, Sputnik, by the Soviet Union on October 4, 1957. On January 31, 1958, the first U.S. satellite, Explorer 1, was launched. In July 1958, Congress passed the National Aeronautics and Space Act, which established the National Aeronautics and Space Administration (NASA) as the civilian arm for U.S. space research and development for peaceful purposes. In August 1960, NASA launched its first communication satellite, Echo I—an inflatable metallic space balloon or "passive satellite," designed only to "reflect" radio signals (like a mirror).

The next period of satellite technology development was marked by an effort to launch "active satellites" that could receive an uplink signal, amplify it, and retransmit it as a downlink signal to an earth station. In July 1962, AT&T successfully launched the first U.S. active satellite, Telstar I. In February 1963, Hughes Aircraft launched the first U.S. geostationary satellite, Syncom, under a contract with NASA.

In tandem with these experimental developments was the creation, via the passage of the Communications Satellite Act of 1962, of the Communication Satellite Corporation (Comsat), a privately owned enterprise overseen by the U.S. government. An international satellite consortium, the International Telecommunications Satellite Organization (Intelsat), was formed in August

1964. Intelsat is a non-profit cooperative organization that coordinates and markets satellite services, while member states are both investors and profit sharers of the revenues. The founding member nations, including the United States, 15 Western European nations, Australia, Canada, and Japan, agreed that Comsat would manage the consortium (Hudson, 1990). The era of global communication satellite services began with the launching of Intelsat-I into geostationary orbit in 1965. Intelsat-I was capable of transmitting 240 simultaneous telephone calls.

Other important issues during this period involved satellite transmission frequency and orbital deployment allocation policy. Both issues were decided at the World Administrative Radio Conference (WARC) in 1963. WARC is a technical arm of the International Telecommunications Union (ITU), an international technical organization founded in 1865 to regulate radio spectrum use and allocation. Given that a geostationary satellite's "footprint" can cover more than one-third of the earth's surface, the ITU divides the world into three regions:

- Region 1 covers Europe, Africa, and the former Soviet Union.
- Region 2 encompasses the Americas.
- Region 3 spans Australia and Asia, including China and Japan.

The ITU also designated radio spectrum into different frequency bands, each of which is used for certain voice, data, and/or video communication services. These frequency bands include:

- L-band (0.5 GHz to 1.7 GHz) is used for digital audio broadcast, personal communication services, global positioning systems, and non-geostationary and business communication services.
- C-band (4/6 GHz) is used for telephone signals, broadcast and cable TV signals, and business communication services.
- Ku-band (11 GHz to 12/14 GHz) is used for direct broadcast TV, telephone signals, and business communication services.
- Ka-band (17 GHz to 31 GHz) is used for direct broadcast satellite TV and business communication services.

With the passage of time, the satellite's status as the most efficient technology to connect the world seems firmly established. However, an array of other challenges remains concerning issues related to national sovereignty and the orbital resource as a shared and limited international commodity. At the 1979 WARC, intense confrontations on the issue of "efficient use" versus "equitable access" to the limited geostationary orbital positions by all ITU member nations took place, pitting Western nations that owned and operated satellite services against the developing nations that did not. For the developed nations, equitable access to orbital positions suggests a waste of resources, as many developing nations lack the economic means, technical know-how, and practical need to own and operate their own individual satellite services. They consider an "efficient use" (first-come/first-served) system to be more technically *and* economically feasible. By contrast, developing nations consider equitable access a must if they intend to protect their rights to launch satellites into orbital positions in the future, before developed nations exhaust the orbital slots. This dispute would carry over to the 1990s before it was fully resolved.

International competition for the share of satellite communication markets was not limited to the division between developed and developing nations. In response to U.S. domination of international satellite services, Western European nations established their own transnational version of NASA in 1964. It was named the European Space Agency in 1973, and was responsible for the research and development of the Ariane satellites and launch programs. Modeled after the Intelsat system, Eutelsat was formed in 1977 to service its Western European member nations with the Ariane communication satellites launched. A parallel development was the establishment of the Intersputnik satellite consortium by the former Soviet Union, which served all of the Eastern European Communist bloc, the former Soviet Union, Mongolia, and Cuba. Other developing nations owned and operated their own national satellite services during this early period as well, including Indonesia's Palap, which was built and launched by the United States in 1976.

The success of Intelsat as an international non-profit organization providing satellite communication services was repeated in another international satellite consortium—the International Maritime Satellite System (Inmarsat)—founded in 1979. Inmarsat now provides communication support for more than 5,000 ships and offshore drilling platforms using C-band and L-band frequencies. Inmarsat is also used for land/mobile communications for emergency relief work during natural disasters such as earthquakes and floods.

Recent Developments

During the 1980s, the satellite industry experienced steady growth in users, types of services offered, and launches of national and regional satellite systems. By the mid-1990s, satellite technology had matured to become an integral part of a global communications network in both technical and commercial respects. The following discussion will highlight a few of these important landmark events that set the stage for satellite communication in 2000 and beyond.

Due to a new allocation of frequencies in the Ku-band and Ka-band, direct-to-home (DTH) satellite broadcasts or direct broadcast satellite (DBS) services became more economical and technically viable, as higher-power satellites can broadcast from these frequencies to smaller size earth receiving stations (or dishes). This development helped stimulate the television-receive-only (TVRO) dish industry and ignite an era of DBS services around the world (see Chapter 5). The best success story of this particular type of satellite service can be found in Europe and Japan. For the Europeans, direct-to-home satellite services are deemed an economical means to both transport and share television programs among the various nations that are covered by the footprint of a single satellite, as the pace of cable television development has been slow due to lack of privatization in national television systems. The Japanese used this service to overcome poor television reception, owing to mountainous terrain throughout the island nation.

As the economic benefits of satellite communication became apparent for those nations that owned and operated national satellite services during the 1970s, many countries followed suit in the 1980s to become active players as well. In particular, developing countries saw satellite services as a relatively cost-efficient means to achieve indigenous economic, educational, and social developmental goals in the vastly-underdeveloped regions outside of their selected urban centers. Following the example of Indonesia (Palap, 1976), India (Insat, 1983), Brazil (Brasilsat, 1985), and Mexico (Morelos, 1985) were among the countries that became national satellite owners. Arab nations also formed their own satellite communication consortium, Arabsat (1985), to serve their domestic and regional needs. Asiasat (1989), a joint venture between Rupert Murdoch's News Corp. and the Chinese (now solely owned by News Corp.), serving both China and southeast Asia, carries Chinese language television programs targeting the vast Chinese population in the region. In the United States, direct-to-home satellite services have traditionally served remote and rural areas not reached by broadcast and cable television signals. The launch of DirecTV by Hughes Communications in 1994 created an alternative to cable television for urban and suburban consumers as well.

This growth in the number of satellite services available in the marketplace also provided the impetus for free market competition in both domestic and international satellite communications markets. Intelsat has been a near monopoly for most of its existence. It did endorse two regional consortia, Eutelsat and Arabsat, as well as Inmarsat, as "acceptable" competitors because they do not present any significant economic harm to Intelsat's status as a global service provider. This monopoly status was successfully challenged by the Federal Communications Commission (FCC) approval of five private satellite systems (including PamAmSat and Orion) to enter the international satellite communication service market in 1985. As a result, international competition for global satellite services began when both PamAmSat and Orion launched their satellite services in 1988 (Reese, 1990). Subsequently, the FCC adopted two separate policies that permit all U.S. satellite systems to offer both domestic and international satellite services and allow private satellite networks authorized by the World Trade Organization (WTO) to provide services to the U.S. market in 1996 and 1997, respectively.

Another aspect of this unstoppable growth trend in the satellite industry involves the expansion of satellite business services. The first satellite system launched for business services was Satellite Business Systems (SBS) in 1980. Its purpose was to provide corporations with high-speed transmission of conventional voice and data communication as well as videoconferences to bypass the public switched network. As earth uplink and downlink station equipment became more affordable, a flurry of corporate satellite communication networks began at companies such as Sears, Wal-Mart, K-Mart, and Ford. These firms leased satellite transponder time from satellite business service providers to perform such tasks as videoconferencing, data relay, inventory updates, and credit information verification (at the point-of-sale) using very small aperture terminals (VSATs)—earth stations (or receiving antennas) that range from 3 feet to 12 feet in diameter. In many areas, large and small satellite earth stations were clustered in "teleports," allowing a concentration of satellite services and a place to interconnect satellite uplinks and downlinks with terrestrial networks.

As the market grows, digital transmission techniques have become increasingly popular because they allow maximum transmission speed and channel capacity. For instance, voice and video signals can be digitized and transmitted as compressed signals to maximize bandwidth efficiency. Internet via satellite is another fast-growing service area (including streaming media services), targeting corporate users by all major satellite services in the United States and Europe.

Other digital transmission methods, such as time division multiple access (TDMA), demand assigned multiple access (DAMA), and code division multiple access (CDMA), can achieve similar efficiency objectives. TDMA is a transmission method that "assigns" each individual earth station a specific "time slot" to uplink and downlink its signal (for example, two seconds). These individual time slots are arranged in sequential order for all earth stations involved (e.g., earth station numbers 1 through 50). Since such time allotment and sequential order is repeated, over time, all stations will be able to complete their signal transmission in these repeated sequential time segments using the same frequency. DAMA is an even more efficient or "intelligent" method. This is because, in addition to the ability to designate time slots for individual earth stations to transmit their signals in a sequential order, this method has the capability to make such assignments based on demand instead of an *a priori* arrangement. CDMA utilizes the spread spectrum transmission method to provide better signal security, as it alternates the frequency at which a signal is transmitted and allows for multiple signals to be transmitted at different frequencies at different times.

Yet another important technological advance during this period involves launching satellites that have greatly-expanded payload capability and multiband antennas (e.g., C-band, Ku-band, and/or Ka-band receiving and transmitting antennas). These multiband antennas enable more versatile services for earth stations transmitting signals in different frequency bands for different communication purposes, as well as more precise polarized spot beam coverage by the satellite. Such refined precision in spot beam coverage allows for targeted satellite signals to be received by designated earth receiving stations, which can be smaller and more economical due to the strength of the more focused spot beam signal. This, in turn, allows the satellite to better serve greater numbers of individuals and facilitates growth in the business service sector.

Current Status

As of 1999, Intelsat lists 143 member countries. This organization remains a financially healthy and technologically innovative entity for global communication services. Intelsat's newest satellite package, the Intelsat-VIII series, carries 22,500 two-way voice circuits and three television channels. Employing digital circuit multiplication equipment, the satellite can also include up to 112,500 two-way voice circuits. Private satellite service competitors, such as PamAmSat among other entrants, are all sharing the arena with relative success in an ever-growing market.

The ITU currently has 189 member nations. It successfully resolved the single-most politically contentious issue ever faced by the organization: balancing equitable access to and efficient use of geostationary orbital positions for all member nations. The World Administrative Radio Conference of 1995 presented regulations to ensure the rights of those member nations. These regulations enable those who have actual usage needs to obtain a designated volume of orbit/spectrum resources on an efficient "first-come/first-served" basis through international coordination. Additionally, these regulations also guarantee all nations a predetermined orbital position associated with free and equitable use of a certain amount of frequency spectrum for the future.

Subsequently, during WARC '97, decisions involving frequency allocation (see Table 20.1) for non-geostationary satellite services were resolved.

Non-geostationary satellites are launched into low-earth orbits (LEOs) or medium-earth orbits (MEOs) to provide two-way business satellite services. An example of this type of service is Teledesic, which proposes to deliver a 288-satellite constellation to provide high-speed global voice, data, and video communication, as well as "Internet-in-the-sky" services. Recent market emergence in cellular radio, personal communication services (PCS), and global positioning systems (GPS) represents another reason why non-geostationary mobile satellite services are on the rise.

Table 20.1
Frequency Allocation for Non-Geostationary Satellite Services

Service	Frequency Band	Frequency Allocation
Fixed Satellite Service	Ka-band	18.9/19.3 GHz and 28.7/29.1 GHz
Mobile Satellite Systems	Ka-band	19.3/19.7 GHz and 29.1/29.5 GHz
Low-Speed Mobile Satellite Systems	L-band	454/455 MHz

Source: C. A. Lin

Cellular radio and PCS both provide wireless mobile telephone services. When radio signals are being relayed via a satellite using digital transmission technologies, these signals can reach a wide area network to accommodate global mobile communication needs (Mirabito, 1997). For instance, Globalstar's LEO satellites can transmit a phone signal originated from a wireless phone unit or a fixed unit to a terrestrial gateway, which then retransmits the signal to existing fixed, cellular telephone, or PCS networks around the world.

GPS is a mobile satellite communication service that interfaces with geographic data stored on CD-ROMs. It is a navigational tool used to pinpoint locations in remote regions and disaster areas, and on vehicles, ships, and aircraft (see Figure 20.2). GPS was originally developed for military use, but it is now moving into the consumer market. The increasing interest in launching business satellites in non-geostationary orbits is an outgrowth of strong market demand and increased congestion in the existing services provided through GEO satellites. The industry experienced an almost exponential market growth around the world toward the end of the 20th century. The projected global outlook for this service looks bright beyond 2000.

By contrast, domestic development of direct-to-home satellite services is experiencing only moderate growth. Even though earth receiving satellite dish size can be as small as 18 inches in diameter, a bull market is not on the immediate horizon. The cable TV industry is responding to this competition through improved channel capacity and program lineups. The arrival of digital television broadcasting paints an even murkier picture for the future of this industry. Worldwide, direct-to-home satellite services are nonetheless expanding at a respectable rate, especially in those countries that still have relatively limited national television fare or cable TV systems due to the lack of privatization of government-owned and/or -operated television systems.

Figure 20.2
Three-Dimensional Triangulation

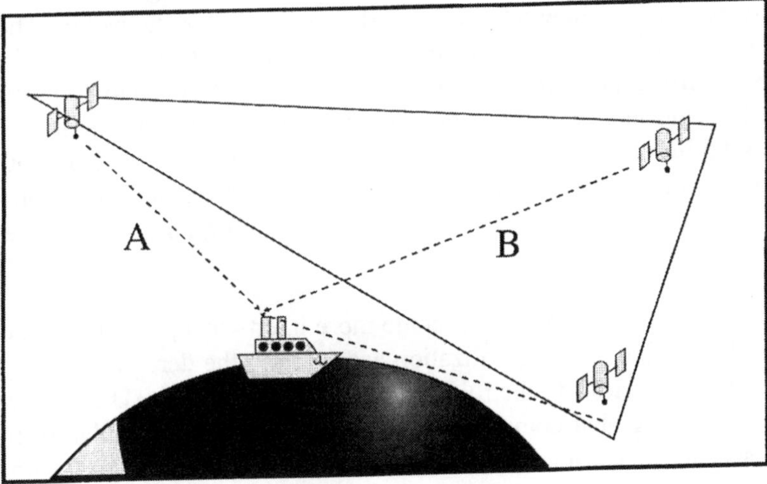

Source: J. Raley

Factors to Watch

Several major developments are expected to take place in the next few years. First and foremost is the continuing trend in deregulation for both national and international satellite services. On the international front, European ministries of Posts, Telephone, & Telegraph (PTT) and their Japanese counterparts, for instance, should speed up their efforts to deregulate government control over the telecom industry and continue privatization efforts. The arrival of the European Union has helped hasten privatization, which includes market mergers as well as shake-ups. More important, the stage has been set for the eventual privatization of Intelsat, the largest global satellite communication service (Cole, 1996), as the U.S. Congress just passed legislation that sets a deadline of April 1, 2001 for the privatization of Intelsat, Inmarsat, and New Skies (Comsat, 2000). Such policies, taken in tandem, should create a vibrant playing field where technological innovations and operational efficiency can help foster a competitive and expanding market.

A related growth event to watch is the Chinese Long March satellite launch program, which is emerging as a major player in the commercial satellite launch industry, on a par with the European Space Agency's Ariane launch program and the Russia-based launch services (including joint-ventures with Germany, France, and Ukraine). Several smaller U.S. commercial launch programs (for example, Boeing's Delta series and Lockheed-Martin's Athena) are also developing, gradually replacing NASA's commercial launch program, which was discontinued after the Challenger explosion in 1986.

The most promising satellite communication sector involves the business applications via mobile satellite systems as well as fixed satellite services. These services, which operate in non-geostationary orbits (i.e., low- or medium-earth orbits), will compete with existing business services operated in geostationary orbits. The advantages of these new services are that they require a smaller and more economical satellite system and launch vehicles, yet they are versatile enough to provide an entire range of data, voice, and video networking services.

Major players entering this market are targeting the wide-open global market for PCS, global positioning systems, satellite Internet access, and wireless virtual networks. These include such well-financed ventures as Loral and Qualcomm's Globalstar (48 satellites), and the much anticipated Craig McCaw/Bill Gates joint venture, Teledesic (ultimately, up to 800 satellites). Nonetheless, the first large-scale service of this kind—Motorola's Iridium (66 satellites)—failed to attract enough mobile satellite phone service subscribers and folded its operation. It was hampered by costly and cumbersome handset units, poor signal reception quality, and high operating costs. Another similar service—ICO Global (12 satellites)—is being absorbed by the up-and-coming Teledesic system.

As the global economy continues to become more intertwined, the ITU is increasingly aware of the rising trend of global trade liberalization accelerating the deregulatory policies in national telecommunications industries. As the role of the WTO grows, the ITU also needs to accommodate the rapid growth of global commerce by making its technical regulatory role a facilitating force instead of an intergovernmental bureaucratic drag. Such an ideological shift is evidenced by the recent Americas Telecom 2000 Conference (held in April 2000), where an advanced interactive multimedia C-band satellite terminal was demonstrated (jointly by France's Alcatel Espace and U.S.'s ATC Teleports). It delivered digital video and Internet traffic in a single data stream (Intelsat, 2000). Internet communication, in many ways, symbolizes the technological freedom of information flow. By the same token, Internet via satellite also comes to showcase the satellite communication industry's entry into an era of free wheeling global market exchange and open competition. Looking inside this "big picture" of an ever more technically connected world beyond 2000, the satellite's unique role as the chosen vehicle to bring forth and link together our global village seems more significant than ever.

Bibliography

Berman, H. (1998, March). LEOs and MEOs: Countdown to launch. *Via Satellite*, 46-53.

Christensen, J. (1998, February). WARC '97 results: The effect on the satellite communications industry. *Via Satellite*, 80-96.

Cole, J. (1996, October 17). Intelsat's privatization beams in rivals' cries of foul. *Wall Street Journal*, B4.

Comsat. (2000, March 10). Comsat applauds congressional passage of satellite reform legislation. *Press Release*, 1-2.

Divine, R. A. (1993). *The Sputnik challenge*. New York: Oxford Press.

Hudson, H. E. (1990). *Communication satellites: Their development and impact*. New York: The Free Press.

Intelsat. (2000, April 9). Intelsat showcases satellite solutions for interactive multimedia applications, rural telephony, and Internet connectivity at Americas Telecom. *Press Release*, 1-2.

Jansky, D. M., & Jeruchim, M. C. (1983). *Communication satellites in the geostationary orbit*. Dedham, MA: Artech House.

Miller, M. J., Vucetic, B., & Berry, L. (1993). *Satellite communications: Mobile and fixed services*. Norwell, MA: Kluwer Academic Publishers.

Mirabito, M. M. A. (1997). Wireless technology and mobile communication. *The new communications technologies*. Boston, MA: Focal Press.

Raley, J. (1994). Global positioning systems. In A. E. Grant (Ed.). *Communication technology update*, 3rd edition. Boston: Focal Press.

Reese, D. W. E. (1990). *Satellite communications: The first quarter century of service*. New York: Wiley Interscience Publications.

21

Distance Learning

John F. Long, Ph.D.*

During the 1990s and into the 21st century, most forms of communication technology have undergone the conversion from analog to digital platforms. The distance learning community has been a direct beneficiary of this metamorphosis. The fundamental rationales for engaging distance learning—retraining, a non-traditional student clientele, and geographic isolation—remain stable. However, the infrastructure, software, and curricular designs have become increasingly sophisticated and adaptive.

Distance learning systems typically employ an amalgam of communication technologies. The actual system architecture is chosen based upon a number of factors. Costs, geographical location of participants, visual quality requirements, sites supported, interactivity, and pedagogical goals jointly define what parameters must be placed on the system design and which technologies should be used for optimal outcomes. Technically, conferencing technologies are systems that provide a multi-user interface, communication and coordination within the group, shared information space, and the support of a heterogeneous, open environment that integrates existing single-user applications. They can be categorized according to a time/location matrix using the distinction between same time (synchronous) and different times (asynchronous), and between same place (face-to-face) and different places (distributed).

In concert with the articulated social needs for distance education, technological developments have provided a means to accomplish and extend directives articulated by education and business. The rapid rise of digital telecommunications has provided a plethora of delivery systems, collaboration tools, and databases that act as a backbone for contemporary distance education efforts. These resources and technologies are no longer restricted to an elite community and are changing the nature of education and training forever as a result (Capell, 1995). Digitization of resources empowers new models of education and an enhanced instructional experience for the traditional classroom. The varied forms of technology in classroom applications have risen dramatically since 1995.

---

* Professor, Communication Design Department, California State University, Chico (Chico, California).

The component processes that comprise distance education include learning, teaching, communication, and instructional character. Characterization of distance education from the systems perspective is critical. Anything that happens in one part of the system affects other parts of the system (Moore & Kearsley, 1996). Decisions to employ a new technology in a learning system must be grounded in the pedagogical soundness of the overall system. It remains imperative that the learning environment incorporates evolving technologies, as opposed to letting emerging technologies drive the learning system.

Background

Historically, distance education has gone through several phases of development. Over a century ago, correspondence study, using mail as the enabling technology, allowed distance learners to obtain degrees and train for professional endeavors. Many of these home study schools are still operating today. For example, International Correspondence School (ICS) and the American Association for Collegiate Independent Study (AACIS) offer courses for nearly four million students annually.

The second development phase could be characterized as an integrated approach to distance education. These systems used multiple media such as correspondence, radio, television, audiotapes, and telephone conferences. The primary examples of these 1960s to 1970s projects were the University of Wisconsin's AIM Project and the Britain's Open University (White, 1996). The Articulated Instructional Media (AIM) Project hypothesized that learners could acquire skills from broadcast media and, at the same time, derive an "interactive" experience through correspondence or telephone conferencing. The original conference for AIM became the basis for the British Open University (OU). The OU model incorporated the public broadcasting resources of the BBC with correspondence materials, offering degree education to any adult. Since 1971, the OU has served over a million students (Witherspoon, 1997).

The emergence of broadcast media and, later, teleconferencing, represented another phase of distance education. By the end of 1922, 74 colleges offered classes on radio (Barnouw, 1966). Spurred by the Ford Foundation, educational TV broadcasts from universities began in the 1930s and were omnipresent by the 1950s. Collegiate consortia developed in 1961 through implementation of the Midwest Program on Airborne Television Instruction (MPATI). This configuration broadcast educational material to six states from transmitters on DC-6 airplanes (Smith, 1961).

The cause for distance learning took a considerable leap forward in the mid-1960s with the launch of Intelsat I. This modest beginning for satellite technology, offering one television channel or 240 telephone circuits, provided the hope for distance insensitive relays of educational material. During the next decade, several universities experimented with the Applications Technology Satellite (ATS). Because of diverse territories and extreme population dispersion, the Universities of Hawaii and Alaska were two of the first to utilize ATS for educational and health training (Rossman, 1992).

As satellite technology improved, its viability in the educational marketplace surged forward. Business, the military, industry, and institutions of higher learning realized the potential that satellites provided. The U.S. military developed numerous distance education networks using satellites.

Using two-way audio and one-way video, the Air Force Technology Network (ATN) reaches 18,000 students in 69 sites (Moore & Kearsley, 1996). To complement its videoconferencing infrastructure, the military has moved to link up 16 sites for fully-interactive exchange of video, voice, and data.

Business and industry have made substantial use of satellites as well. Nearly half of Fortune 500 companies engage in corporate training using this technology (Irwin, 1992). IBM's Interactive Satellite Educational Network (ISEN) broadcasts courses to over 50 locations. The system is configured as a one-way video/two-way audio system with digital keypads for testing responses. Wang Laboratories uses a similar system for international training, as does Ford Motor Company and Aetna Life and Casualty. Currently, there are over 14,000 receive sites for corporate education (Dominick, Sherman, & Messere, 2000).

Because of expertise in the academic fields and corresponding needs in private industry, a virtual university, similar in concept to the British Open University, was founded in 1985. The National Technological University (NTU) offers advanced degrees by distributing digitally compressed video by satellite. Offering more than 400 courses, its clients are corporations whose employees require advanced training (such as AT&T, GE, and Kodak). Because of its digitally compressed signal, NTU was able to substantially reduce transmission costs and increase channel capacity (Witherspoon, 1997).

Other higher education consortia have developed to defray institutional costs and share expertise. The Adult Learning Satellite Service (ALSS), operated by the Public Broadcasting System (PBS), offers courses to public and private entities also using satellite-delivered compressed video. Through generous funding from Congress, the Star Schools Program initiated regional satellite consortia designed to enhance educational opportunities in grades K-12. Most programs were advanced to educate audiences not served by traditional methods.

The next phase of distance learning technologies uses computer networking and multimedia. The Internet has provided the opportunity to offer courses and complete degree programs directly through a student's computer. Unlike satellite conferencing, the student would not have to attend a remote site to receive instruction and interact with the instructor. Computer networking also provides asynchronous learning alternatives through the use of electronic mail, bulletin boards systems, and the like. Groups such as the Open University, University On-Line, and CalCampus transform regular university classes into digital format. These classes can then be replayed live or delayed through a Web browser. Graphics, digital images, and computer simulations can be added to the presentation (Open University, 2000). Most programs using the Internet have faced technological limitations due to modem speeds. The narrow bandwidth available through the public Internet cannot yet simulate virtual classrooms as in satellite teleconferencing. The emerging Internet 2 may help address these bandwidth challenges.

Other ambitious distance education arrangements came about in the late 1990s. The Western Governors University (WGU) brought together 17 western states to create a virtual university. An interactive "Smart Catalog" and technological nerve center of WGU identifies the media delivery system for each class. The principal means of delivery are satellites using compressed digital video and interactive Web-based designs (Blumenstyk, 1998).

A similar design emerged with the California Virtual University (CVU). In this model, over 300 California public and private higher education institutions would provide access to the expected 500,000 new students during the next decade. At the center of this cooperative is an interactive Internet catalog of technology-mediated distance learning offerings combined with a Website and Intranet tools available to all campus Websites (Design team, 1997). By April 1999, the CVU lay in ruins due to faculty opposition and lack of funding. The WGU, although still operational, is a mere shell of its expectations (Downes, 1999).

Other consortia remain solvent. Serving students, teachers, and adult learners in all 50 states, the Star Schools networks reach over one million students and 5,000 schools annually. For the past 10 years, these networks have used interactive satellite conferencing. In 1999, the STEP-Star network expanded to direct broadcast satellite (DBS) using the Ka-band. This move inexpensively expands the network scope and allows access directly to homes. In addition, current satellite learning programs in grades K-12 and higher integrate Web and Internet resources and create virtual online classrooms that expand the scope of the satellite delivered activities.

Distributed learning environments using convergent technologies are the mainstay of the International University (IU). This virtual university offers business degrees by combining the media of the Internet, videotape, telephone, and print. Instructors and students interact online using a virtual classroom. Starting in 1999, Ka-band satellites began delivering multimedia directly to homes, enabling areas without high-bandwidth connections to access an array of materials and resources.

One of the more significant projects of the 1990s that influenced the scope of distance learning was the creation of Internet 2. In October 1996, President Clinton adopted the central goals as the Next Generation Internet (NGI) Initiative and committed the administration to a system 1,000 times faster than today's. A very elaborate Internet 2 partnership was developed that includes industry, universities, and government. As of mid-2000, over 150 universities were involved in R&D on collaborative projects (Olsen, 2000b).

In order to facilitate development and use of Internet 2 infrastructure, the National Science Foundation (NSF) and private industry have funded various phases for state or regional internetworking. System architectures require a gigaPOP (a network point of presence capable of transmitting speeds exceeding a gigabyte). In contemporary design and experimentation, geographically proximate institutions share the gigaPOPs. These provide for high-bandwidth desktop-to-desktop communication between institutions. One prototype, VITALnet, interconnects Duke, UNC–Chapel Hill, North Carolina State, and an Internet service provider (ISP), MCNC, at 2.4 Gb/s. System advocates recognize its value in distance collaborative learning and telemedicine (NC Giganet, 1997). The California Research and Education Network-2 (CalREN-2) has initiated a venture involving research campuses in two major high-speed clusters or aggregation points (APs) forming distributed gigaPOPs. The APs connect to each other using OC12 lines (high-speed lines capable of transmitting at 655 Mb/s), and OC3 or DS3 lines to regional universities. The OC12 lines then connect APs to Internet 2 and vBNS for national collaborative efforts.

With universal high-bandwidth networking capability on the horizon, the issue of instructional software that would serve as the basis of distributed instruction becomes an issue. The harnessing of information in a comprehensible fashion must be addressed. In groundbreaking experimentation, Sun Microsystems has developed a Distributed Tutored Video Instruction (DTVI). This form of desktop videoconferencing (DVC) allows for "social presence" and interaction among visually present learners. Sun's digital version of DTVI is called "Kansas." This virtual collaborative learning environment allows participants to move about in 2-D space, to make or break connections with others to collaborate, and have independent discussions. The "Kansas" window establishes the instructional feasibility of a network-based virtual space (SunLabs, 2000).

Figure 21.1
"Kansas" DTVI System

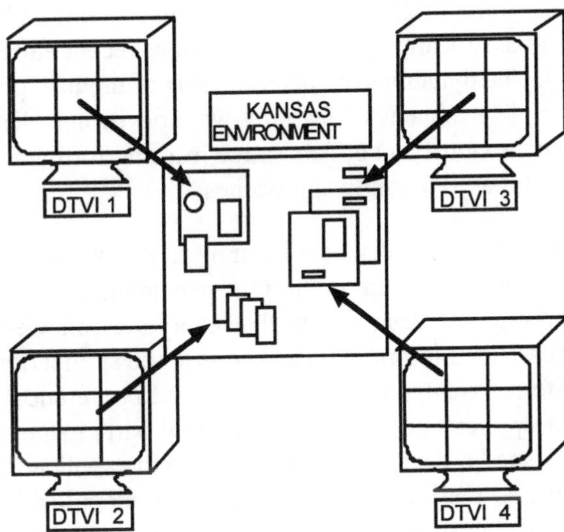

Source: J. F. Long

High-bandwidth access with Internet 2 provides opportunities for a wide variety of collaborative research projects. One potential application is tele-immersion. This permits individuals at different locations to share a single virtual environment. These systems would emulate the "Kansas" classroom environment, using a third dimension of facilitated communication.

Similar to tele-immersion is the virtual laboratory. This climate enables a distributed problem-solving condition wherein researchers interact and direct simulations of virtual projects. Sometimes called a "collaboratorium," these projects allow for simultaneous exploration and manipulation of physical properties from various sites and perspectives. The NSF is planning to support software standards for virtual laboratories in the near future (NGI, 1998).

Recent Developments

Beginning in 1999, there was a continued buildout of high-speed terrestrial networks, a greater involvement of technology manufacturers to provide network solutions, and an overwhelming array of software designed for these learning networks. An increased clientele ranging from business, K-12, higher education, and research consortia have created demands on network providers and software developers to create more sophisticated and virtual environments for teaching and learning. In higher education alone, the number of students enrolled in distance education courses is predicted to jump from 710,000 in 1998 to 2.2 million in 2002. Additionally, data released by the International Data Corporation suggest that 85% of two-year and 84% of four-year colleges will offer distance education programs by 2002 (ITPE, 1999). Colleges are expected to spend $1.2 billion on academic hardware in 2000, up 28% over 1999 (Olsen, 2000a). The distance learning revenue is projected to reach approximately $3 billion a year in the United States by 2005. The system designs in terms of network configuration and software varies according to the pedagogical requirements of the business or institution.

As a result of increasing processor speeds, video compression, and implementation of standards such as H.323, Web-based learning has received more recent attention than any other form of information technology. The dynamics of specific fields of inquiry, performance requirements, and cost factors are typical determinants of learning-ware decisions. Generally, these are categorized as asynchronous, synchronous, and collaborative learning formats. Proprietary software has addressed each of these Web-based learning approaches.

Application sharing is one type of distance learning tool that allows a conference with users who may not have the same application programs. One person launches the application, and it runs simultaneously on all participants' desktops so that everyone can view the application screen simultaneously. All users can input information and control the application using the keyboard and mouse. Files can be transferred easily, and the results of the conference are available to all users immediately. This kind of application is designed to allow participants at multiple sites to view and sometimes control it. Usually, the person who launched the application can lock out other persons from making changes, so the locked-out person sees the application, but cannot control it. Application conferencing is almost never used as a stand-alone technology; rather, it is usually used jointly with one of the other conferencing technologies discussed in this section.

Another application is data conferencing, which is a term used to describe the placement of shared documents or images in an on-screen "shared notebook" or "whiteboard," which allows a document or image to be viewed simultaneously by two or more participants. All participants can then view the document, while making annotations using the drawing or text capabilities of the specific whiteboard program. Most of the shared whiteboard programs use different colors to indicate whose annotations are whose. Some software includes "snapshot" tools that enable the user to capture entire windows or portions of windows and place them on the whiteboard (Center for Technology, 2000).

In a data conference, users can type, draw a chart, or paste images, Website addresses, or anything else onto an electronic whiteboard on the computer screen that other users can see simultaneously on their screens—no matter where they are. A chairman or host typically organizes and controls the conference, displaying the initial image or document. Authorized participants join at the appointed time by clicking a button, which allows the whiteboard's contents to be displayed in a browser on their screens. Initially, the host has the floor, controlling the cursor and what's drawn or placed on the whiteboard. But a participant can request control, typically by choosing a command from a menu. When the current participant is finished, the host passes control to the next person, who can then place or annotate something on the whiteboard.

Whiteboards usually have a toolbar for drawing, adding text, highlighting, or otherwise marking what's being viewed. To change an image or document, its owner typically edits the original document and pastes the result on the whiteboard. Most products also have a file transfer feature that allows the user to copy a file to one or more users' machines without requesting cursor control and interrupting the conference.

Outside of the whiteboard, participants can type comments in a chat window, directing them to an individual or to the group. When the meeting is over, chat text can be saved in a file and distributed as meeting notes. Typing comments back-and-forth isn't always expedient for involved conversations among a large group, but it works surprisingly well for most small groups.

The asynchronous software suites, as a rule, keep track of who is in the class, organizes their threads of discussion, and shares relevant information with group members. This type of "groupware" does not require the instructor or students to be connected at the same time. There are currently about 35 of these products for lease or purchase, each directed to a specific clientele. Some are developed for particular applications in business, industry, and education and exhibit varying degrees of scalability and reliability. The standard products in higher education have been "WEBCT" and "Topclass." Because of aggressive pricing and scalability, Blackboard's "Courseinfo" has become the contemporary groupware of choice (Barone & Luker, 1999).

Alternate versions of groupware support synchronous Web-based environments. These allow students to participate in real time over the network, instantly receiving information that is text-based, audio/visual (video streaming), or a combination. Some products are voice interactive over the Internet; others are instructor-led training systems that enable interaction through polling and text-based chat. At this point, the two most advanced products are "Centra Symposium" and "Lotus Learningspace." Both have scalability to hundreds of users per session and retain a digital replay feature. This category of synchronous software is most appropriate for formal instructor-led sessions with limited, but possible, student interaction. Other interactive synchronous systems have developed within the context of the H.323 Multiuser Control Unit (MCU) protocol. Driven by Microsoft's "Netmeeting," companies such as White Pine with "Meetingpoint" have developed software to facilitate desktop videoconferencing. Because of connectivity issues, only small numbers of users can interact simultaneously. As a result, the utilitarian value of this and similar software is limited to small group data collaboration and two-way audio sessions (Jackson, 2000). Intel, Cisco, and Microsoft are currently engaged in developing compression techniques that will alleviate this drawback.

Other collaborative synchronous environments have emerged without the assistance of proprietary software. Cyber-venues known as MUDs, MOOs, and MUSEs consider educational topics and can provide virtual environments for group discussions in real time. MOOs are object-based, thus permitting the creation of virtual rooms and objects that become permanent elements. Students can also create objects for online learning projects.

Virtual learning environments (VLEs), similar to MUDs and MOOs, have become functional in all levels of education. Elementary school students in Nova Scotia evaluated products in a simulated Quaker Oats Company (Follows, 1999). Students in a North Carolina college are confronted with deductive reasoning problems as they enter a virtual cell. The cell was composed using three technologies: VRML (virtual reality markup language) visualization, Java simulation software, and a text-based MOO server (Slator, 1999). Penn State medical students perform surgery on virtual patients networked to powerful computers that grade their effectiveness (Mangan, 2000).

In Alabama, sophisticated networks link rural areas with medical specialists to "virtually" examine patients (Teleconference, 2000). Similar projects in Arizona using the vBNS network multimedia capabilities allow for interactive consultations with doctors at rural facilities. Researchers in New Mexico collaborate with data from remote telescopes to engage in a wide range of network-based experimentation (3Com, 1999). Literally hundreds of projects at all levels of education are currently engaging the potential of virtual learning environments.

Another development is the application of video streaming to distance learning. Streaming video is a sequence of "moving images" that are sent in compressed form over the Internet and displayed by the viewer as they arrive. Streaming media has become an effective tool for one-way distance education. With streaming video or streaming media, a Web user does not have to wait to download a large file before seeing the video or hearing the sound. Instead, the media is sent in a continuous stream and played as it arrives. The user needs a player, which is a special program that decompresses and sends video data to the display and audio data to speakers. A player can either be an integral part of a browser or downloaded from the software maker's Website.

Streaming video is usually sent from prerecorded video files, but it can be distributed as part of a live broadcast "feed." In a live broadcast, the video signal is converted into a compressed digital signal and transmitted from a special Web server that is able to do multicasting (sending the same file to multiple users at the same time).

Current Status

The major telecommunications infrastructure and software corporations continue to recognize that distance or distributed learning systems are an important market segment. The Cisco IP/TV product line is a comprehensive network video streaming solution for businesses and universities. Cisco's video streaming applications, in concert with Microsoft's Windows Media Technologies, create an "anytime, anywhere" environment directly to the desktop (Cisco, 2000). In addition, the Cisco IP/VC MCU allows multiple participants in multiple locations to attend the same meeting with full real-time interactivity.

3Com has formed alliances with businesses and universities internationally to provide enterprise network solutions. High-speed ATM distributed learning networks are now deployed in Mexico to deliver telemedicine applications to media students (3Com, 2000). At Suffolk University, 3Com's 1,000M Gb/s Ethernet network provides the most advanced legal research and training system yet devised (3Com, 2000). In an effort to support growing needs for bandwidth-intensive academic application, the University of Pennsylvania is now collaborating with 3Com. The major purpose of this venture is to access Internet 2 for reasons such as genetic modeling, biostatistics, and other medical research designs (3Com, 2000). 3Com is currently engaged in similar network solutions for many institutions involved in Internet 2 connections.

The issue of satellite versus terrestrial-based distance learning still exists. The high-bandwidth buildout is not yet universal. Asian, Latin American, and European populations have increasing distributed learning needs, many of which are met through a synthesis of satellite and IP (Internet protocol) connections (Puetz, 2000). Fiber cables, however, continue to be put in place across the Atlantic and Pacific to meet the demand of distance learners. By early 2000, the total transatlantic bandwidth was 3 terabits per second, compared to just 100 Mb/s one year earlier. By 2001, the rate should exceed 6 Tb/s (Foley, 2000). The proliferation of international fiber connection certainly does not mean an end to the satellite industry. Financing and regulation will hamper fiber buildout in underdeveloped countries. Moreover, the entertainment niche for DBS will only increase its market viability.

Factors To Watch

The high-bandwidth infrastructure spurred by Internet 2 development will continue to engage all constituents of the distance learning community. Software standards enabling virtual classrooms, collaborative research, and tele-immersion will likely develop to provide more opportunities for sophisticated learning environments. With transoceanic fiber buildout, the long-term impact will likely be international in scope.

The issue of wireless broadband will definitely receive considerable attention. With the recent FCC ruling, the use of ITFS for two-way services is now a reality. These fixed broadband wireless solutions will provide the necessary flexibility to geographic locations without the recurring costs of T1 lines. Spike Technologies, MCI WorldCom, and Lucent Technologies are some of the major players in the "anytime, anywhere" distance education continuum (Spike Technology, 2000).

As the demand for distance learning grows, the need for better, faster, and more advanced distance learning technologies will continue. Advances in streaming media, broadband networks, wireless technology, and the Internet will spur further development in distance learning technologies. These technologies provide the hope that more people can obtain the education they need and desire, regardless of place and time.

Bibliography

3Com. (1999). *University of New Mexico partners with 3COM to create ultra-fast ATM backbone.* [Online]. Available: http://education.3com.com/internet2/projects.html.

3Com. (2000a). *UNAM set up the largest ATM network in Mexico.* [Online]. Available: http://education.3com.com/internet2/upennrel.html.

3COM. (2000b). *Suffolk University heralds a new era in legal training.* [Online]. Available: http://www.3com.com/news/releases/pr00/feb.html.

Barnouw, E. (1966). A *Tower in Babel: A history of broadcasting in the United States*. New York: Oxford Press.

Barone, C., & Luker, M. (1999). *The role of advanced networks in the education of the future.* [Online]. Available: http://www.educause.edu/ir/library/erm9968.html.

Blumenstyk, G. (1998, February 6). Western Governors U. takes shape as a new model for higher education. *Chronicle of Higher Education, 44,* (8), 21-24.

Capell, P. (1995). *Report on distance learning technologies.* Pittsburgh: Carnegie Mellon Software Engineering Institute.

Center for Technology in Government Universal Interface Project. (2000). *Group collaboration: Synchronous groupware.* [Online]. Available: http://www.ctg.albany.edu/projects/inettb/univ/Reports/CH4/grpware.html.

Cisco. (2000). *Cisco IP/VC quick product overview.* [Online]. Available: http://www.cisco.com/warp/public/cisco/mkt/video.html.

Design team of the California Virtual University. (1997). *California Academic Plan.* California Virtual University, Draft 3.0.

Dominick, J., Sherman, B., & Messere, F. G. (2000). *Broadcasting, cable, the Internet, and beyond: An introduction to modern electronic media,* 4th Ed. New York: McGraw-Hill.

Downes, S. (1999). *What happened at California Virtual University.* [Online]. Available: http://www.atl.ualberta.ca/downes/threads/column041499.html.

Foley, T. (2000). Facing the fiber threat. *Via Satellite, 15,* (1), *42-48.*

Follows, S. (1999). Virtual learning environments. *T.H.E. Journal, 27,* (4), 100.

Geary, S. (1997). Educators take a trip on the Internet. *Teleconference, 16,* (8), 21.

Irwin, S. (1992). *The business television directory.* Washington, DC: Warren Publishing.

ITPE. (1999). *Distance education students expected to hit 2 million by 2002.* Higher Education Washington.

Jackson, R. (2000). *An overview of Web-based learning.* [Online]. Available: http://www.outreach.utk.edu/weblearning/#Personal Collaborative Environments.html.

Krebs, A. (1997, November). Star Schools: Approaching a decade of accomplishments. *Via Satellite, 12,* (11), 56-65.

Mangan, K. (2000). *Teaching surgery without a patient.* [Online]. Available: http://www.chronicle.com/weekly/v46/i25/25a09901.html.

Moore, M., & Kearsley, G. (1996). *Distance education: A systems view.* Belmont, CA: Wadsworth.

NC Giganet. (1997). *Background information.* [Online]. Available: http://www.ncgni.org/background.html.

NGI. (1998). *Netamorphosis demonstration.* [Online]. Available: http://www.ngi.gov/events/netamorphosis/press.html.

Olsen, F. (2000a). *Colleges are on an I. T. spending spree.* [Online]. Available: http://www.chronicle.com/free/2000/03/2000031701t.html.

Olsen, F. (2000b). *Internet 2 aims to build digital-video network for higher education.* [Online]. Available: http://www.chronicle.com/free/v46/i33/33a04901.html.

Open University. (1997). [Online]. Available: http:www.online.edu.

Puetz, J. (2000). The Internet epoch: Not just idle bandwidth. *Satellite Communications, 24,* (1), 22-27.

Rossman, P. (1992). *The emerging worldwide electronic university.* Westport, CT: Greenwood Press.

Slator, B. (1999). Virtual environments for education at NDSU. *Proceedings from the World Conference on Educational Media, Hypermedia, and Telecommunications.* Seattle, WA.

Smith, M. (1961). *Using television in the classroom.* New York: McGraw-Hill.

Spike Technology. (2000). *Spike Technology to discuss developing broadband market.* [Online]. Available: http://www.spike.com/press/prelease/press5.html.

SunLabs. (2000). *The Kansas project.* [Online]. Available: http://www.sunlabs.com/researcg/distancelearning/kansas.html.

Teleconference. (2000). *University of South Alabama and BellSouth team up.* [Online]. Available: http://www.teleconferencemagazine.com/2000/janfeb.html.

vBNS. (1998). *vBNS University case studies.* [Online]. Available: http://www.vbns.net/press/case.

Wedenmeyer, C. (1981). *Learning at the back door: Reflections on non-traditional learning in the lifespan.* Madison, WI: University of Wisconsin Press.

White, J. (1996). Britain's open road. *The Distance Educator, 2,* (2), 8-11.

Witherspoon, J. (1997). *Distance education: A planner's casebook.* Boulder, CO: Western Interstate Commission for Higher Education.

22

Wireless Telephony*

Shubho Ghosh**

The wireless telephone industry has enjoyed continued and unprecedented growth in the past decade. This trend is indicative of its future potential in a market clearly on a high-growth trajectory. The industry's growth over the past 20 years has resulted in a market capitalization of over $1.7 trillion in the wireless industry as a whole. This figure translates to a compound annual growth rate of 13%, indicative of how dynamic this industry is, especially when compared to aggregate economic growth over the same period (Fertig, et al., 1999).

Wireless telephone technology can be divided into three components: cellular, PCS (personal communications services), and satellite. The sections below outline the specifics of each and their role in the future growth of the industry. Indeed, wireless telephone technology is here to stay, and the one thing we can be sure of is its unprecedented development as a medium that will transform, forever, the way in which we talk and do business with each other.

Numerous factors have contributed to the dramatic increase in wireless telecommunications. Development of new digital technologies and the allocation of additional spectrum have facilitated wireless growth, but four additional fundamental factors merit further elaboration:

(1) There has been an increase in the "mobile" population of developed countries as a whole. Such a population has a natural propensity to demand wireless services as a convenient way of communicating while away from the office or home.

(2) Economic factors have begun to favor the growth of the wireless sector. Landlines require a higher initial investment and lead-time as compared with the cheaper and quicker wireless alternative. The cost of connecting a basic copper wire to a telephone subscriber often exceeds $2,000 for a company, and delays (bureaucratic and otherwise) are common even in mature economies (Housel & Skopec, 2000). In contrast, new wireless connections can

* Portions excerpted from R. L. Hodges, *Telecommunications Access Technologies: Overview and Competitive Assessment* (Austin, TX: Technology Futures, Inc., 1999).

** Information & Operations Management, The Marshall School of Business, University of Southern California (Los Angeles, California).

be set up almost instantaneously, and without the investment and deployment issues involved in physically connecting a phone to the local loop and the local telephone exchange.

(3) Global deregulation has facilitated development of wireless telecommunications. Regulators in many developing countries—for reasons outlined above and with the aim of achieving connectivity as quickly as possible—favor wireless deployment. Added to this is the perceived value of wireless devices: consumers who were dependent on national phone companies in developing countries perceive wireless systems as unique, high-value services (Housel & Skopec, 2000).

(4) The promise of the Internet in delivering information anytime, anywhere through wireless connections from phones and laptops has seen the stirrings of a major trend in the "portable information" market. All indicators point to the fact that the wireless industry will see phenomenal growth in the years to come.

Background

The wireless industry has come a long way since Marconi transmitted the first wireless telegraph message in 1899. The contemporary background of the industry was shaped between 1984 and 1995, when the U.S. government created a duopoly consisting of one local exchange operating company and another independent wireless operator in each regional market. This decision resulted in similar service offerings and prices from both players in a region, and high earnings for the players in return. Once viewed as a luxury, wireless telephony moved down-market through the 1990s, and demand increased substantially over the years. Penetration rates increased from 6% of the U.S. population in 1993 to 18% in 1997. According to the Cellular Telecommunications Industry Association (CTIA), at the end of 1999 there were 86 million U.S. cellular subscribers (including PCS) or 31% of the population. Figure 22.1 shows historical and forecast trends of cellular subscribers as a percentage of the U.S. population. By the end of 2000, the United States will cross the 100-million subscriber threshold, and 50% penetration should be reached by 2003. The United States is not the leader in this respect. Finland, for example, has already reached 60% adoption (Hodges, 1999).

Growth rates, however, declined from 30% in 1996 to a projected rate of between 7% and 12% by 2001. In response to declining revenues per customer, wireless providers undertook process reengineering and operational efficiency measures, while simultaneously increasing prices for low-use segments (Arnold, et al., 1998). Revenue growth, as a result, continues to be positive and unabated.

The 1990s were the age of cellular technology. However, PCS is expected to become the dominant technology in the first half of the current decade, while analysts predict that, beyond 2010, there is a possibility of satellite or even a new technology (third-generation) penetrating and eventually dominating the wireless market. Thus, it can be assumed that wireless will remain the realm of cellular and PCS technologies in the near future, but satellite and third-generation technologies will become a significant force in telecommunications in the years beyond 2010.

Exhibit 22.1
U.S. Cellular/PCS Subscribers—Percentage of U.S. Population

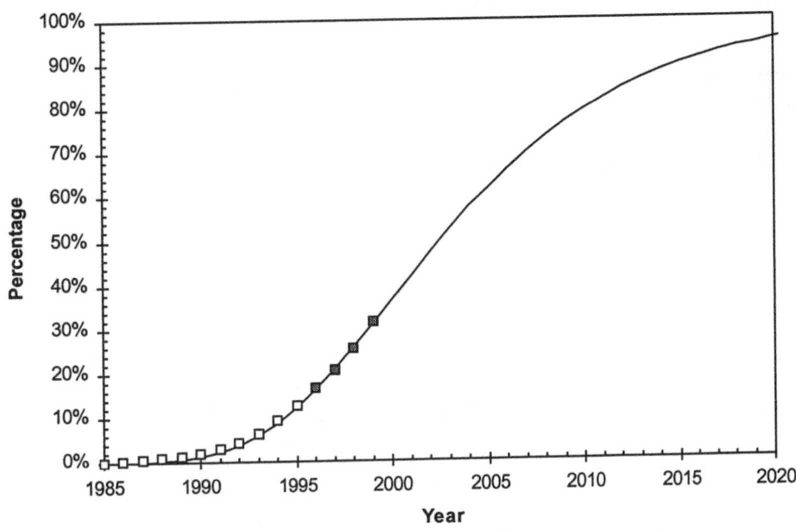

Source: Technology Futures, Inc.

Cellular, PCS, and Satellite

In 1971, the Federal Communications Commission (FCC) opened the way for cellular by allocating spectrum in the 800 MHz to 900 MHz band. Seven years later, the industry initiated trial service, and, in 1980, the FCC settled operational issues and set technical standards. Finally, in 1983, the first commercial mobile cellular telephone service began. Far exceeding traditional mobile service in capacity, quality, and cost, cellular adoption has been phenomenal, with U.S. subscribership reaching one million within three years. According to the CTIA, U.S. subscribership surpassed 68 million in January 1999 (Hodges, 1999).

The term cellular is key to the operational concept of these communications systems. Contiguous cells for radio communication are established within a given area so that the frequencies available for mobile communication can be re-used repeatedly as the mobile unit moves through the clustered cells. Typically, none of the cells adjacent to any given cell use the same frequencies, but cells one cell or more from any given cell may "re-use" that cell's frequencies. This type of cellular pattern is shown in Figure 22.2. Actual cell boundaries are not sharply defined as indicated by the diagram, but are established by the gradual weakening of power received by the mobile unit to the point that power from an adjacent cell is greater. Theoretically, there is no limit to the overall size of a cell cluster or to the number of cells that can be included in such a cluster.

Each cell is served by a base station called a "cell-site" that communicates by radio transmission with each mobile telephone while it is within the area served by that cell-site. This is the first step in the access of the mobile telephone to the telecommunications network, as shown in Figure

22.3. The mobile telephone unit signals the cell-site concerning the connection desired, just as the landline telephone signals the central office or remote switching unit (RSU). The cell-site locates the mobile unit, sets up and supervises the call, monitors signal strength, and terminates the call. Overall control of the cluster of cell-sites is accomplished by a mobile telephone switching office (MTSO), which is equivalent to a telephone company central office. The MTSO receives the signal from the cell-site and makes the switching connections.

Figure 22.2

Diagrammatic Cellular Cluster Frequency Pattern

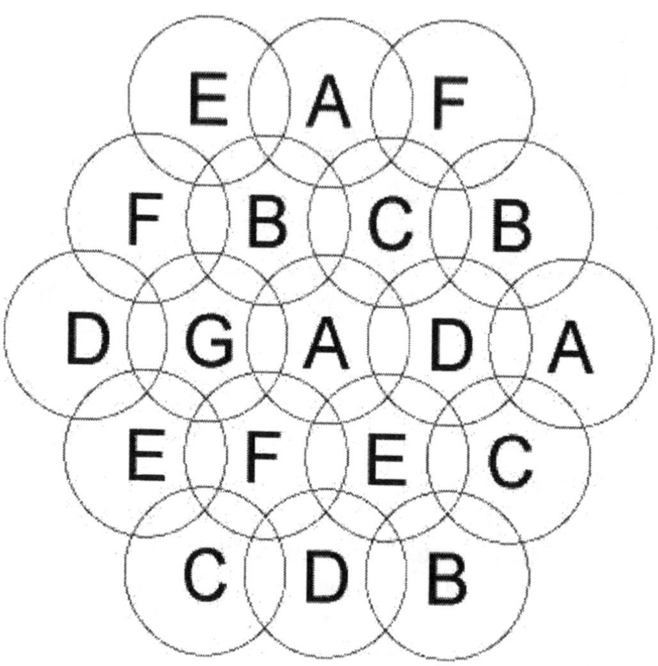

Source: Technology Futures, Inc.

Access to telephones connected to the landline network is by trunk connections between the MTSO and the local telephone company's central office. This architecture is also shown in Figure 22.3. As the mobile unit moves from its original calling point toward other cell-sites in the cell cluster, the initial cell-site reports to the MTSO that the signal level being received from the mobile unit is dropping below an acceptable level. The MTSO then connects the next cell-site to the mobile unit. Transportable and portable cellular phones operate in the same manner and at the same frequencies as mobile units. The only significant differences are in unit size, power, and handoff needs.

MTSO switches are designed to provide all the basic switching functions and features that characterize the telephone company central office switching equipment, including signaling, data transmission, call recording, call forwarding, no-answer transfer, call waiting, three-party conferencing, and many others. These MTSO switches are technologically indistinguishable in form, function, and architecture from their telephone company central office equivalents and, with appropriate interface equipment, may be used interchangeably.

The U.S. Advanced Mobile Phone System (AMPS), as described above, is an analog transmission cellular system with digital switching. This system is now in the process of being converted to digital transmission capabilities, which will include time division multiple access (TDMA) and code division multiple access (CDMA) systems. TDMA is the digital cellular standard (IS-54) of the Telecommunications Industry Association (TIA). It calls for the division of an existing 30 KHz radio channel into size time slots yielding three equivalent voice channels. In theory, TDMA can be implemented one channel at a time. CDMA is a direct spread-spectrum technique developed by Qualcomm and supported as an interim standard by the TIA. Their system calls for a channel bandwidth 1.25 MHz wide, or forty-two 30-KHz channels.

Figure 22.3
Cellular Telephone Network Architecture

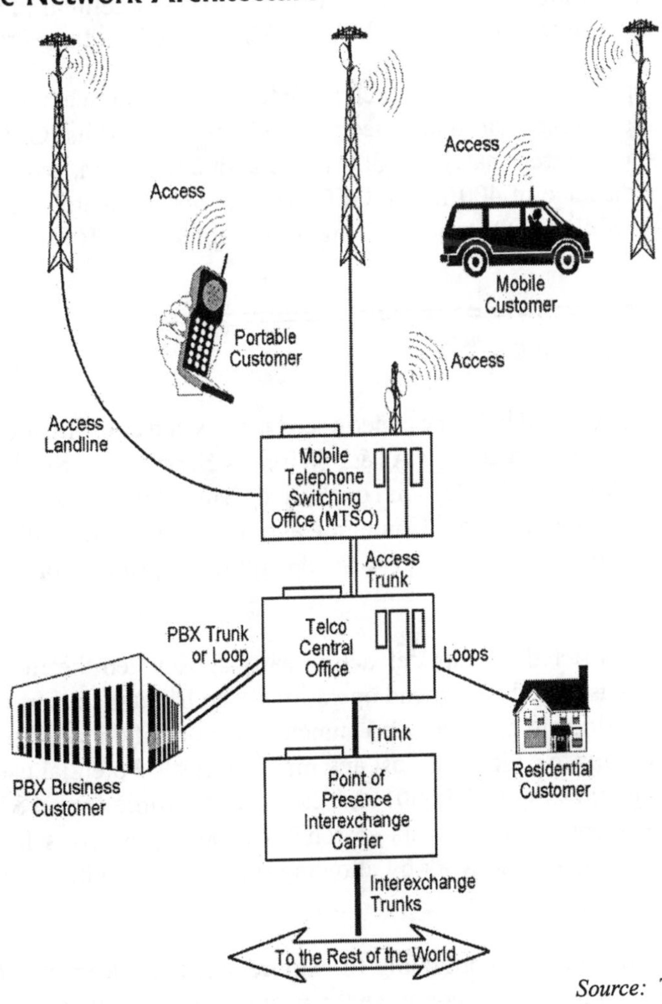

Source: Technology Futures, Inc.

Personal Communication Services

Traditional cellular operates at 800 MHz to 900 MHz, while PCS operates at 1.8 GHz to 2.0 GHz. Both systems can offer the same end-user features, and at least one service provider (AT&T) refers to their service in both frequency bands as Digital PCS. As a result of propagation characteristics of the higher PCS frequencies, the maximum cell size is smaller than for cellular for a given power level. The most important distinction for the purposes of this chapter is the fact that the FCC has allocated over three times the amount of spectrum for PCS than the original cellular allocation. This is important since it directly translates to more capacity and more competitors. Other differences are:

(1) All PCS terminals are small, low-power units, while cellular has both low-power portables and high-power mobile units.

(2) All PCS systems are multiple access digital technology, while cellular is in transition from single access analog to multiple access digital.

(3) Two digital technologies are used in the United States for cellular (TDMA and CDMA), while three are utilized for PCS (TDMA, CDMA, and GSM, the European standard).

Satellite-based approaches can cover much more territory with fewer cells by employing satellites in varying altitudes: geosynchronous, medium, and low earth orbits (GEOs, MEOs and LEOs, respectively). GEOs are approximately 22,300 miles above the earth, while MEOs are at 5,000 miles and LEOs at a distance of 400 miles. GEOs require approximately four satellites for complete global coverage, while MEOs require 12 and LEOs at least 48 (Housel & Skopec, 2000).

Recent Developments

The U.S. wireless telephone industry experienced large-scale consolidation from 1998 through 2000, with four main players emerging: AT&T Wireless Services, Sprint PCS, Verizon (a joint venture between Bell Atlantic Mobile, AirTouch, and PrimeCo), and an alliance between SBC Communications and BellSouth. Nextel Communications is also a leading player in the market. As of this writing, MCI WorldCom is in the process of acquiring Sprint to form a new conglomerate, WorldCom PCS.

Satellite telephony suffered a major setback in early 2000 when the first major satellite telephony effort, Iridium, was forced to suspend operations. The Iridium consortium, led by Motorola, invested more than $6 billion to construct and launch a network of 66 satellites that would allow a person to make phone calls to or from almost any place on earth. Potential users, however, balked at the cost of the telephones (up to $3,000) and the cost of airtime (up to $3 per minute). As of mid-2000, Iridium was unsuccessful in its search for a buyer, and was facing the prospect of destroying its $6 billion satellite network by allowing the satellites to burn up in the earth's atmosphere.

Other satellite companies are continuing to explore satellite telephony. The leading satellite company players that can potentially comprise the next wave of wireless communications include

Hughes Electronics and EchoStar Communications. Telecommunications chip makers Conexant Systems and Broadcom are fueling the growth of this market.

Wireless Data

After years of bubbling below the surface, wireless data has finally arrived as a significant force in the market. It is estimated that, by 2003, there could be 60 million users of wireless data services in North America, and the market could be worth up to $10 billion (Arnold, et al., 1998). Technology companies are rushing to take advantage of this opportunity. Qualcomm and Ericsson have jointly agreed to support and license a single world standard for the third generation of wireless communications. Cisco and Motorola have announced plans to develop and build wireless networks based on the Internet protocol (IP). Two other wireless data standards are the Wireless Applications Protocol (WAP) and Bluetooth. (For more on these standards, see Chapter 19.)

Wireless Application Protocol. WAP is a global standard communications protocol (as TCP/IP is for the Internet), which aims to define the standard for how content from the Internet is filtered for mobile communications. It is thus the definitive protocol for application, session, transaction, security, and transport layers for wireless communications. Similar to the Internet model, WAP can operate when a wireless device contains a microbrowser, and when content and applications are hosted and communicated from Web servers (International Engineering Consortium, 2000).

The advantage of WAP lies in the promise of a range of value-added services for subscribers. Moreover, WAP provides access to the same desktop information from a wireless pocket-sized device, thereby adding a new dimension of connectivity. Thus, this technology bridges the gap between the Internet, corporate Intranets, and easy access to "anytime, anywhere" information.

WAP is poised as the future of third-generation wireless applications. Not controlled by any single company, WAP comprises over 100 members that represent terminal and infrastructure manufacturers, operators, carriers, service providers, software houses, content providers, and companies developing services and applications for mobile devices.

Bluetooth. Bluetooth is a diverse group of industry leaders promoting technology to enable a common wireless environment. The five founding companies of the Bluetooth Special Interest Group (SIG) are Ericsson, IBM Corporation, Intel Corporation, Nokia, and Toshiba Corporation. Since then, 3Com Corporation, Lucent Technologies, Microsoft Corporation, and Motorola have joined the alliance as partners. Bluetooth wireless technology has broad industry support, with more than 1,200 companies already having joined the SIG as adopters since its inception in 1998 (Bluetooth, 2000).

Behind the deals and alliances taking place in the wireless industry as a whole, and Bluetooth in particular, is a growing awareness that no single player has all of the assets or technology to dominate this emerging market. Indeed, there is a synergy between content providers, software companies, wireless service providers, and equipment and device manufacturers, who all have to work together in the drive for new wireless connectivity.

Factors to Watch

Generations 2.5 and 3

Today, especially internationally, it is becoming more common to refer to all cellular type systems by technology generation rather than by frequency band. The first generation consists of the analog single access per channel systems and narrowband analog multiple access systems (N-AMPS). The second generation is the multiple access digital systems (CDMA, GSM, and TDMA), with narrowband data rates and limited or no intersystem roaming. The 2.5 generation has been planned to address the near-term data needs by trading-off some mobility for higher-speed data up to 384 Kb/s). The third generation is a coordinated international effort for quantum leap improvements. Some of the objectives are:

- Wireline quality.
- Security comparable to the traditional telephone network.
- Interconnection to other mobile/fixed users.
- National and international roaming.
- Service to fixed users.
- Packet and circuit data.
- Higher spectral efficiency.
- Coexistence/interconnection with satellite systems.
- Data rates up to 2 Mb/s.

While a single international standard was sought, it now looks as if at least three will be adopted. This makes the objective of international roaming more costly and complex to accomplish. The main issue favoring multiple standards is achieving a smooth migration path from second-generation technologies. Anticipated migration paths and standards are shown in Figure 22.4. Forecasts of how each generation will evolve are shown in Figure 22.5.

Other Technologies

It is obvious that cellular technology is still evolving and improving as evidenced by the continuous progression from first generation to third generation. While radio technology itself may be mature, advances in integrated circuit (IC) technology, computers, and software enable more economical and feature-rich systems. Commercialization of previously classified defense technology, such as spread-spectrum (which resulted in CDMA) and smart antennas, is resulting in systems with much more capacity. Microcells and picocells provide more and improved coverage. Continued implementation of these technologies will make it possible for essentially everyone to use wireless access for almost all voice and most data communications.

Figure 22.4
Next-Generation Wireless Systems

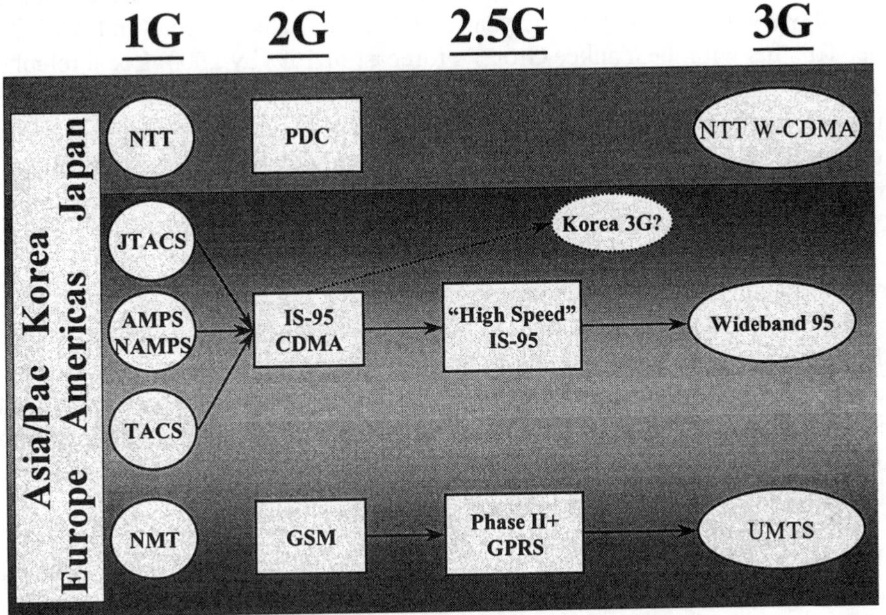

Source: Motorola

Figure 22.5
Wireless Subscribers by Technology

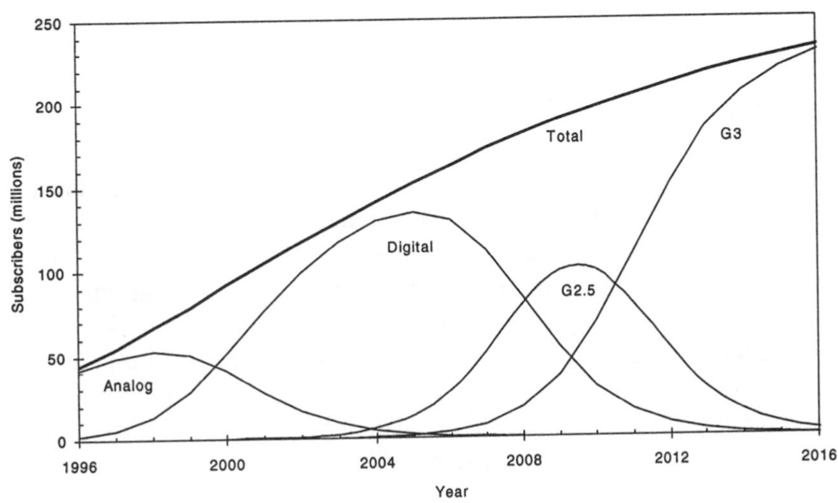

Source: Technology Futures, Inc.

Wireless vs. Wireline

Wireless communications present a significant threat to existing wireline voice services in terms of both access and usage. According to Technology Futures, Inc., wireless access will supplant 5% of today's wired access lines by year-end 2001 and 25% by year-end 2005 in the United States alone. This fits with the Yankee Group's forecast of 15% by 2003. Local telephone companies receive about half their revenue from usage charges, mostly involving long distance service. Thus, a shift in usage from wireline to wireless, regardless of whether monthly wireline service is retained, can cost local wireline providers. In terms of usage-sensitive revenue, Technology Futures predicts that nearly 20% of the local telephone company market share could be lost by year-end 2001 and 50% by year-end 2005 (Hodges, 1999).

So far, in the United States, wireline versus wireless competition has been from cellular and PCS customers using their portable phones in place of wireline phones. A concerted wireless local loop (WLL) strategy would accelerate losses. With WLL, wireless technology is used to provide standard telephone service to a fixed location. FCC Chairman William Kennard made the following statements during a speech at PCS 98: "We must bring competition to the local loop.... WLL is a critical and important step to making that happen" (Karissa, 1998, p. 26). These statements signal regulatory support for the effort.

In mid-2000, there were several trial or small-scale offerings in the United States. Bell Atlantic was conducting a trial of CDMA WLL in New Jersey, while Wireless North, Western Wireless, and Pioneer were directly targeting the local service market in rural America (Karissa, 1998). AT&T also had a trial in Plano, Texas where they were offering unlimited airtime for $40 per month bundled with Extended Area Service (EAS). This price substantially undercuts SBC's comparable wireline rates (Cut the cord, 1999).

It is interesting to note that the most severe impact may first be felt in small towns. For example, in a town of 10,000 population, 90% likely live within a five-mile radius of the center and can easily be reached from a single tower. There is ample spectrum available in most of these locations for someone to build a digital system to provide unlimited usage and compete with the local telephone company.

Spectrum Allocation

The growth of wireless telecommunications will depend, in large part, on the amount of bandwidth available to the expanding wireless industry. The FCC has used and may continue to use allocation and auctioning of spectrum as methods for facilitating competition in the telecommunications marketplace. Opinions are sharply divided on this issue. A large group (56%) believes that spectrum auctioning represents a good method for allocating spectrum. Another smaller group (25%) raised objections, comparing the airwaves to national parks, beaches, and lakes that belong to everyone and which should thus not be permanently sold. The third group (25%) believes that auctions are acceptable, but that changes to the existing system of auction allocation to ensure fairer distribution is necessary (Housel & Skopec, 2000).

Satellites

In terms of technologies, both PCS and cellular are seeing increased growth in the current market. Satellite-based systems have seen initial snags in their deployment. However, it is estimated that, after 2010, their use will become increasingly prominent. Factors that can be attributed to their future success are:

- Provision of services to areas without adequate telecommunications. An FCC industry panel recently noted that, while the majority of the world's population currently lives in areas with cellular service, 85% of the earth's land mass will remain without adequate telecommunications infrastructure.
- Increased convenience for consumers who will be able to carry one phone number anywhere, avoid hotel surcharges, and bypass local networks.
- The prospect of dramatically reduced costs, with expected satellite communication charges ranging from $0.25 to $3 per minute.
- Expanded bandwidth with data rates on the order of 1 Mb/s to 1.5 Mb/s to support voice, data, video, and multimedia services (Housel & Skopec, 2000).

All this points to the phenomenal growth that is yet to be witnessed in the wireless sector. It is estimated that, by 2003, there could be 60 million users of wireless data services in North America, and the market could be worth up to $10 billion (Daugherty, et al., 1999). Implications for corporate players are obvious; what is not so obvious is how the industry and standards will ultimately align to provide for a stable wireless environment in a bold new world of instant integrated anytime information and communication.

Bibliography

Arnold, S., Auguste, B. G., Knickrehm, M., & Roche, P. J. (1998). *The McKinsey Quarterly, 2*, 19-31.*

Bluetooth. (2000). [Online]. Available: http://www.bluetooth.com.

Brown, R. (1995, May 1). Wireless cable: A capital idea. *Broadcasting & Cable*, 16.

Cellular Telecommunications INdustry Association. (2000). [Online]. Available: http://www.wow-com.com.

Cut the cord. (1999, January 25). *Forbes*.

Daugherty, W., Eugster, C., Roche, P., & Stovall, T. (1999). *The McKinsey Quarterly, 2*, 88-99.*

Fertig, D., Prince, C. H., & Walrod, D. (1999). *The McKinsey Quarterly, 4*, 144-148.*

Hodges, R. (1999). *Telecommunications access technologies: Overview and competitive assessment*. Austin, TX: Technology Futures, Inc.

Housel, T. J., & Skopec, E. (2000). *Global telecommunications revolution: The business perspective*. Boston: McGraw-Hill.

International Engineering Consortium. (2000). *Web ProForums*. [Online]. Available: http://www.webpro-forum.com.

Karissa, T. (1998, November 15). The road to local competition. *Wireless Review*, 26.

Learn cellular's lingo before you buy. (1997, September 10). *Investor's Business Daily*, A1.

23

Personal Communication: Computing on the Go

Philip J. Auter, Ph.D.*

It's a typical Wednesday afternoon in October of the early 21st century.

In Cancun, Mexico, a middle-management executive sits on the deck of a beachside bar sipping a beer and skipping rocks into the bay. Back in San Francisco, her division of the corporation is involved in a multimillion-dollar sales deal. She's had a very productive, but stressful year and is on vacation—but her boss has asked her to stay in contact via pager. Suddenly, the pager goes off—for the twelfth time in two hours. Her junior associate wants to clarify yet another minute point about the negotiations. After reading the text message, she consults her full-color personal data assistant (PDA) (e.g., http://www.palm.com/*) or PocketPC (*http://www.pocketpc.com*)—a handheld computer where she stores all her appointments, to do's, contacts, and memos. She opens her pager like a makeup compact and types a quick message on the tiny QWERTY keyboard and presses the send button (e.g.,* http://www.motorola.com/MIMS/MSPG/SmartPagers*). The page zips off through the ether and arrives in San Francisco a few moments later. The executive settles back into her chair and sips her beer, skipping rocks. Fifteen minutes later, the pager goes off again. She reaches over and grabs it, instead of a rock, and tosses it into the bay. Picking up her full-color PDA, she plays a classic game of* Ms. PacMan *for a few minutes until the ghosts finally catch up with her. Resigned in her defeat, she slips the PDA into her bag, focuses on the picturesque sunset, and goes back to sipping her beer....*

In Evansville, Indiana, a second-grader, and his older sister, a fifth-grader, arrive home on the bus from school. They say hi to their dad who's cleaning house and surfing the Internet, and immediately head their separate ways. The fifth-grader goes down the street to a friend's house. Her friend's mother drops them both off at the mall. The second-grader plays with a "toy laptop computer" in his room (e.g., http://ww.vtechkids.com*). It's not a real computer like mom and dad's, but it beeps and buzzes and you can type on it. A few hours later, dad looks up from his work and realizes*

* Assistant Professor of Telecommunications, Department of Communication Arts, University of West Florida (Pensacola, Florida).

that it's getting close to dinnertime. Mom will be home from work soon. He logs onto the pager company's Website and sends a text page to his daughter (e.g., http://epage.arch.com/*). "Time to come home. Dinner in an hour." He goes to check on his son, confident that his daughter will get the message on the "raspberry-ice" alpha pager she keeps clipped to her bag and be home on time. His daughter reads the message and replies using the two-way capabilities of her pager. He discovers that the boy has connected the toy PC to the second phone line and is sending e-mail to strangers again. The boy thinks these people are his friends, but as far as mom and dad are concerned, he doesn't really know these people, as they've never met face-to-face. The e-mail account was originally set up as a fun learning experience so the boy could send messages to his grandparents and friends from the neighborhood. Dad logs on to the e-mail system and adds access restrictions to the account (e.g.,* http://www.vtechworld.com/qa/*). His son can now only send and receive e-mails with people approved by mom and dad in advance....*

In Atlanta, a professor of communication is just arriving at Hartsfield International Airport from Pensacola, Florida. He's on his way to a conference in Las Vegas, but has a two-hour layover before his next flight. He exits the plane, and finds the concourse for his next flight. Thirsty, he goes to the nearest soda dispenser. Realizing he has no change handy, he pulls out his PDA, aims its infrared port at the one on the soft drink machine, presses a few buttons, and buys a drink that's charged to his credit card (e.g., http://www.televend.com/*). Taking the drink to his gate, he sits down to send and receive a few e-mails while he waits for his flight. The professor extends the antenna of his Internet-ready PDA, logs on to the Internet, and connects to his e-mail account. At about that time, his Internet connection breaks off and he cannot log back on. The line is busy....*

In Boston, an MIT graduate student sits in a bohemian coffeehouse eliciting fewer stares than usual. He used to be glanced at quite often because of the odd headgear he wears—and the wires running from those spectacles to the black box on his belt (e.g., http://www.media.mit.edu/projects/wearables/*). But people have apparently gotten used to his appearance, his almost vacant gaze, and his occasional predilection to speak to ethereal associates. Somewhat disappointed at the lack of attention he's drawing, he goes back to work on his thesis project....*

Whether these stories sound like dreams, nightmares, or descriptions of the stuff in your bag right now, they are definitely not a vision of the future. All of these technologies and services are available today, although some are still in the "early adopter" stages. Personal, portable communication and computers are not only here, they are here to stay—and virtually impossible to avoid.

Background

"Personal communication" and "portable computing" devices have developed along five lines: cellular telephones, paging devices, palmtop computers or PDAs, children's toys, and wearable computers. (Others might also include laptop and vehicle-based computers in these categories.) Each has its own unique history.

Paging

The concept of paging was born in 1949 when a hospitalized radio engineer, Charles Neergard, was annoyed at the public address system used to notify doctors that they had messages. He felt that there had to be a quieter way to inform physicians, and shortly thereafter, the first internal radio-paging network was developed. In the 1970s, voice circuitry was added to pagers. Digital displays became common in the early 1980s, and alphanumeric paging technology soon followed (Hon, 1996).

All paging systems work on the same basic concept. A caller dials a phone number that reaches a paging terminal, which then transmits a tone, voice, numeric, or alpha message. The paging terminal converts the message into a pager code and relays it to transmitters throughout the coverage area. The transmitters send out the code to all pagers in the coverage area tuned to a specific frequency, but only the pager with the proper address (cap code) will be alerted (Motorola, 1996b).

Alphanumeric paging has become the fastest-growing segment of the industry, in part due to services such as e-mail-to-pager gateways and information services that send news updates and stock quotes directly to pagers. Motorola "flex" technology has also allowed for greater reliability in two-way paging (Motorola, 1999). Fewer pages are lost with flex-equipped units, because a message is continuously rebroadcast from the tower until the pager confirms that it has been received. Messages sent while a pager is off or out of the service area are thus received once the device is active and within range.

Palmtops/PDAs

The ever-shrinking laptop computer evolved into a sub-notebook and finally the palmtop—a PC that would fit into the palm of your hand. Almost simultaneously, personal data assistants (PDAs) evolved from what were originally 32K+ organizers produced by Sharp and other companies. Initially, they were rudimentary ways of digitizing lists of contacts, appointments, and things to do. Most palmtops can be differentiated from PDAs because they have a tiny keyboard and no handwriting recognition. However, more and more palmtops are sporting unique data-entry abilities (including voice and handwriting recognition) as the two merge. Meanwhile, a very prevalent accessory to the world's most popular PDA—the PalmPilot—is an attachable keyboard. Like digital watches a decade ago, the early basic organizers—which offer limited but functional address books, calendars, to-do lists, and calculators—are still available today at remarkably low prices. They are now targeted toward low-end users, and even marketed to children as "digital diaries."

PDA operating systems and user interfaces have varied, but a general attempt has been made to mimic the Windows-based desktop computer operating system (OS). Apple's Newton Message Pad, introduced in 1993, was one of the first PDAs and, for a while, held over 75% of the then small PDA market. Although large and expensive by today's standards, it incorporated some of the current PDA features including handwriting recognition, a graphical interface, and synchronization of data to a desktop computer (Dor, 1996). Current PDAs and palmtop computers have become almost indistinguishable. Both operate with one of two dominant operating systems: WindowsCE or the Palm OS. Both now have color capability and a surprising amount of storage, which currently ranges from 64K up to 2 MB or more for palmtops. (By comparison, the original

Commodore desktop computers had only 64K.) Also, today, PDAs consistently come packaged with 8K to 16K of RAM.

As with desktop computers, one of the "handheld" PC's biggest battlegrounds was over operating systems. Although Microsoft's WinPad OS was doomed to failure, their newer WindowsCE operating system has been a strong entrant in the field. Originally a subsidiary of U.S. Robotics and then 3Com, Palm, Inc.'s PalmPilot OS was a stronger, more versatile predecessor to WinCE. Unlike Apple in the desktop area, Palm licensed their OS early in the game to increase market share and become a stronger competitor against Microsoft. Currently, the Palm OS has been licensed to TRG, Handspring, and Kyocera/Qualcomm, the latter of which is incorporating a Palm OS-based PDA into their newest "smart cell phone" (e.g., http://www.kyocera-wireless.com/pdq/).

One of the newer competitors in the PDA market is Handspring Technologies (www.handspring.com). Formed by the people originally responsible for developing and marketing the PalmPilot, the company has taken the device to the next level. Handspring's "Visor" is built on the Palm operating system, but includes enhancements such as a world clock, a more advanced calculator, and an integrated datebook and to-do list. It also comes with four times the memory of a basic PalmPilot and a math coprocessor. But most important, the Visor has a patented "Springboard" expansion module that allows for quick upgrades and easy integration with global positioning system (GPS), pagers, cell phones, and other additions.

Certainly one of the biggest advantages of today's PDAs and palmtops is the ability to "easily" synchronize data with a desktop personal computer (PC). Early attempts to do this via infrared were cumbersome and expensive. Today, PDAs such as the PalmPilot are placed in a cradle that connects to a data port on any PC or Macintosh. After an initial configuration, the device can synchronize data with as little effort as pressing a single button. Although this system works almost flawlessly most of the time, duplicate records are sometimes created when the same record is modified on both the PDA and the PC.

New operating system software can be credited for synchronization and other advancements in PDAs and palmtops. The open architecture of the Palm OS has allowed for the development of many commercial and shareware applications that stretch this device's flexibility to the limit. Games, calculators (of many types), drawing applications, and Internet access are but a few of the functions that can make the PDA seem more like a pocket-sized desktop computer.

Not to be outdone, Microsoft has finally developed a stable palmtop/PDA operating system with their WinCE. Although it operates in an entirely different way than Windows for the desktop (a program that is tens of megabytes larger), this stripped-down OS mimics the Windows operating environment, making life simpler for some users. Publishers of many of the more popular desktop programs are developing palmtop versions of their software for these portable devices (http://www.microsoft.com/windowsce/).

Another major advancement in PDAs is a more stable and forgiving handwriting recognition technology. Here again, the PalmPilot has received rave reviews. Ironically, one of the first PDAs with handwriting recognition, the costly Apple Newton (and its younger relative, the eMate), have finally died after a lingering illness (Apple, 1998; Dor, 1996). "Writing" on a PDA is accomplished by using a digital stylus—a not-too-sharp, pointed object—on an electronic screen. Regular pens, pencils, and similar items are *not* recommended. Forward thinking pen companies such as

Cross (http://www.cross-pcg.com/), however, have developed inserts for their traditional pens, effectively turning them into digital styli. The Cross Pen Computing Group has also developed their own version of the PDA, the CrossPad, which allows you to write onto paper, but simultaneously transcribe the data as a picture file into the clipboard-style PDA.

Connectivity has advanced dramatically for PDAs and palmtops. Streamlined Internet software, credit-card-sized modems, and wireless options have allowed the PDA and palmtop users *almost* seamless access to e-mail and the Web. Wireless cell modems make it possible for almost universal access for the financially well off. For those of us on a more limited budget, tiny modems allow for access wherever a dataport can be found.

In actuality, most of the PDA and palmtop offerings are becoming indistinguishable from each other—and from other electronic devices. As the technology advances, pagers, PDAs, palmtops, and cell phones are moving toward one hybrid communication unit. With color screens, increased data storage, Internet connectivity, GPS, voice recognition, MP3 playback ability, and upgradability, today's PDAs operate more like desktop computers than travel organizers.

Children's Portable Computing Toys

Since the dawn of time, children's toys have mimicked their parents' tools. From toy kitchens and tools to today's toy remote controls and video cameras, some of these replicas "really work," while others do not. The latest batch of toy replicas includes pagers, computers, cell phones, and PDAs. While most of these do not really work, there are a significant and growing number of "real" computers with Internet access masquerading as toys. Today, thanks to advances in microchip quality and dramatic reductions in cost, many toys have some sort of rudimentary computer chip inside. Children's diaries are going digital, and many contain features once considered advanced in adult organizers such as calendars, memo pads, and to-do lists (e.g., Tiger Toys' digital diaries; see http://www.game.com/tigertoys/precamera.html).

Wearable Computers

As computing devices grow more portable, smaller, and more powerful, they have ventured well beyond the realm of science and into science fiction. Today's palmtops and PDAs are approaching the power and functionality of *Star Trek's* original tricorder in the 1960s TV show. Despite these advances, computers and their cousins have not achieved the expectations of the most imaginative visionaries—the wearable computer: a powerful, portable, comfortable, connected, hands-free, truly interactive device.

We've all seen the prototype wearable computer in science fiction and even in an IBM concept commercial. It consists of a non-obstructing, eyeglasses-based visual display, wireless connectivity to the Internet and telephone systems, an easy-to-operate voice command interface, and a lightweight yet powerful central processing unit (CPU) and hard drive, usually mounted to a belt. While academic and corporate research groups are still hard at work on such a device, the origins of the wearable computer go back almost four decades.

In 1966, Ed Thorp and Claude Shannon of MIT revealed their invention of the first wearable computer, which was originally designed in 1961. The cigarette-pack-sized device was an analog machine designed to predict the speed of a roulette wheel. The same year, a Harvard professor and his graduate assistant introduced the first head-mounted visual display. Nicknamed "the Sword of Damocles," the device was mounted to the ceiling and suspended forbiddingly over the wearer. Development of the wearable computer—and even the PC, for that matter—was slow. The highlight of wearable computing in the 1980s was Steve Roberts' bicycle-mounted system, the Winnebiko II, which was built in 1986. The system included a laptop computer and a packet data communication system for e-mail via ham radio (Rhodes, 1997). In the ensuing years, researchers at the MIT Wearable Computing Lab and other academic think-tanks slowly brought the wearable computer of science fiction into reality. Today, IBM and other manufacturers are spearheading the development and commercialization of these powerful devices, which may one day replace cell phones, pagers, and PDAs (IBM, 1998).

Recent Developments

Paging

One of the paging industry's recent triumphs has been two-way paging. Technology has advanced and come down in price, and soon these devices will be affordable to most customers. Despite this and other technological advances, however, paging services (often sold by franchises that also carry a cellular phone option) are now facing direct competition from digital cell phones that can also send and receive text messages (e.g., DigiphPCS [http://www.digiph.com]). In an attempt to gain market diversification, paging companies have started targeting younger segments of the market—and their parents—suggesting that this is an inexpensive way for families to stay in contact (Motorola, 1996a). Additionally, hardware and service companies have marketed pagers inside watches (e.g., http://www.beepwear.com) and pager card hardware upgrades for PDAs (Shapirio, 1998) to capture a segment of the market that does not want to carry multiple devices.

Palmtops and PDAs

Despite their advances in capabilities, portable computing in the form of palmtops and PDAs cannot realistically get much smaller and still be usable. Thus, most of the manufacturers have concentrated on improving performance and making the devices more and more like pocket-sized PCs. Color, dramatically increased memory sizes, and a plethora of PC-like software have all become standard. Additionally, newer PDAs and palmtops use permanent batteries that recharge while the device sits in its synchronization cradle, rather than disposable batteries that must be replaced at an alarming rate.

The latest frontier is wireless Internet connectivity. Hardware manufacturers are working hand-in-hand with Internet service providers and cellular companies to provide wireless access to a lean and mean version of the Web. A number of services offer "Web clippings" or proxy server services, but connectivity and compatibility problems abound. One industry group, Bluetooth (www.bluetooth.com) is working on a way to deal with connectivity issues. The Bluetooth technology concept consists of placing a low-power radio transmitter/receiver chip in PDAs, cell phones, laptops, etc. so that they can communicate and share data without wires—and across platforms—in a more seamless manner. The chip also allows for wireless Internet connectivity. A consortium of major technology providers, including Palm, Intel, and IBM, are backing Bluetooth, but the technology is still untested. And, although this technology may help solve connectivity issues, it still does not overcome the hurdles of squeezing the Internet into such a tiny device.

Although traditional PDAs and palmtops have probably gotten as small as they can and still be usable, a few entries in the watch-sized category have surfaced (e.g., http://www.casio.com/timepieces/). These "wrist data devices" synchronize with PCs and PDAs via infrared, but due to their size do not support direct data entry and modification. Still, the watch-sized PDAs pack many features into their tiny containers, including calendars, to-do lists, address books, text browsers, and a world clock. In fact, a new category of portable computing may be on the horizon. Casio has created PDA watches, watches with MP3 players, wireless Internet cameras, and even global positioning systems. These watches might just be another step in the evolution toward wearable PCs. Motorola is working backward from the pager to create a two-way paging device with a built-in PDA that has access to the "Web without wires" (e.g., www.motorola.com/www).

Children's Portable Computing Toys

Perhaps one of the most surprising advances in "toy computers" is the attempt to teach children how to become truly PC-compatible. For just $90, you can buy a small, text-only laptop PC for your eight-year-old with free e-mail connection to a service that allows worldwide e-mail access, with parental accessibility control, for about $7.50 per month.

Wearable Computers

Wearable computers are still in their infancy, but several companies are working hard to make them viable. Like cell phones and PDAs, they will initially be distributed in industry, predominantly where individuals will need hands-free data access in unique locations. For example, the telephone company employee, working on phone lines 30-feet off the ground, could access important technical data through her wearable PC without coming down—or falling—off the pole. As these devices become smaller, sleeker, and more stylish, they could become the cool tech gadget of the next decade.

Which person in this picture is not wearing a computer? It's *not* the Professor. Steven Mann, inventor of the WearComp teaches his computing class wearing the latest version of his invention, which looks like normal bifocal glasses (see www.wearcam.org).

Current Status

Paging

In 1998, there were 51 million individual paging service subscribers in the United States or 19% of the potential market. The vast majority of that, 99%, used one-way (receive-only) pagers. Although at one time the fastest-growing segment of the telecommunications industry, projections for the growth of paging have slowed. By 2004, market penetration is projected to be only 22.3% or 62.5 million pagers in service, an increase of only 10 million new accounts in six years. Projections suggest a steady decline in the market for one-way pagers as two-way paging technology improves and becomes less expensive. New markets may not be opening up in the United States: two-way pagers may simply be replacing one-way devices, unless the users switch to text-messaging digital cell phones (Brad Dye's paging, 1999).

Palmtops and PDAs

Sales in palmtops and PDAs continue to grow exponentially, and the devices are spreading across the globe. As of 1997, the estimated market in 2000 for all palmtop and PDAs was five million units—a number that has been far surpassed. Palm alone has sold over five million of its units as of early 2000 (Palm operating, 2000). Currently, the Palm OS has the lion's share of the U.S. market, over 75% (News briefs, 1999), and Palm has licensed its OS to a number of major players including IBM, Sony, Qualcomm, Motorola, and Nokia (Spangler, 2000).

Although PDAs are becoming more and more popular, pundits are already ringing their death knell. The newest wave of device is the multifunctional phone/PDA such as the Quartz. Not only will it have all the capabilities of today's cell phones and PDAs, such a device may be "free" with phone service. Already moving into European markets, these devices may take hold in the United States, but only if a connectivity standard is developed and coverage improves (Orlowski, 2000).

Today, PDAs range from about $150 to well over $800. What do you get at the various levels? Today's basic, $150 PDA packs all the advanced features of the original high-end models. They usually come with handwriting recognition or a small touch keypad, gray-scale screen, 1 MB to 2 MB of memory, and a small suite of applications. Additional memory, increased functionality, and cutting-edge technology come at a higher price. These include color screens, 64 MB or more of memory, wireless capability, and upgradeablilty. (For a comparison of PDA prices versus features, check out http://www.eshop.msn.com/giftfinder/giftfinder.asp, and do a search on the term PDA.)

Children's Portable Computing Toys

Vtech has carved out a specialty niche in the Internet-ready toy market, but they will soon be facing stiff competition by latecomers to the game. Tech-based toys are such a fast growing segment of the toy market that they received their own pavilion for the first time at the toy industry's 97th annual trade show in 1999 (Krochmal, 2000a). One new entry, the Cybiko, which debuted in April 2000, is targeted directly to teenagers. The device is a combination PDA, game machine, and low-power text communicator—all for about $170 (www.cybiko.com). Kids can chat with from one to 100 other Cybikos within range of the transmitter/receiver (up to 300 feet outdoors). The extremely powerful "toy" sports 256 MB of RAM and expansion ports for MP3 players, GPS receivers, and other add-ons (Krochmal, 2000b).

Wearable Computers

Wearable PCs are somewhere in the early adopter stage of development and distribution. Wearable PC think tanks exist at a number of universities, including MIT, Georgia Tech, Columbia, and Stanford, but the devices are also being manufactured for commercial use. Xybernaut (www.xybernaut.com) is just one of a number of commercial businesses that is marketing wearable devices to commercial and industrial clients that need hands-free access to data in the field (Horowitz, 2000). Interest is so high, in fact, that a conference was built around wearable computers back in 1995 (http://www.xybernaut.com/Icwc2000/index_icwc.html). Although very few wearable computers have made it into the mainstream, the industry appears be poised to take off, and it could possibly overtake palmtop/PDA sales in the not-too-distant future

Factors to Watch

Convergence and power continue to be the watchwords for the future of portable computing. It may take many months—more likely years—before cross-technology standards are set. Still, the multi-function individual appliance appears to be on the way. And a fairly advanced and colorful version will be available for your children. Market size for the new appliances will probably increase, even as demand for single-function PDAs and pagers levels off. The primary stumbling blocks are wireless Internet services using low-frequency radio broadcasts at less than 2.2 Gigahertz that are still too expensive, unreliable, and lack coverage. As these factors are minimized, the result will be a take off in wireless data minutes, putting pressure on available bandwidth.

The future of wearable computing is surprisingly bright, but we're still on the cutting edge of this new technology. One of the earliest wearables was a cigarette-pack sized calculating machine. Today, a matchbox sized wireless server might just be the perfect compliment to your heads-up display and voice-recognition wearable (Sci/tech, 1999). But the true stumbling block to the wearable PC is making it cool—an acceptable fashion statement for the average individual. While most of us might not be inclined to head out of the house with awkward, high-tech headgear and a black box strapped to our belts, we might not mind leaving home dressed in our most trendy PC clothing. Researchers are working right now on weaving the components of wearables into actual clothing, and fashion designers have begun adding their input. The 21st century's next technology and fashion trend might just be PC clothing. Dry-clean only, of course.

Bibliography

Apple, Inc. (1998, February). *Apple discontinues development of Newton OS.* Apple Media & Analyst Information Press Releases. [Online]. Available: http://www.apple.com/pr/library/1998/feb/27newton.html.

Arch: To send an e-page. [Online]. Available: http://epage.arch.com.

Beepwear: One beeping great pager watch from Timex and Motorola. [Online]. Available: http://www.beepwear.com.

Brad Dye's paging information resource. (1999). [Online]. Available: http://www.refreq.com/braddye/DLJForecast.htm.

Casio timepieces. [Online]. Available: http://www.casio.com/timepieces/.

Cross Pen Computing Group. [Online]. Available: http://www.cross-pcg.com.

Cybiko. [Online]. Available: http://www.cybiko.com.

DigiphPCS. [Online]. Available: http://www.digiph.com.

Dor, S. (1996). *Newton, where goest thou?* [Online]. Available: http://info.berkeley.edu/courses/is296a-3/f96/Projects/sdoi/pda.html.

Hon, A. S. (1996, 1993). *An introduction to paging: What it is and how it works.* Motorola Electronics Pte Ltd. [Online]. Available: http://www.mot.com/MIMS/MSPG/Special/explain_paging/ptoc.html.

Horowitz, A. (2000). *Wearable computers: Silicon Valley meets Seventh Avenue?* [Online]. Available: http://www.planetit.com/techcenters/docs/mobile_computing/technology/PIT20000305S0001/.

International Business Machines. (1998). *Wired for wear: IBM researchers demonstrate a wearable ThinkPad prototype.* [Online]. Available: http://www.ibm.com/news/ls/1998/09/jp_3.phtml.

International Conference on Wearable Computing 2000. [Online]. Available: http://www.xybernaut.com/Icwc2000/index_icwc.html.

Krochmal, M. (2000a). *Intel/Mattel lab adds to high-tech toy box.* [Online]. Available: http://www.techweb.com/wire/story/TWB20000215S0007.

Krochmal, M. (2000b). *Wireless is included.* [Online]. Available: http://www.techweb.com/wire/story/TWB20000330S0015.

Kyocera: pdQ smartphone. [Online]. Available: http://www.kyocera-wireless.com/pdq/.

Microsoft WindowsCE. (2000). [Online]. Available: http://www.microsoft.com/windowsce/.

Motorola Electronics Pte. Ltd. (1996a). *100 uses for your Motorola pager: For family.* [Online]. Available: http://www.mot.com/MIMS/MSPG/Special/what100.html#ss2.

Motorola Electronics Pte. Ltd. (1996b). *What is paging?* [Online]. Available: http://www.mot.com/MIMS/MSPG/Special/explain_paging/overv.html.

Motorola Electronics Pte Ltd. (1999). *Motorola Flex Technologies.* [Online]. Available: http://www.mot.com/MIMS/MSPG/FLEX/.

Motorola Web w/o Wires. [Online]. Available: http://www.motorola.com/www.

News briefs: Palm slaps CE. (1999). [Online]. Available: http://mobilecomputing.com/mobilecomputingn2/mcc200002/news1.htm.

Orlowski, A. (2000). *Quartz: The Palm-killing PDA?* [Online]. Available: http://www.zdnet.com/zdnn/stories/news/0,4586,2445796,00.html.

Palm, Inc.: Handheld computing solutions. [Online]. Available: http://www.palm.com.

Palm operating net tops forecasts. (2000, March). [Online]. Available: http://www.techweb.com/wire/story/reuters-finance/REU20000328S0008.

Rhodes, B. (1997). *A brief history of wearable computing.* [Online]. Available: http://wearables.www.media.mit.edu/projects/wearables/timeline.html.

Sci/tech: Surfing on a matchbox. (1999). [Online]. Available: http://news.bbc.co.uk/hi/english/sci/tech/newsid_276000/276762.stm.

Shapiro, S. (1998). Pager card for the PalmPilot. *Palm Power Magazine.* [Online]. Available: http://www.palm-power.com/issues/issue199804/pagercard001.html.

Smart pagers by Motorola. [Online]. Available: http://www.motorola.com/MIMS/MSPG/SmartPagers/.

Spangler, T. (2000). *Palm's future in its own hands.* [Online]. Available: http://www4.zdnet.com/intweek/stories/news/0,4164,2449224,00.html.

Televend, Inc. [Online]. Available: http://www.televend.com.

The MIT wearable computing Web page. [Online]. Available: http://www.media.mit.edu/projects/wearables/.

The official Bluetooth Website. [Online]. Available: http://www.bluetooth.com.

Tiger Electronics: Cameras, organizers & diaries. [Online]. Available: http://www.game.com/tigertoys/precamera.html.

Vtech: Award-winning toys designed to develop minds. [Online]. Available: http://www.vtechkids.com.

Vtechworld.com: Questions? We've got answers. [Online]. Available: http://www.vtechworld.com/qa/.

Xybernaut. [Online]. Available: http://www.xybernaut.com.

24

Videoconferencing

Kyle Nicholas*

The mystique of two-way interactive video is being supplanted by high user expectations as videoconferencing establishes a beachhead on the World Wide Web. In this chapter, we define a videoconference as when two or more individuals "meet" via a two-way, interactive, audio/visual link. For most users, videoconferencing acts and feels like two-way television, joining them across distance in real time. The connection is made possible by sending audio and video signals from cameras through the air or down phone lines, generally with a computer in between to compress and filter the signal for greater speed and clarity. Videoconferences and audioconferences facilitate large groups such as corporate conventions, small groups such as work groups, and one-on-one conferences. Users employ different hardware and software depending on the scale of the conference, the level of interactivity required, and locations of participating conferees.

Large group videoconferences are generally held in convention halls or theatres, where hundreds of spectators can watch a speaker, and the speaker gets a view of the crowd. The system resembles broadcast television, and generally utilizes satellite or fiber optic transmission networks. Such conferences are generally less interactive; often, only an audio link is provided for the crowd to respond to the speaker. These presentations are also known as business TV (BTV). Small group conferences may be held in specially-built studios or in small rooms utilizing "roll-around" monitors with attached video cameras. By using omnidirectional microphones and voice tracking cameras, small group videoconferences allow a great deal of interactivity, much like a work group or small seminar.

Both small groups and one-on-one videoconferences may use desktop computer technology, and this technology is driving the future of videoconferencing. Desktop systems use a small video camera and microphone attached to the computer. Incoming video is visible in a frame on the user's monitor. Information has traditionally been transmitted via a dedicated local area network (LAN) or wide area network (WAN) using an all-digital transmission medium such as Integrated

* Assistant Professor, Old Dominion University (Norfolk, Virginia).

Services Digital Network (ISDN). Increasingly, however, videoconferences use cable or high-speed phone lines to communicate via the Internet. If a group is using a desktop videoconference system, users can be connected through a "bridge" or a "portal," which can either be a hardware switch or software that allows a single user's transmission to reach the desk of each member of the group. For a desktop videoconference participant, the process is much like a conference phone call, only with video, graphics, and other computer applications included. Desktop video is the fastest-growing segment of videoconferencing technologies, but wireless videoconferencing is on the horizon.

Background

In 1935, the German Post Office linked four locations via a studio-based audio/visual system. Shortly thereafter, an Iowa school district employed audioconferencing to teach homebound students. By 1958, AT&T had linked 10 U.S. Department of Defense sites through a fast-connecting audio system. A few years later, the first private videoconferencing system was installed by First National City Bank (Olgren & Parker, 1983). Despite these pioneering efforts, and at least two concentrated marketing efforts by AT&T, videoconferencing failed to catch on with consumers. AT&T demonstrated a two-way videophone at the 1934 World's Fair and again at the 1964 fair. Participants in these demonstrations were impressed with the technology, but considered it more novelty than necessity (Noll, 1992). The technology continued to improve, and color video images approaching broadcast quality were available by the early 1980s. High-speed phone lines and the emergence of digital ISDN lines, combined with improving compression techniques, enabled more reliable and functional videoconferences. Innovators in medicine, education, and industry experimented with new applications (Olgren & Parker, 1983).

By the late 1980s, however, videoconferencing was still waiting for widespread acceptance. One important reason was the cost; the special studios required for a videoconference were prohibitively expensive. Another critical factor still inhibiting videoconference acceptance may be human behavior. Because of the technical demands of computerized video compression, users must move and speak more slowly to avoid a fuzzy or jerky picture (Noll, 1992). Underlying many of the critiques of videoconferencing are the rising expectations of users. Video representations on computers (video games), in television (high-definition television), and in the theater (digital video) have dramatically improved in the past decade.

Despite the factors mentioned above, videoconferencing is slowly catching on (King, 1996). Adoption of videoconferencing in business is being driven by an emerging need for routine dissemination of multimedia information (Cukor & Coppock, 1995). While many companies still employ LANs for videoconferencing, improvements in the CODEC—the computer processor that uses software to *co*mpress and *dec*ompress digital information in videoconferencing—allows small companies to videoconference via the public phone network with increasingly robust results (Testa, 2000). Desktop videoconferencing is improving due, in part, to higher computer processing speeds and high-bandwidth networks. Key inhibiting factors have been a lack of interoperability between systems, impaired audio/visual quality due to limited bandwidth, and a failure to generate a "killer application" that would attract a wide variety of users.

Vendors have employed proprietary algorithms and packaged hardware and software so that one system will not "communicate" with another. This has led to two problems: First, proprietary systems can quickly become obsolete, as new innovations are not always "backward compatible" with old systems. Second, consumers are accustomed to technological transparency in communication media. Phones and faxes "talk" to each other regardless of brand. Videoconference users expect the same from conferencing technologies (Videoconferencing, 1996.) Standardization of software and operating protocols has helped the industry overcome some of these problems.

Standards

The development and acceptance of technological standards for compression and transmission of audio and video signals is a key development in videoconferencing. The International Telecommunications Union (ITU) has adopted several related standards, and vendors are rapidly building compliant software and hardware. The three most important standards for the implementation and adoption of videoconferencing are known as H.320, H.324, and H.323.

H.320. The ITU's H.320 standard covers direct connections between conference sites. The connection is usually made using a dedicated, circuit-switched network, such as the ISDN employed by the majority of large group videoconferences (ITU, 1996). International videoconferencing may get a boost from the adoption of this standard, particularly in Europe and the Pacific Rim where ISDN is widespread. Prices for ISDN desktop systems using H.320 are well under $2,000.

H.324. The number of telecommuters in the United States is expected to grow to 18 million by 2005 (ITCA, 1998). The H.324 standard, ratified in 1996, was developed to facilitate desktop videoconferencing in the home via twisted pair copper lines. With the advent of H.324, vendors can exploit the home videoconferencing and videophone markets, including the market for telecommuters. Videoconferencing capabilities are often part of a standard personal computer package (Intel, 1999). With the introduction of home videoconferencing packages between $100 and $1,000, small non-profit groups and individual users can afford to try videoconferencing, dramatically increasing the number of video-ready endpoints (Davis, 1997; Essex, 1999). Standards are particularly important in this market because they allow individuals to shop for price and features, without worrying about compatibility with other systems.

H.323. The Internet uses packet switching: Digitized messages are broken up into small chunks and routed through various pathways to their final destination, a method that makes videoconferencing over the Internet particularly problematic (Labriola, 1997). The third ITU standard, H.323, applies to networks using Internet protocol (IP), generally LANs and WANs. Data conferencing via the Internet is already appearing as an embedded capability in Internet browsers. H.323-compliant videoconferencing capabilities are being embedded into the latest versions of Microsoft Explorer and Netscape Communicator Web browsers.

The move to desktop videoconferencing for organizations is also enabled by multicast software that allows a source to send a data stream (video, audio, and text) that several destinations can receive by subscribing to a designated address (Krapf, 1997).

Some vendors have introduced videoconferencing products designed for home users and handheld devices (generally referred to as videophones) compatible with H.323 (Intel, 1999;

Moch, 2000). However, few of the products utilizing Internet connections from the home support frame rates comparable to the 30 frames per second of full-motion video

Three New Standards

Although ITU standards have facilitated development of hardware, a series of industry standards in routing and switching video over the Internet may prove to be the key to next-generation videoconferencing (Veil, 2000). The differentiated service (DiffServ) and multi-protocol label switching (MPLS) initiatives are designed to assist network management of video streaming applications, including real-time videoconferencing. DiffServ is an initiative of the Internet Engineering Task Force (IETF) that assigns high-bandwidth sessions, such as videoconferencing, priority routing. It speeds up the videoconferencing process by moving critical routing services to the edge of the network. This process helps balance the flow of traffic through the Internet, easing clogs and potentially providing clearer, more reliable videoconferencing. The alternative protocol, MPLS, works within a different layer of the network to simplify and streamline the routing of "data packets." The MPLS packet labeling system works with various kinds of videoconferencing systems, reducing the need for sending information along specific pathways in the Internet and making transmission more efficient (Degnan, 1999).

Current Status

Emerging social and organizational factors are driving the diffusion of videoconferencing. Collaboration and productivity are the watchwords for enterprises in communications, and videoconferencing is increasingly accepted as a tool that enables both. Enterprise networking—connecting the computers in an organization and with the resources of vendors, suppliers, and collaborators—is the *sine qua non* of the decentralized organization. Videoconferencing facilitates enterprise networking to the extent that it allows users to share information efficiently and collaborate effectively (Morrison & Liu Sheng, 1992). The introduction of high-speed networks and easy-to-use video portals enables more efficient and more "life-like" videoconferencing. There has been significant movement from hardware to software solutions for videoconferencing (Hibner, et al., 1997).

Bandwidth—the speed and capacity of a telecommunications network—has long been a key deterrent to videoconferencing diffusion. Enterprise consumers have relied on ISDN lines to provide adequate capacity for reliable service. A significant factor in adoption may be the cost of ISDN, still the preferred transmission medium. ISDN prices in the United States vary widely by provider and region. Costs associated with ISDN remain a considerable part of the videoconferencing price equation, especially when multiple lines are required for videoconferences. But, for many businesses, ISDN enhances communication speeds and efficiencies across the board, including accelerated Internet access and faster processing throughout the LAN. The portion of ISDN costs accorded specifically to videoconferencing will vary from business to business or home to home.

As videoconferencing continues to move to the desktop, however, "prosumers" are looking to two different high-bandwidth pipelines. Digital subscriber line (DSL) is the option touted by telephone companies. DSL is affordable for small companies and other enterprises and has proved to be fairly reliable (Cohen, 1999). There are a few of drawbacks to DSL, however. Incumbent telephone companies have been accused of slowing the diffusion of DSL by delaying implementation for competing Internet service providers (ISPs), so customers are unable to get the service through their local ISP (New Networks Institute, 2000). Second, the technology is distance-sensitive: The farther a consumer is from the central telephone switch, the lower the bandwidth. Finally, the most popular DSL configuration is asynchronous—bandwidth is generally much higher going to the consumer than back out. That downstream bias will negatively affect videoconferences from homes or remote locations. Downstream bias is a problem for cable modems as well; the upstream bandwidth is about 10% of the downstream. Cable modems are also limited by the ISP's connection to the Internet, rarely more than 1.5 Mb/s (Megabits per second).

Both cable modems and DSL are faster than typical ISDN lines, and both can utilize at least some of the existing infrastructure. Provider investment in high-bandwidth infrastructure is outpacing demand in many areas. Some experts predict a bandwidth glut that may force providers to step up their marketing efforts of high bandwidth services, especially videoconferencing (King, 2000). New videoconferencing products delivered via the Internet will drive the industry in the near future (Schober, 1999).

Factors to Watch

New Products

Videoconferencing revenues have not approached even the most cautious growth predictions (Nicholas, 1998). PictureTel and VTel, two of the largest players in the industry, suffered combined losses of more than $100 million on sharply lower revenues in 1999. System service and support have been the revenue producers in the industry lately, while international sales have fallen. The industry will have to find new markets and capitalize on Internet-based services in order to grow and survive (Culver, 1999). Videoconferencing companies note customers are taking a wait-and-see attitude, as they anticipate multi-functional units that operate on a variety of delivery "pipelines" (VTel, 1999). Internet-based systems promise to connect more people at cheaper prices with more videoconferencing options. Nevertheless, a variety of videoconferencing systems are available. Desktop systems range from less than $100 to around $1,500 for high-quality systems such as the PictureTel 550. Portable, integrated systems weighing less than 10 pounds and including a

Photo courtesy of VTel Corporation.

power supply can be had for under $8,000. The Intel TeamStation 5.0, a conference room workstation, costs closer to $11,000. Most industry watchers agree that the future will depend on how well companies can serve the growing demand for portability and affordability, two chief attributes of IP-based systems. Although history cautions against making predictions, a few trends will be worth watching.

The concepts of standards, infrastructure, and services come together in the videoconferencing portal. These portals, launched by video veterans such as Lucent, Polycom, and Microsoft, offer a "place" on the Web to organize one-on-one videophone calls or multipoint conferences (Schober, 1999; Vonder Haar, 2000). Portals also sell videoconferencing software online and offer some basic services for free (see below). Internet portals—such as Yahoo!—were considered to be the key attractor by early Internet investors, but they did not catch on with consumers. It remains to be seen if the videoconferencing aspect will provide the added appeal to attract enterprises and other key markets.

Past experience with Internet-based communications indicates that the "killer app" for videoconferencing may have less to do with the quality of the video image than with the ease of communicating. Video pioneers are piggybacking on instant messaging systems, a clear favorite of Internet communicators (Quinton, 2000). Like e-mail and instant messaging, the video messaging offered through White Pine's CUseeMe software is widely available, easy-to-use, and inexpensive. A complete video messaging kit costs less than $100, and users can log-on and chat within a variety of communities before they make even that monetary commitment (www.cuseeme.com). "Videochats" may prove to be particularly popular with younger users who have grown up with instant messaging systems.

Instant video messaging may prove to be the first viable video application for wireless communications devices as well. Although enterprises are not particularly interested in wireless videoconferencing, experiments with personal wireless video messaging will escalate as cellular and PCS devices continue their rapid diffusion into the mainstream (Wireless to be led, 1999). Streaming video applications may come first, but Sanyo plans to market a two-way mobile videophone in the United States in 2001 (Moch, 2000). VisualPhone by Kyocera debuted in 1999 and is expected to be available in Japan in 2000 (www.kyocera.com).

Social Issues

As Internet videoconferencing advances, both video and telecommunications companies continue to roll out new products to create new revenue streams. However, cultural factors, such as language barriers and cultural norms, continue to inhibit the effectiveness of international videoconferencing (Cukor, 1995). Increasing expectations of quality and mobility in telecommunications will shape the destiny of these firms.

It is also worth noting that the geographical position of a user on the network can affect their ability to videoconference. Those near the fringes of the network have limited access to high-bandwidth technologies, and experience more delays and breakups in video transmission (Weber, 2000). Network economics dictate that any action by one user on the network affects all other users. The resulting network externalities can be positive as in the increased connectivity for each additional user, or negative as when increased demand on the network slows communication for all users. Since the days of the experimental videophone, videoconferencing has been envisioned as a publicly accessible service. However, active videoconferencers such as schools, businesses, and medical facilities have generally relied upon dedicated networks.

As videoconferencing applications move to the Internet, Web video portals and other services are essentially privatizing sections of the public network. This privatization can have deleterious effects on network users, especially those whose position on the network is most precarious (Sassen, 1999). Codecs and software protocols that identify and prioritize video packets reverse the old Internet adage that "bits are bits" (Negroponte, 1996). The same packet discrimination that allows videoconferencing to function smoothly could potentially be used to prioritize messages from particular senders. Just as first class, business, and economy passengers all use the same airline network but receive different priority and service, Internet users could have their messages sorted according to their ability to pay.

The videoconferencing market is moving on parallel tracks. The ISDN-based standard has the advantage of reliability, but the disadvantage of limited access. The IP-based standard has the advantage of almost universal connectivity, but does not match the quality of ISDN systems. These tracks are not mutually exclusive. Depending on the nature of the conference, ISDN may be appropriate in some instances, and IP in others. Companies are likely to keep their options open, at least over the next few years, until the efficacy of the H.323 standard can be demonstrated.

It's inevitable that desktops will migrate to IP; however, the road to IP will be evolutionary, not revolutionary. IP conferencing provides the opportunities for low-cost, point-to-point, and multipoint conferencing (when using multicast). However, the network infrastructure will have to be upgraded before we see any meaningful movement toward IP.

Most corporations already have a sizable investment in their existing networks, and many either have upgraded or are in the process of upgrading their networks. The corporate Intranet can support IP videoconferencing (or soon will be able to), but this isn't true for the public Internet.

The principal obstacle to IP conferencing is the WAN. The public Internet does not provide the quality of service required for business-quality video. This is in contrast to ISDN, where the public network does provide a guaranteed quality of service. There are other options for IP videoconferencing over the WAN, but these options are not universally available, have performance limitations, or add to the operating costs.

So, is it ISDN or IP? ISDN desktops are a proven technology and will be around for a while. These products are best utilized when deployed in small numbers and used to communicate with other ISDN (group or desktop) products over the WAN. On the other hand, IP products are ideal for collaborative communications when used in a campus environment over high-speed LANs or in an environment where existing IP networks are already linked by a high-speed enterprise WAN. It's just another application riding on the IP network.

It is safe to say that IP and ISDN desktops will coexist for the foreseeable future. The two networks will interoperate thanks to the H.323 gateway—the "glue" tying the two together.

Bibliography

Borthick, S. (1997, November/December). DVC market: Are we there yet? *Desktop Videoconferencing Magazine*. [Online]. Available: http://www.bcr.com/dvcmag/novdec/dvc6p10.htm.

Bracker, Jr., W., & Sarch, R. (1995). *Cases in network implementation: Enterprise networking*. New York: Van Nostrand Rheinhold.

Cohen, A. (1999, November 19). Bionic browsing. *PC Magazine*. [Online]. Available: www.zdnet.com/pcmag/stories/trends/0,7607,2397451,00.html.

Cukor, P., & Coppock, K. (1995, April). International videoconferencing: A user survey. *Telecommunications*, 33-34.

Culver, D. (1999, November 15). IP videoconferencing: Hope & hype. *Interactive Week*. [Online]. Available: http://www.zdnet.com/intweek/stories/news/0,4164,2396451,00.html.

Davis, A. (1997, September/October). Videoconferencing markets and strategies. *Desktop Videoconferencing Magazine*. [Online]. Available: http://www.bcr.com/dvcmag/septoct/dvc5p23.htm.

Degnan, C. (1999, June 25). PictureTel to unveil eVideo application server. *PCWeek Online*. [Online] Available: www.zdnet.com/pcweek/stories/news/0,4153,1015267,00.html.

Forward Concepts. (1997). *Teleconferencing videoconferencing markets and strategies: From novelty to necessity*. [Online]. Available: http://www.fwdconcepts.com/studindx.htm.

Hibner, D., Noufer, M., Ahuja, Y., Reed, G., Mangia, K., & Kovac, R. (1997). *Desktop videoconferencing systems get better*. Applied Research Institute, Center for Information and Communication Sciences,

Intel Internet Videophone. (1999, January 12). *Internet product watch*. [Online]. Available: http://ipw.internet.com/communication/internet_telephone/916161459.html.

International Telecommunications Union. (1997). *Summary of H.320, H.323 and H.324 standards*. [Online]. Available: http://www.itu.int/itudoc/itu-t/rec/h/s_h320_e_43422.html.

ITCA. (1997). Teleconferencing revenues passed $5 billion in North America in 1996. News release from the International Teleconferencing Videoconferencing Association. [Online]. Available: http://www.itca.org/web/memberservices/pressitca/970627.html.

King, R. (1996, December). A work (still) in progress. *tele.com*, 85-88.

King, R. (2000, February 24). V-Span seeks life beyond videoconferences. *Interactive Week*. [Online]. Available: www.zdnet.com/intweek/stories/news/0,4164,2445325,00.html.

Korostoff, K. (1996, September 16). Desktop video is coming, really! *Computerworld*, 35.

Krapf, E. (1997, November/December). The role of multicast in videoconferencing over IP networks. *Desktop Videoconferencing Magazine*. [Online]. Available: http://www.bcr.com/dvcmag/novdec/dvc6p32.htm.

Kyocera. *Visual Phone*. [Online]. Available: http://www.kyocera.co.jp/frame/product/telecom/english/vp210_e/index.htm.

Labriola, D. (1997, December 16). Videoconferencing on the net. *PC Week Online*. [Online]. Available: http://www.zdnet.com/pcmag/pclabs/nettools/1622/tools/tools1.htm.

Lapolla, S. (1997, March 17). Vendors bolster Web-based videoconferencing efforts. *PC Week*, 123.

Moch, C. (2000, March 20). Video streams to handheld devices. *Internet Telephony*. [Online]. Available: www.internettelephony.com/asp/ItemDisplay.asp?ItemID=8464.

Morrison, J., & Liu Sheng, O. (1992). Communication technologies and collaboration systems: Common domains, problems and solutions. *Information and Management*, *23*, 93-112.

New Networks Institute. (2000). [Online]. Available: http://www.internetnews.com/isp-news/article/0,2171,8_339351,00.html.

Nicholas, K. (1998). Teleconferencing. In A. E. Grant & J. H. Meadows (Eds.). *Communication technology update*, 6th ed. Boston: Focal Press.

Noll, A. M. (1992). Anatomy of a failure: Picturephone revisted. *Telecommunications Policy*, 307-317.

Olgren, C., & Parker, L. (1983). *Teleconferencing videoconferencing technology and applications*. Dedham, MA: Artech House.

Pacific Bell. (1998). ISDN: Prices quoted by PacBell. [Online]. Available: http://www.pacbell.com/products/index.html.

PictureTel. (2000). *Annual report*. [Online]. Available: http://www.picturetel.com/info/default.htm.

Polycom. (1997). [Online]. Available: http://www.polycom.com.

Quinton, B. (2000, March 20). Ericsson's instant messaging is a sight. *Internet Telephony*. [Online]. Available: www.internettelephony.com/asp/ItemDisplay.asp?ItemID=8611.

Schober, D. (1999, October 12). Lucent, Polycom team on videoconferencing. *Internet Telephony.* [Online]. Available: www.internettelephony.com/asp/ItemDisplay.asp?ItemID=6795.

Testa, B. M. (2000, January 3). Video free flow. *Internet Telephony.* [Online]. Available: www.internettelephony.com/asp/ItemDisplay.asp?ItemID=7429.

Veil, M. (2000, January 24). The solution for IP service quality. *Internet Telephony.* [Online]. Available: www.internettelephony.com/asp/ItemDisplay.asp?ItemID=7627.

Videoconferencing: Ready for prime time? (1996, May). *Managing Office Technology,* 52-53.

Vonder Haar, S. (2000, April 7). Let's talk video on the Web. *Interactive Week.* [Online]. Available: www.zdnet.com/intweek/stories/news/0,4164,2522405,00.html.

VTel. (1997). [Online]. Available: http://www.vtel.com.

VTel. (1999). *Annual report.* [Online]. Available: http://www.vtel.com/company/invest/Default.htm.

Weber, T. (2000, March 28). Lucent launches Net video venture. *Internet Telephony.* [Online]. Available: www.internettelephony.com/asp/ItemDisplay.asp?ItemID=8587.

Wireless to be led by same old apps. (1999, August 3). *Cahners In-Stat.* [Online]. Available: http://cyberatlas.internet.com/big_picturehardware/article/0.1323.5921_173781.00.html.

V

CONCLUSIONS

25

Trends in Selected U.S. Communications Media

Dan Brown, Ph.D.*

This chapter updates three previous efforts (Brown & Bryant, 1989; Brown, 1996; Brown, 1998) to chart the progress of various communications media from their outset. The philosophy of reporting in all four efforts focuses on media units and penetration (i.e., the percentage of marketplace use such as households), rather than on dollar expenditures. The example of box office receipts from motion pictures offers a notable exception. Even here, however, the data include unit attendance figures with the dollar amounts. Considering the dynamic value of the dollar, more meaningful media consumption trends emerge from examining changes in non-monetary units.

Another premise of this chapter holds that government sources should provide as much of the data as possible. Researching the growth or sales figures of various media over time quickly reveals conflicts in both dollar figures and units shipped or consumed. Such conflicts exist even among government reports. For example, marked differences have existed for some years between the *Statistical Abstract* tables and *Industrial Outlook* tables in reporting motion picture box office receipts. Although private sources frequently form the basis for government reports, emphasizing government sources provides at least some consistency to the data-gathering process. Readers should use caution in interpreting data for individual years and instead emphasize the trend over several years.

Print Media

The American print industry is the largest such industry among the printing countries of the world, consisting of approximately 70,000 firms that generated $211 billion in domestic revenues

* Associate Dean of Arts & Sciences, East Tennessee State University (Johnson City, Tennessee).

in 1998 (U.S. Department of Commerce/International Trade Association, 1999, p. 25-1). A printing firm or newspaper publisher exists in almost every American community. Foreign investment in printing is bringing a more global characteristic to the industry, with half of the 20 largest American book publishers having foreign ownership.

Despite some minor fluctuations, both the number of American newspaper firms and newspaper circulation (Table 25.1 and Figure 25.1) remain lower than the figures posted in the early 1990s. Circulation among American newspapers inched downward throughout the decade but actually rose in many developing nations. Heavy volume in buying and selling newspaper firms occurred in the United States during the latter half of the 1990s. As large newspaper chains bought smaller ones, only 77 independent American newspapers with circulation exceeding 30,000 remained in business in 1997. The number of daily newspapers fell, but newspaper firms frequently expanded to offer online versions. More than 3,600 newspapers around the world offer a World Wide Web presence (U. S. Department of Commerce/International Trade Association, 1999, p. 25-5), and more than 500 American daily newspapers offer online versions (p. 25-7).

Figure 25.1
Newspaper Firms and Daily Newspaper Circulation

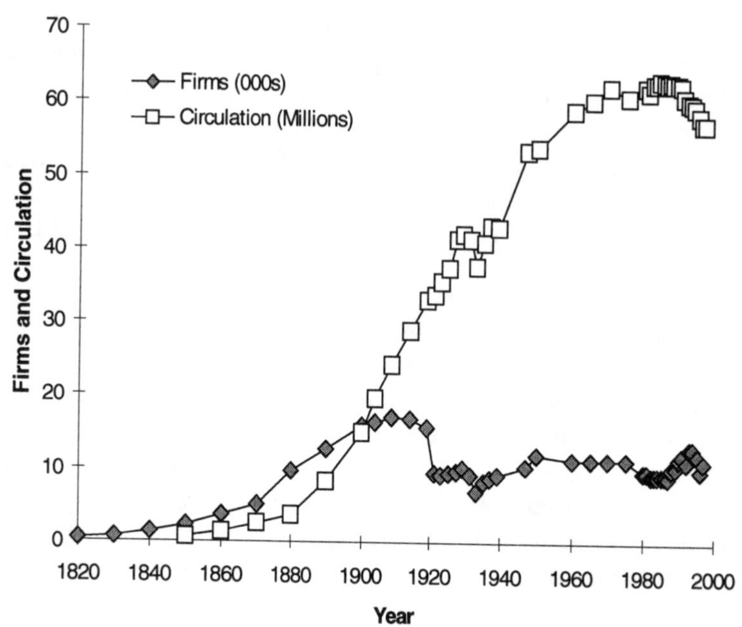

Table 25.1

Newspaper Firms and Daily Newspaper Circulation

Year	Firms	Circulation (millions)	Year	Firms	Circulation (millions)
1704	1		1931	9,299	41.3
1710	1		1933	6,884	37.6
1720	3		1935	8,266	40.9
1730	7		1937	8,826	43.3
1740	12		1939	9,173	43.0
1750	14		1947	10,282	53.3
1760	18		1950	12,115	53.8
1770	30		1960	11,315	58.9
1780	39		1965	11,383	60.4
1790	92		1970	11,383	62.1
1800	235		1975	11,400	60.7
1810	371		1980	9,620	62.2
1820	512		1981	9,676	61.4
1830	715		1982	9,183	62.5
1840	1,404		1983	9,205	62.6
1850	2,302	0.8	1984	9,151	63.1
1860	3,725	1.5	1985	9,134	62.8
1870	5,091	2.6	1986	9,144	62.5
1880	9,810	3.6	1987	9,031	62.8
1890	12,652	8.4	1988	10,088	62.7
1900	15,904	15.1	1989	10,457	62.6
1904	16,459	19.6	1990	11,471	62.0
1909	17,023	24.2	1991	11,689	60.0
1914	16,944	28.8	1992	11,339	60.0
1919	15,697	33.0	1993	12,597	60.0
1921	9,419	33.7	1994	12,513	59.0
1923	9,248	35.5	1995	12,246	57.0
1925	9,569	37.4	1996	10,466	57.0
1927	9,693	41.4	1997	10,042	57.0
1929	10,176	42.0	1998	10,508	56.2

Note: The data from 1704 through 1900 are from Lee (1973). The data from 1904 through 1947 are from U.S. Bureau of the Census (1976). The data between 1947 and 1986 are from U.S. Bureau of the Census (1986). The data from 1987 through 1988 are from U.S. Bureau of the Census (1995c). The data after 1988 are from U.S. Bureau of the Census (1999).

Forces driving newspapers to develop Websites include the proliferation of non-newspaper sites on the Web that offer classified advertising. Newspapers will face critical business decisions in the short term in deciding whether to move to directly offer online news content to end users or to take advantage of their ability to provide content to other concerns that need news for their sites.

Except for a dip in 1996 and 1997, the number of periodical titles (Table 25.2 and Figure 25.2) remained fairly stable. Many American firms rapidly expanded international operations in the mid-1990s. Concern about paper and distribution costs, together with new marketing opportunities, sparked higher per-copy prices among periodicals and considerable corporate interest in new media. Single copy newsstand sales of magazines fell below 20% of total magazine circulation. In

1995, half of the 10 largest-selling American magazines reduced circulation to focus on target demographic groups and cut marketing costs on circulation (U.S. Department of Commerce, 1998).

The American book publishing industry accounts for about 30% of worldwide demand for books, and 90% of the American industry product goes to domestic trade (U.S. Department of Commerce, 1998, p. 25-10). During the 1990s, the number of book titles printed (Table 25.3 and Figure 25.3) experienced sharp growth after a period of steady increase, with the 1995 total reflecting nearly one-third more titles than the 1990 output.

Table 25.2
Published Periodical Titles

Year	Titles	Year	Titles
1904	1,493	1975	9,657
1909	1,194	1980	10,236
1914	1,379	1981	10,873
1919	4,796	1982	10,688
1921	3,747	1983	10,952
1923	3,829	1984	10,809
1925	4,496	1985	11,090
1927	4,659	1986	11,328
1929	5,157	1987	11,593
1931	4,887	1988	11,229
1933	3,459	1989	11,556
1935	4,019	1990	11,092
1937	4,202	1991	11,239
1939	4,985	1992	11,143
1947	4,610	1993	11,863
1954	3,427	1994	12,136
1958	4,455	1995	11,179
1960	8,422	1996	9,843
1965	8,990	1997	8,350
1970	9,573	1998	12,036

Note: Data from 1904 through 1958 are from U.S. Bureau of the Census (1976). The data from 1960 through 1985 are from U.S. Bureau of the Census (1986). Data from 1987 through 1988 are from U.S. Bureau of the Census (1995c). Data for 1988 through 1991 are from U.S. Bureau of the Census (1997), and data after 1991 are from U.S. Bureau of the Census (1999).

Figure 25.2
Published Periodical Titles

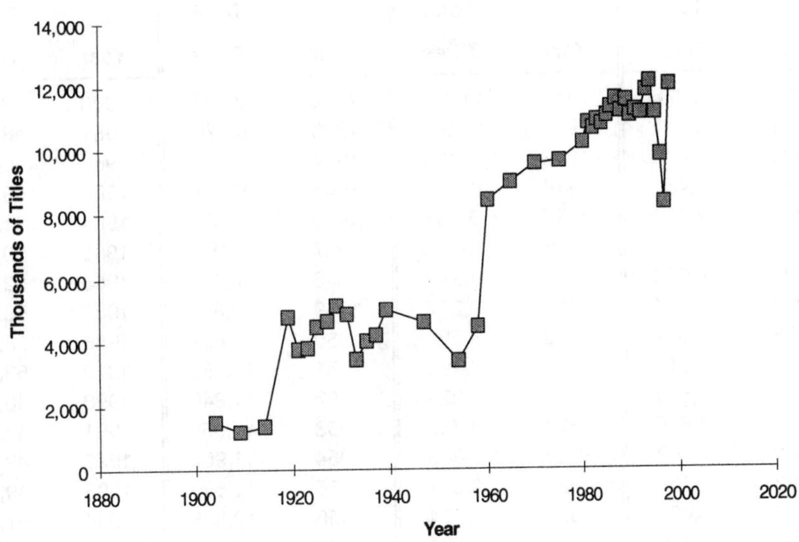

Figure 25.3
Published Book Titles

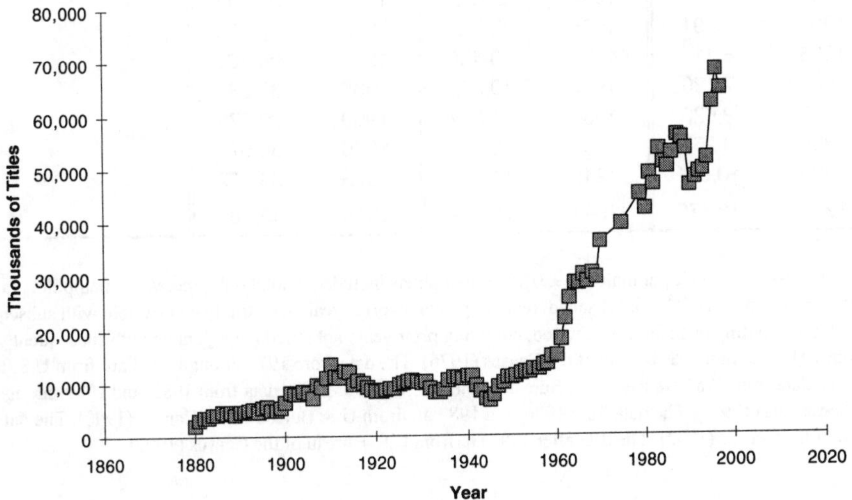

Table 25.3
Published Book Titles

Year	Total Titles	Year	Total Titles	Year	Total Titles	Year	Total Titles
1880	2,076	1911	11,123	1942	9,525	1980	42,377
1881	2,991	1912	10,903	1943	8,325	1981	48,793
1882	3,472	1913	12,230	1944	6,970	1982	46,935
1883	3,481	1914	12,010	1945	6,548	1983	53,380
1884	4,088	1915	9,734	1946	7,735	1984	51,058
1885	4,030	1916	10,445	1947	9,182	1985	50,070
1886	4,676	1917	10,060	1948	9,897	1986	52,637
1887	4,437	1918	9,237	1949	10,892	1987	56,057
1888	4,631	1919	8,594	1950	11,022	1988	55,483
1889	4,014	1920	8,422	1951	11,255	1989	53,446
1890	4,559	1921	8,329	1952	11,840	1990	46,738
1891	4,665	1922	8,638	1953	12,050	1991	48,146
1892	4,862	1923	8,863	1954	11,901	1992	49,276
1893	5,134	1924	9,012	1955	12,589	1993	49,756
1894	4,484	1925	9,574	1956	12,538	1994	51,663
1895	5,469	1926	9,925	1957	13,142	1995	62,039
1896	5,703	1927	10,153	1958	13,462	1996	68,175
1897	4,928	1928	10,354	1959	14,876	1997	64,711
1898	4,886	1929	10,187	1960	15,012		
1899	5,321	1930	10,027	1961	18,060		
1900	6,356	1931	10,307	1962	21,904		
1901	8,141	1932	9,035	1963	25,784		
1902	7,833	1933	8,092	1964	28,451		
1903	7,865	1934	8,198	1965	28,595		
1904	8,291	1935	8,766	1966	30,050		
1905	8,112	1936	10,436	1967	28,762		
1906	7,139	1937	10,912	1968	30,387		
1907	9,620	1938	11,067	1969	29,579		
1908	9,254	1939	10,640	1970	36,071		
1909	10,901	1940	11,328	1975	39,372		
1910	13,470	1941	11,112	1979	45,182		

Note: The data for 1880-1919 include pamphlets; 1920-1928, pamphlets included in total only; thereafter, pamphlets excluded entirely. Beginning in 1959, the definition of "book" changed, rendering data on prior years not strictly comparable with subsequent years. Beginning in 1967, the counting methods were revised, rendering prior years not strictly comparable with subsequent years. The data from 1904 through 1947 are from U.S. Bureau of the Census (1976). The data from 1975 through 1983 are from U.S. Bureau of the Census (1984). The data from 1984 are from U.S. Bureau of the Census (1985). The data from 1985 and 1989 through 1992 are from U.S. Bureau of the Census (1995c). The data from 1986 and 1987 are from U.S. Bureau of the Census (1990). The data from 1988 are from U.S. Bureau of the Census (1992). The data after 1993 are from U.S. Bureau of the Census (1999).

Telephone

The Telecommunications Act of 1996 encouraged competition between providers of electronic communications, and the Federal Communications Commission (1999b) publicly announced its intention to encourage the development of new telecommunications technologies and reduce regulatory barriers to their entry into the marketplace. The telephone continued as an American staple (Table 25.4 and Figure 25.4), and the exploding growth of wireless telephones accelerated in the late 1990s. Following the sevenfold increase in the number of cellular phones from 1990 to 1996, the rate of increase in cellular telephone subscribers exceeded 25% from 1997 to 1998.

Figure 25.4
Telephone Penetration and Cellular Telephone Subscribers

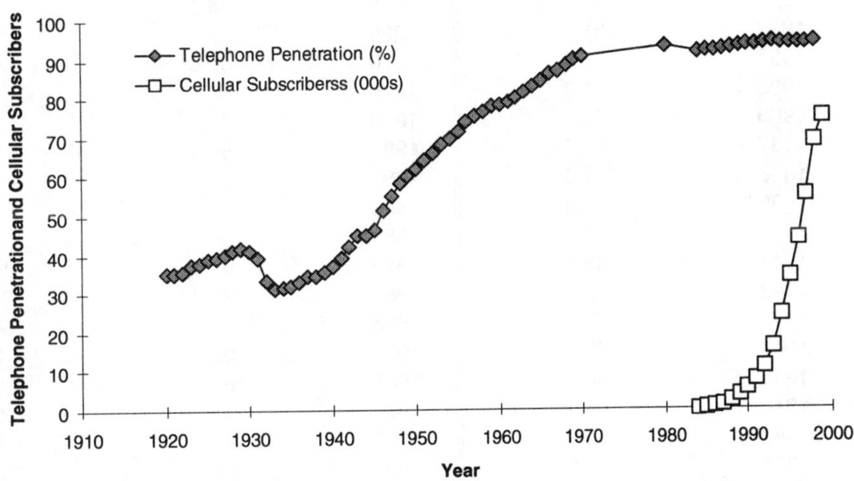

Table 25.4

Telephone Penetration and Cellular Telephone Subscribers

Year	Households with Telephones (%)	Year	Households with Telephones (%)	Cellular Subscribers (thousands)
1920	35.0	1956	73.8	
1921	35.3	1957	75.5	
1922	35.6	1958	76.4	
1923	37.3	1959	78.0	
1924	37.8	1960	78.3	
1925	38.7	1961	78.9	
1926	39.2	1962	80.2	
1927	39.7	1963	81.4	
1928	40.8	1964	82.8	
1929	41.6	1965	84.6	
1930	40.9	1966	86.3	
1931	39.2	1967	87.1	
1932	33.5	1968	88.5	
1933	31.3	1969	89.8	
1934	31.4	1970	90.5	
1935	31.8	1975		
1936	33.1	1979		
1937	34.3	1980	93.0	
1938	34.6	1981		
1939	35.6	1982		
1940	36.9	1983		1
1941	39.3	1984	91.8	100
1942	42.2	1985	92.2	350
1943	45.0	1986	92.2	682
1944	45.1	1987	92.5	1,231
1945	46.2	1988	92.9	2,069
1946	51.4	1989	93.0	3,509
1947	54.9	1990	93.3	5,283
1948	58.2	1991	93.6	7,557
1949	60.2	1992	93.9	11,033
1950	61.8	1993	94.2	16,009
1951	64.0	1994	93.9	24,134
1952	66.0	1995	93.9	33,786
1953	68.0	1996	93.8	44,043
1954	69.6	1997	93.9	55,312
1955	71.5	1998	94.1	69,209
		1999		75,000

Note: 1950-1982 data applied to principal earners filing reports with FCC; earlier data applies to Bell and independent companies. Beginning in 1959, data includes figures from Alaska and Hawaii. The data for 1986 and 1987 are estimates. The data to 1970 are from U.S. Bureau of the Census (1976). The data from 1970 through 1982 are from U.S. Bureau of the Census (1986). The data after 1982 are from U.S. Department of Commerce (1987). The data from 1986 and 1987 are from U.S. Bureau of the Census (1987, 1988). The data for 1987 through 1989 are from U.S. Bureau of the Census (1997). The data from 1989 through 1998 are from U.S. Bureau of the Census (1999), except that households with telephones for 1998 are from the FCC (2000). Cellular telephone subscribership for 1999 is from the FCC (2000b).

Motion Pictures

Table 25.5 and Figure 25.5 show motion picture box office receipts and average weekly attendance at motion picture theaters. The government *Statistical Abstract* and *Industry Outlook* stopped reporting average weekly attendance for recent years. However, a new figure replaced the older one to show trends in hours per person per year that people spent viewing motion pictures in theaters. The latter figure reported 11 hours per person per year for 1992, which increased by slightly more than 15% to 13 hours in 1998, with no change in 1999. Admissions increased every year after 1992 except in 1995, and the total number of admissions grew by about 2% from 1.37 billion in 1996 to 1.39 billion a year later (U.S. Department of Commerce/International Trade Association, 1999, p. 32-2). Overall, the stability of attendance reflects little change for more than three decades. However, box office receipts increased by about 17% during the 1990s, perhaps reflecting changes in ticket prices. Such increases may also reflect the increasing average production and distribution costs that characterized the industry in the late 1990s (U. S. Department of Commerce/International Trade Association, 1999).

Figure 25.5
Motion Picture Attendance and Box Office Receipts

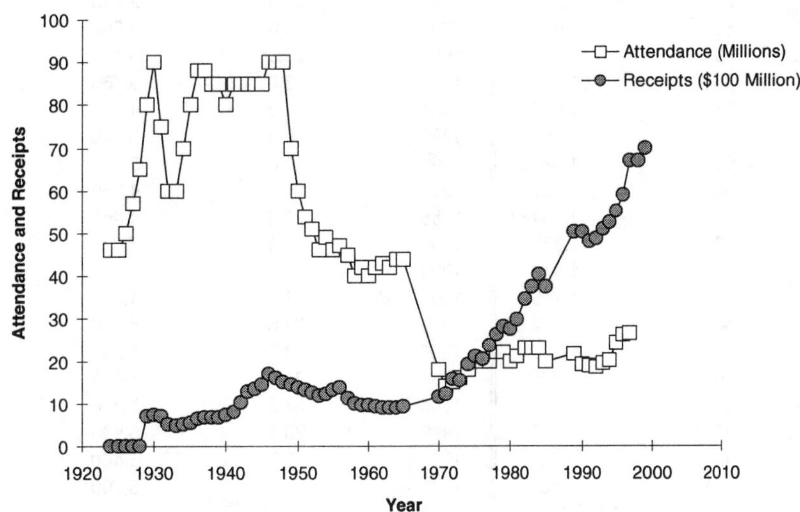

Table 25.5
Motion Picture Attendance and Box Office Receipts

Year	Average Weekly Attendance*	Receipts ($ million)	Year	Average Weekly Attendance*	Receipts ($ million)	Hours Per Person Per Year
1922	40		1961	42.0	921	
1923	43		1962	43.0	903	
1924	46	1.71	1963	42.0	904	
1925	46		1964	44.0	913	
1926	50		1965	44.0	927	
1927	57		1966		964	
1928	65		1967		989	
1929	80	720	1968		1,045	
1930	90	732	1969		1,099	
1931	75	719	1970	18.0	1,162	
1932	60	527	1971	14.0	1,214	
1933	60	482	1972	15.0	1,583	
1934	70	518	1973	16.0	1,524	
1935	80	556	1974	18.0	1,909	
1936	88	626	1975	20.0	2,115	
1937	88	676	1976	20.0	2,036	
1938	85	663	1977	20.0	2,372	
1939	85	659	1978	22.0	2,643	
1940	80	735	1979	22.0	2,821	
1941	85	809	1980	20.0	2,749	
1942	85	1,022	1981	21.0	2,966	
1943	85	1,275	1982	23.0	3,453	
1944	85	1,341	1983	23.0	3,766	
1945	85	1,450	1984	23.0	4,030	
1946	90	1,692	1985	20.3	3,749	
1947	90	1,594	1986	19.6	3,780	
1948	90	1,506	1987	20.9	4,250	
1949	70	1,451	1988	20.9	4,460	
1950	60	1,376	1989	21.8	5,030	
1951	54	1,310	1990	22.8	5,020	
1952	51	1,246	1991	21.9	4,800	
1953	46	1,187	1992	22.6	4,870	11
1954	49	1,228	1993	23.9	5,200	12
1955	46	1,326	1994	24.8	5,400	12
1956	47	1,394	1995	24.3	5,500	12
1957	45	1,126	1996	26.3	5,900	12
1958	40	992	1997	26.7	6,400	12
1959	42.0	958	1998		6,700	13
1960	40.0	951	1999		7,000	13

* In millions.

Note: The data to 1970 are from U.S. Bureau of the Census (1976). The data from 1970, 1975, and 1979 through 1985 are from U.S. Bureau of the Census (1986). The box office data from 1971 are from U.S. Bureau of the Census (1975). The box office receipts data from 1972 through 1988 are from U.S. Department of Commerce (1988), and the data for 1988 through 1992 are from U.S. Department of Commerce (1994). The 1991 attendance came from U.S. Bureau of the Census (1996). The data for 1993 through 1996 are from U.S. Department of Commerce (1998). The data for 1997 through 1999 are from U.S. Department of Commerce (1999). All data for hours per person per year are from U.S. Department of Commerce (1999), and the figures after 1997 represent projected estimates.

Recording

Overall recorded music shipments, excluding CDs, steadily declined through the 1990s, as shown in Table 25.6. Figure 25.6 illustrates the trends in various music types other than compact discs, and Figure 25.7 tracks overall music sales of non-CD types. The first decline in the number of CDs shipped (Table 25.7 and Figure 25.8) since CDs were introduced in 1983 occurred in 1997. However, the number of units shipped rebounded by more than 12% in 1998. Sales revenues for the industry remained strong, and several albums registered vigorous sales. Good and bad news arose from estimates that music sales on the Internet would reach as much as 7.5% of total sales by 2002 (U.S. Department of Commerce/International Trade Association, 1999). The new market opportunities were balanced by increased concerns about piracy.

Table 25.6
Recorded Music Unit Shipments (Excluding CDs)

Year	Singles	LPs/EPs	Cassettes	8-Tracks	Total
1973	228.0	280.0	15.0	91.0	614.0
1974	204.0	276.0	*15.3*	96.7	592.0
1975	164.0	257.0	*16.2*	94.6	531.8
1976	190.0	273.0	21.8	106.1	590.9
1977	190.0	344.0	*36.9*	127.3	698.2
1978	190.0	341.3	61.3	133.6	726.2
1979	195.5	318.3	82.8	104.7	701.1
1980	164.3	322.8	110.2	86.4	683.7
1981	154.7	295.2	137.0	48.5	635.4
1982	137.2	243.9	182.3	14.3	577.7
1983	125.0	210.0	237.0	6.0	578.0
1984	132.0	205.0	332.0	6.0	675.0
1985	121.0	167.0	339.0	4.0	631.0
1986	93.9	125.2	344.5		658.0
1987	82.0	107.0	410.0		683.3
1988	65.6	72.4	450.1		588.1
1989	36.6	34.6	446.2		517.4
1990	27.6	11.7	442.2		481.5
1991	22.0	4.8	360.1		386.9
1992	19.8	2.3	336.4		358.5
1993	15.1	1.2	339.5		355.8
1994	11.7	1.9	345.4		359.0
1995	10.2	2.2	272.6		285.0
1996	10.1	2.9	225.3		238.3
1997	7.5	2.7	172.6		182.8
1998	5.4	3.4	158.5		167.3

Note: The data for all years prior to 1983 are from U.S. Department of Commerce (1986). The data from 1983 through 1985 are from U.S. Bureau of the Census (1986). The data for 1986 through 1989 are from U.S. Bureau of the Census (1995c). The data after 1989 are from U.S. Bureau of the Census (1999). "Singles" refer to vinyl singles.

Figure 25.6
Recorded Music Unit Shipments (Excluding CDs)

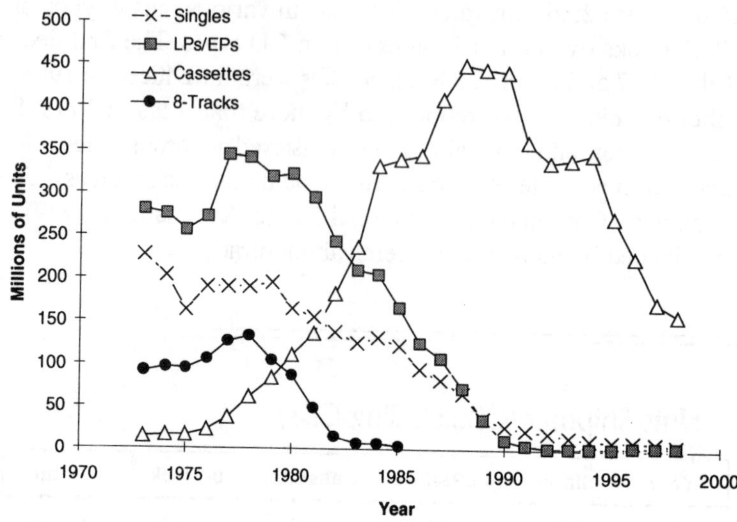

Figure 25.7
Total Recorded Music Sales (Excluding CDs)

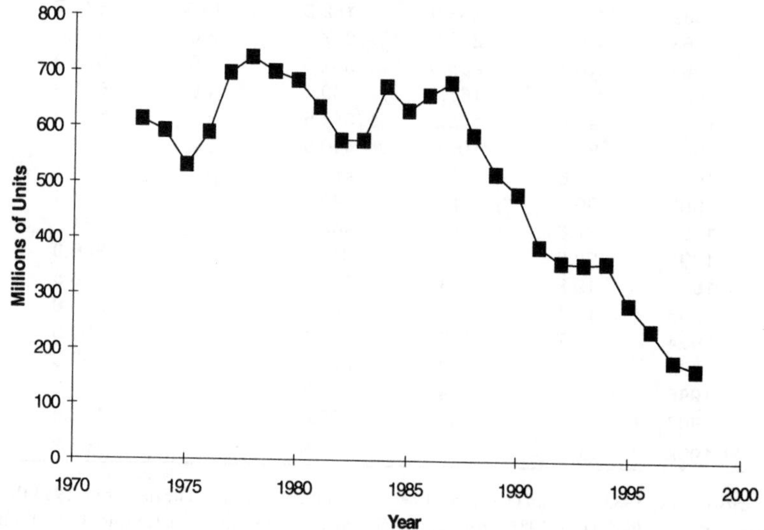

Table 25.7
Compact Discs and Players

Year	Disc Shipments (millions)	Player Shipments (thousands)	Penetration (percent)
1983	1	45	0
1984	6	200	1.60
1985	23	850	6.40
1986	53	1,500	7.00
1987	102.1	1,384	
1988	149.7	4,800	7-8
1989	207.2	4,280	12-16
1990	286.5	6,900	19
1991	333.3	9,260	28
1992	407.5		35
1993	495.4	21,500	42
1994	662.1	26,500	
1995	722.9		
1996	778.9		
1997	753.1		
1998	847.0		

Note: The data for discs through 1985 are from U.S. Bureau of the Census (1986), and after 1986 are from U.S. Bureau of the Census (1995c). The data for players through 1985 are from U.S. Bureau of the Census (1986), for 1986 from U.S. Bureau of the Census (1988), and 1986-1991 from respective issues of U.S. Department of Commerce (1986, 1987, 1988, 1989, 1991, and 1992). The data from 1993 and 1994 are from Trachtenberg (1995). The data for penetration for 1984 and 1985 are from Graham (1986). The data for penetration for 1986 are from Lewyn (1987) and from 1992-1993 from U.S. Bureau of the Census (1995c). The shipment figure for 1994 is from U.S. Bureau of the Census (1997). Other data after 1993 are from U.S. Bureau of the Census (1999).

Figure 25.8
Compact Discs and Players

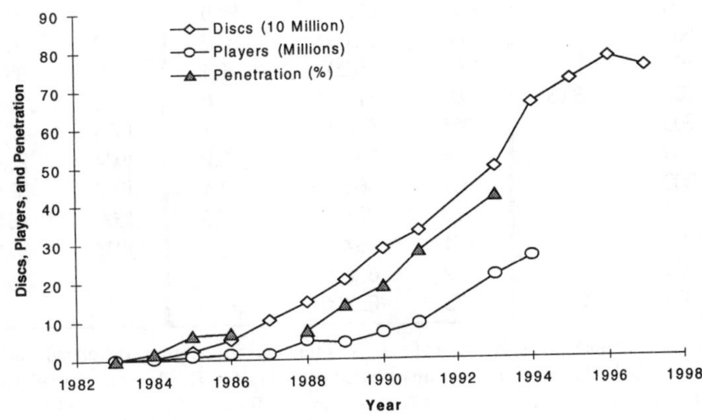

Radio

As of November 1997, 10,470 commercial radio stations held licenses in the United States. That figure swelled to 12,582 in 1999 (FCC, 1999), but the concentration in ownership increased. The 15 largest broadcasters accounted for 42% of 1998 industry advertising revenues (U.S. Department of Commerce/International Trade Association, 1999, p. 32-8). Ownership restrictions have been relaxed, allowing single entities to own up to eight radio stations in one market. Ownership consolidation created cost savings and marketing efficiencies, benefiting economic strength in the industry that has manifested itself in advertising growth since 1993 (U.S. Department of Commerce/International Trade Association, 1999). Table 25.8 and Figure 25.9 show the penetration of radio in American households.

Table 25.8

Radio Households and Penetration

Year	HHs with Sets (000s)	% Penetration	Year	HHs with Sets (000s)	% Penetration	Year	HHs with Sets (000s)	% Penetration
1922	60		1948	37,623		1974	70,800	
1923	400		1949	39,300	93.4	1975	72,600	98.6
1924	1,250		1950	40,700		1976	74,000	
1925	2,750		1951	41,900		1977	75,800	
1926	4,500		1952	42,800		1978	77,800	
1927	6,750		1953	44,800		1979	79,300	
1928	8,000		1954	45,100		1980	79,968	99.0
1929	10,250		1955	45,900	95.9	1981	81,600	99.0
1930	13,750	40.3	1956	46,800	95.7	1982	82,691	99.0
1931	16,700		1957	47,600	95.8	1983	83,078	99.0
1932	18,450		1958	48,500	96.1	1984	84,553	99.0
1933	19,250		1959	49,450	96.1	1985	85,921	99.0
1934	20,400		1960	50,193	95.1	1986		99.0
1935	21,456		1961	50,695	94.7	1987		99.0
1936	22,869		1962	51,305	93.7	1988	91,100	99.0
1937	24,500		1963	52,300	94.6	1989	92,800	99.0
1938	26,667		1964	54,000	96.2	1990	94,400	99.0
1939	27,500		1965	55,200	96.1	1991	95,500	99.0
1940	28,500	80.3	1966	57,200	97.6	1992	96,600	99.0
1941	29,300		1967	57,500	97.1	1993	97,300	99.0
1942	30,600		1968	58,500	96.2	1994	98,000	99.0
1943	30,800		1969	60,600	97.4	1995	98,000	99.0
1944	32,500		1970	62,000	97.8	1996	98,000	99.0
1945	33,100		1971	65,400		1997	98,000	99.0
1946	33,998		1972	67,200				
1947	35,900	91.8	1973	69,400				

Note: Authorization of new radio stations and production of radio sets for commercial use was stopped from April 1942 until October 1945. 1959 is the first year for which Alaska and Hawaii are included in the figures. The data prior to 1970 are from U.S. Bureau of the Census (1976). The households with sets data from 1970-1972 are from U.S. Bureau of the Census (1972). The households with sets data from 1973 and 1974 are from U.S. Bureau of the Census (1975). The households with sets data from 1975 through 1977 are from U.S. Bureau of the Census (1978). The households with sets data from 1978 and 1979 are from U.S. Bureau of the Census (1981). All data from 1988 through 1994 are from U.S. Bureau of the Census (1995c). The data after 1994 are from U.S. Bureau of the Census (1999).

Figure 25.9
Radio and TV Households

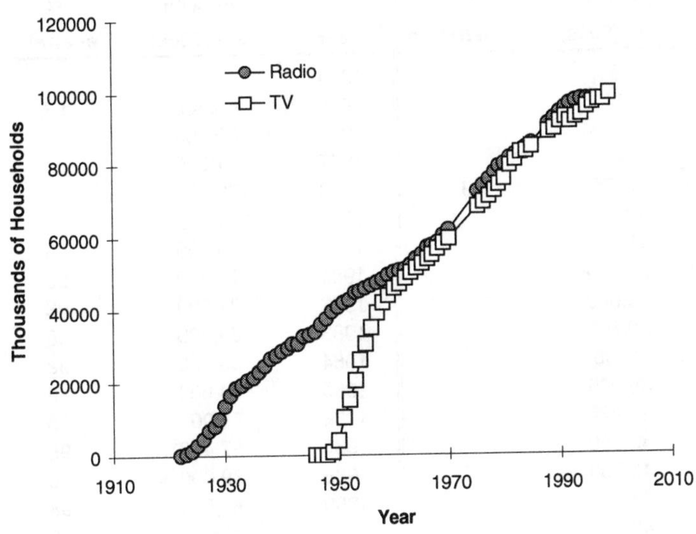

Television

Table 25.9 shows the steady rate of penetration and continuing growth in the number of households with TV sets. Figure 25.9 charts the close parallel growth in households with both television and radio. The number of broadcast television stations increased from 1,561 in 1997 to 1,599 in 1999 (FCC, 1999).

Progress toward digital television broadcasting fell short of the FCC requirement that all of the affiliates of the top four networks in the top 10 markets transmit digital signals by May 1, 1999. As of December 1999, 32 of these stations in the top 10 markets were transmitting with these signals, and 111 stations in the country were transmitting digital signals by the end of 1999. Sales of digital TVs set a fast early pace, with demand apparently exceeding supply. By August 1999, 27,000 such sets were sold, and the first digital VCRs entered the market in the summer 1999. As the penetration of digital broadcasting increases and compatible equipment becomes available, consumer benefits of digital broadcasting will accrue. Broadcasters will compete more effectively with cable operations and satellite vendors through the broadcaster offering high-definition signals, multicasting, and Internet access (FCC, 2000a).

Table 25.9
Television Households and Penetration

Year	HHs with Sets (000s)	% Penetration	Year	HHs with Sets (000s)	% Penetration
1946	8		1974		
1947	14		1975	68,500	97
1948	172		1976	69,600	
1949	940		1977	71,200	
1950	3,875	9	1978	72,900	
1951	10,320		1979	74,500	
1952	15,300		1980	76,300	98
1953	20,400		1981	79,900	98
1954	26,000		1982	81,500	98
1955	30,700		1983	83,300	98
1956	34.90		1984	83,800	98
1957	38,900		1985	84,900	98
1958	41,924		1986	85,900	98
1959	43,950		1987	87,400	98
1960	45,750	87	1988	89,000	98
1961	47,200		1989	90,000	98
1962	48,855		1990	92,000	98
1963	50,300		1991	93,000	98
1964	51,600		1992	92,000	98
1965	52,700		1993	93,000	98
1966	53,850		1994	94,000	98
1967	55,130		1995	95,000	98
1968	56,670		1996	96,000	98
1969	58,250		1997	98,000	98
1970	59,550	95	1998	98,000	
1971	61,600		1999	99,400	
1972					
1973					

Note: 1959 is the first year for which Alaska and Hawaii are included in the figures. The data dealing with households with television to 1971 are from U.S. Bureau of the Census (1976). The data dealing with households with television from 1980 through 1984 are from U.S. Bureau of the Census (1985). The data about penetration for all other pre-1987 years and all data for 1985 and 1986 are from U.S. Bureau of the Census (1986), and data for 1987-1991 are from U.S. Bureau of the Census (1995c). The data for 1991 through 1996 are from FCC (1999). The data after 1996 are from FCC (200a).

Cable Television

Table 25.10 shows the number of cable systems and subscribers and the proportion of homes using cable. Figure 25.10 portrays the growth of subscribers and penetration. Between the mid-1990s and now, the cable industry increased most measures of reach and capacity, including homes with cable available, subscribers, penetration, premium channel subscriptions, proportion of overall television viewing, and channel capacity (FCC, 1998a). Of the 99.4 million homes with televisions in 1999, cable passed 96.7% (96.1 million), up from 96.5% at the end of 1997. Among all cable systems, 22.4% of the 66,690,000 subscribers offered 54 or more channels, and 64.2% of cable subscribers could access 54 or more channels (FCC, 2000a).

Table 25.10
Cable TV Systems, Subscribers, and Penetration

Year	Systems	Subscribers (thousands)	Penetration (Percent)
1952	70	14	
1955	400	150	
1960	640	650	
1965	1,325	1,275	
1967	1,770	2,100	
1968	2,000	2,800	
1969	2,260	3,600	
1970	2,490	4,500	
1971	2,639	5,300	
1972	2,841	6,000	
1973	2,991	7,300	
1974	3,158	8,700	
1975	3,506	9,800	
1976	3,681	10,800	
1977	3,832	11,900	
1978	3,875	13,000	
1979	4,150	14,100	
1980	4,225	16,000	
1981	4,375	18,300	25.3
1982	4,825	21,000	29.0
1983	5,600	25,000	37.2
1984	6,200	30,000	41.2
1985	6,600	31,275	44.6
1986	7,600	36,933	46.8
1987	7,900	41,100	48.7
1988	8,500	41,100	49.4
1989	9,050	48,600	52.8
1990	9,575	52,600	56.4
1991	10,704	54,900	58.9
1992	11,075	55,800	60.2
1993	11,217	56,400	61.4
1994	11,230	57,200	62.4
1995	11,126	58,000	63.4
1996	11,119	60,280	67.8
1997	10,950	64,050	68.2
1998	10,845	64,170	69.1
1999	10,466	69,000	69.4

Note: The systems and subscribers data are from U.S. Bureau of the Census (1986). The penetration data through 1986 are from U.S. Bureau of the Census (1986). The penetration data through 1986 are from U.S. Bureau of the Census (1986) except for 1970, 1980, and 1987 through 1995. Of the latter, 1987 data are from U.S. Bureau of the Census (1988), and data from 1970, 1980, and 1988-1995 are from U.S. Bureau of the Census (1995c). The penetration data for 1995-1997 are from FCC (1998). The number of systems after 1995 is from U.S. Bureau of the Census (1999). The number of subscribers from 1996-1998 are from U.S. Bureau of the Census (1999). The number of subscribers from 1999 is from U.S. Department of Commerce/International Trade Association (1999). Penetration data for 1998 and 1999 are from FCC (2000). The number of cable systems in 1999 is from National Cable Television Association (2000).

Figure 25.10
Cable TV Subscribers and Penetration

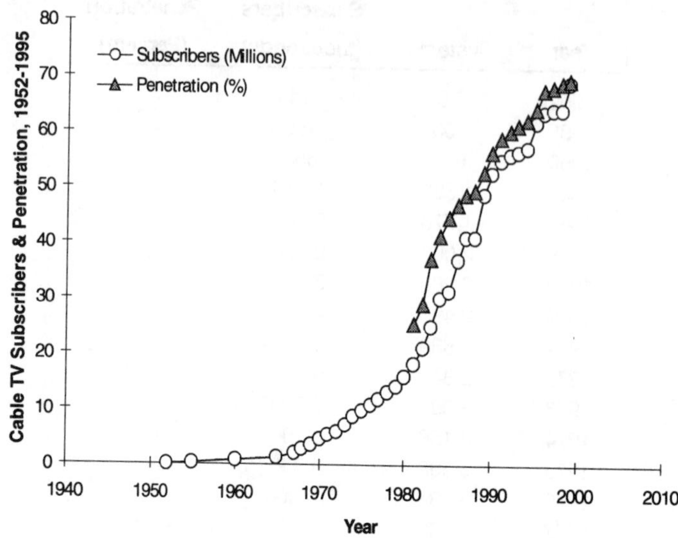

Movement at the end of the 1990s from analog to digital signals constituted a growing trend in cable. As of December 1998, the number of digital cable subscribers exceeded 1.2 million, and estimates projected the figure to surpass four million subscribers by the end of 1999 (FCC, 2000a). Digital transmissions can provide superior video picture quality through the ability to correct signal distortion that inevitably happens as signals move along transmission lines. Digital transmission also allows for signal compression that increases channel capacity and supports additional services in telephony and Internet access.

Broadband Internet connection by cable modem was available to more than 15 million homes by August 1998, and 300,000 subscribers took advantage of the opportunity. By July 1999, cable modems were available to more than 32 million households in the United States and Canada. More than one million households subscribed. Among the subscribers, more than 90% received two-way connection, and the remaining subscribers used telephone connections to send data back to cable operators (FCC, 2000a).

Direct Broadcast Satellite and Other Cable TV Competitors

Competitors for the cable TV industry include a variety of technologies, such as direct broadcast satellite (DBS), home satellite dish (HSD) services, multichannel multipoint distribution services (MMDS), and satellite master antenna television (SMATV). These technologies and cable TV make up a group known as multichannel video programming distributors (MVPDs). Table 25.11 and Figure 25.11 display trends for non-cable categories. As of June 1999, the total number of households receiving MVPD programming reached 80,882,411 or 81.4% of television households, an increase of 5.5% over the previous year (FCC, 2000a).

Cable television accounted for 82.5% of MVPD subscribership in June 1999, down from 90.7% in 1995. Cable has experienced annual declines since that year. The number of home satellite dish subscribers continued to decline in 1999, as it has every year after the last increase in 1995.

MMDS, also known as wireless cable, transmits line-of-sight signals to boosters or receiver systems and offers both analog and digital versions. Analog systems compete poorly with cable because of their limitation of 33 channels. Digital MMDS adds, in addition to increased channel capacity, features such as Internet access and telephony. For the third consecutive year, MMDS subscribers declined in 1999, falling to 821,000 subscribers. Among these subscribers, only 100,000 received digital services.

However, direct broadcast satellite subscriptions mushroomed to more than four times the number of subscribers recorded in the highest year of home satellite dish subscriptions (1995). SMATV systems deliver signals to adjacent buildings located mostly in urban areas. Their subscribers increased in 1999 by 54.3% over the 1998 total to 1,450,000. SMATV systems can join with direct broadcast satellite operations to offer combinations of services to residents of multiple dwelling units. This combination avoids the necessity for each family to set up a separate satellite dish and appears to offer good growth prospects.

Figure 25.11
Multichannel Video Program Distribution

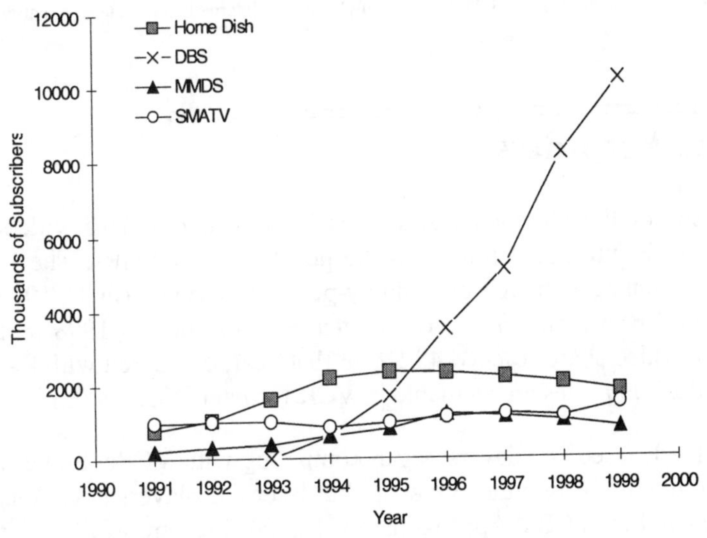

Table 25.11
Multichannel Video Program Distribution

Year	Satellite Dishes Sold	Home Satellite Dish Subscribers	DBS Subscribers	MMDS Subscribers	SMATV Subscribers
1980	4				
1981	20				
1982	60				
1983	225				
1984	525				
1985	700				
1986	280				
1987	400				
1988	330				
1989	335				
1990	380				
1991	146	764		180	965
1992	153	1,023		323	984
1993	201	1,612	<70	397	1,004
1994	436	2,178	602	600	850
1995	222	2,341	1,675	800	950
1996		2,300	3,500	1,200	1,100
1997		2,200	5,100	1,100	1,200
1998		2,028	8,200	1,000	1,100
1999		1,783	10,200	821	1,450

Note: Home satellite dishes represent the four- to eight-foot dishes. DBS uses smaller dishes. For dish sales, the 1986 figure is estimated, and the data from 1980 through 1983 are from Lewyn (1986). The dish sales for 1984 through 1986 are from U.S. Department of Commerce (1987), and the dish sales for 1987 through 1993 are from U.S. Bureau of the Census (1995c). All subscriber data through 1995 are from FCC (1995). Dish sales for 1995 exclude sales after August of that year. Subscriber data for 1996 and 1997 are from FCC (1998). Data for 1997 and 1998 are from U.S. Department of Commerce/International Trade Association (1999). Data for 1999 are from FCC (2000a).

Videocassette Recorders

Table 25.12 shows the shipments and household penetration of videocassette recorders (VCRs) by year, and Figure 25.12 illustrates the penetration of VCRs. The FCC sees VCRs as competition for premium cable television and pay-per-view services (FCC, 1995). Cassette rentals and sales accounted for $6.7 billion of an estimated $13.8 billion in 1998 motion picture studio revenue. This home video share reached 48.5% in that year, compared with $4.5 billion (45%) in 1996. An estimated 25,000 titles are available in VCR format (FCC, 2000a).

Beginning in 1981, laserdisc players began competing with VCRs. Some observers liked the digital features and quality of laserdisc viewing, but laserdisc players never challenged the levels of penetration achieved by VCRs. Approximately 18,000 titles are available in that format, and about two million households own laserdisc players (FCC, 2000a).

Table 25.12
VCR Shipments and Penetration

Year	VCRs (000s)	VCR Penetration (%)
1978	402	
1979	478	
1980	804	1.1
1981	1,330	1.8
1982	2,030	3.1
1983	4,020	5.5
1984	7,143	10.6
1985	18,000	20.8
1986		36.0
1987	43,000	48.7
1988	51,000	58.0
1989	58,000	64.6
1990	63,000	68.6
1991	67,000	71.9
1992	69,000	75.0
1993	72,000	77.1
1994	74,000	79.0
1995	77,000	81.0
1996	79,000	82.2
1997	82,000	84.2

Note: The data from 1978 are from U.S. Bureau of the Census (1982). The data from 1979 through 1984 are from U.S. Bureau of the Census (1984). The data from 1985 are from U.S. Bureau of the Census (1986). The VCR penetration data are from U.S. Bureau of the Census (1986). The VCR sales data for 1985 and 1988-1994 are from the U.S. Bureau of the Census (1995c), and data from 1987 are from the U.S. Bureau of the Census (1990). The VCR penetration data for 1988-1994 are from U.S. Bureau of the Census (1995c), and the data for 1993 and 1997 are from U.S. Bureau of the Census (1999).

About the same number of 1999 households owned digital videodisc (DVD) players, a technology that entered the market in February 1997. Although only about 2,800 DVD titles were available in mid-1999, about 5,000 titles were anticipated by the end of the year. Forecasts predicted that four million homes would own DVD players by the end of 1999. These machines suffer from shortcomings of high costs relative to VCRs and the inability to record programs (FCC, 2000a).

New competitors for the home video market, known as personal video recorders (PVRs), are actually digital video recorders. These devices can pause, rewind, and perform slow motion and instant replay of a live program. They are designed for use with programming guides through telephone connections to create personal menus of recording programs. Some services carry monthly fees, and some rely on advertising support. Unfortunately, these PVRs cannot play videocassettes or videodiscs. Price ranges from $499 to $1,500 varied with the amount of storage capacity. The

FCC reported that the number of units sold and programming service subscriptions taken were small (FCC, 2000a).

Figure 25.12
VCR Shipments and Penetration

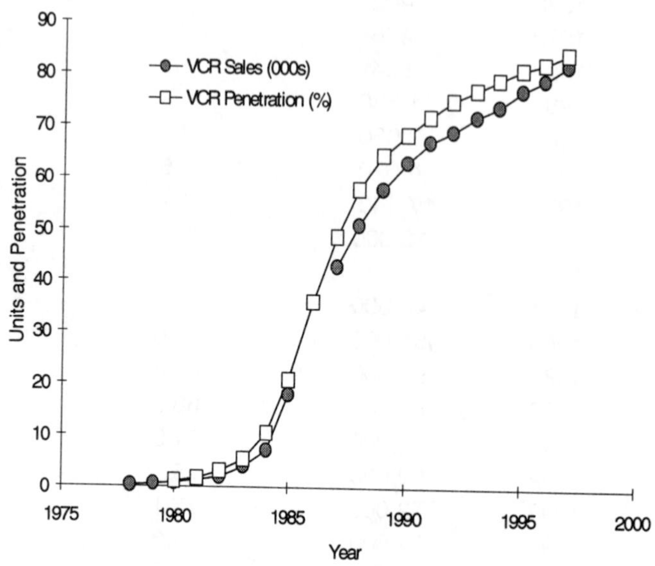

Personal Computers

Shipments of personal computers, including desktops, servers, and notebooks, surpassed 25 million units in 1996, a 14% increase over figures from the previous year. By 1997, shipments grew by another 20% to 30 million personal computers, creating an installed base of 110 million computers in 1997. This represents a 21% increase from the previous year.

Falling prices of computers, growth of Internet use, and convergence of computers with other media stimulated demand, and the proportion of American households with computers rose to 42% in 1998. Forecasts of annual shipments exceeding 60 million computers by 2002 demonstrated the enthusiasm with which Americans made computers a common home appliance (U S. Department of Commerce/International Trade Association, 1999). Table 25.13 and Figure 25.13 trace the steady rise in sales and penetration of personal computers.

Table 25.13
Personal Computer Shipments and Home Use

Year	PCs Shipped	Homes with PCs	% of Homes with PCs*
1,978	212,000		
1,979	246,000		
1,980	371,000		
1,981	1,110,000	750,000	0
1,982	3,530,000	3,000,000	0
1,983	6,900,000	7,640,000	0
1,984	7,610,000	11,990,000	0
1,985	6,750,000	14,960,000	0
1,986	7,040,000	17,380,000	0
1,987	8,340,000	19,970,000	0
1,988	9,500,000	22,380,000	0
1,989	9,329,970		0
1,990	9,848,593		0
1,991	10,903,000	25,000,000	0
1,992	12,544,374	25,000,000	0
1,993	14,775,000	31,000,000	0
1,994	18,605,000	32,010,000	0
1,995	22,582,900		
1,996	25,000,000		
1,997	30,000,000	35,020,000	0
1,998			0

* The percentages of households with computers before 1988 are calculated by dividing the number of homes with computers by the number of homes with TVs from Table 25.9. The figures after 1988 are from Electronic Industries Association (1995). The percentage of homes with PCs for 1997 is from Newburger (1999). The homes with PCs for 1997 are calculated by multiplying the percentage of homes with PCs times the number of homes with TVs from Table 25.9. The percentage of homes with PCs for 1998 is from FCC (2000a).

Note: Shipments for 1978 are from U.S. Bureau of the Census (1983); from 1979 and 1980 are from U.S. Bureau of the Census (1984). The data for 1981-1988 are from U.S. Bureau of the Census (1992). Shipments for 1989-1990 are from U.S. Bureau of the Census (1993), 1991-1993 data are from U.S. Bureau of the Census (1995c), and 1994-1995 data are from U.S. Bureau of the Census (1997). The computer shipments for 1996 are from U.S. Department of Commerce (1998), and shipments for 1997 are from U.S. Department of Commerce/International Trade Association (1999).

Figure 25.13
Personal Computer Shipments and Home Use

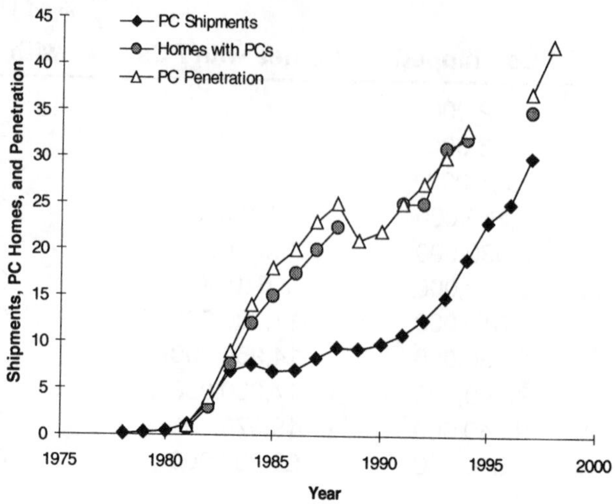

Synthesis

This chapter documents trends from government records involving traditional media. This method suffers from uneven availability of data reported in readily-available government sources. Book titles, VCR shipments, and radio usage have not been updated since 1997. Newspaper figures, periodical titles, music sales, and computer shipments were updated to 1998. Usage of broadcast television, cable television, satellite dish connection, MMDS, SMATV, motion pictures, and cellular phones were updated to 1999.

During the 1990s, media forms showing declines included newspaper circulation and the number of newspaper firm owners, although the number of newspaper firms rebounded in 1998. Published book titles decreased slightly, the first decline since 1990.

Most forms of recorded music showed declines in the 1990s, including the first decline in CD sales since their debut. Shipments of total non-CD music units declined noticeably with dropping sales of cassettes. Sales of music singles continued to plummet drastically, but LPs reclaimed some lost popularity. LP shipments increased every year after 1993, except for a drop in 1997. The 1998 LP gain exceeded 25% of the 1997 levels. The recording industry began to cope with emerging tools for distributing digital copies of recordings via the Internet. Radio households remained steady, and television households increased slightly. Cable television penetration and total subscribers increased slightly.

Media showing increases in recent years included the number of published periodical titles, which reversed a three-year decline with a gain of more than 44% in 1998 over 1997. Motion picture attendance and box office receipts increased somewhat in the decade, and box office receipts increased every year after 1991. However, the number of hours per person per year that people spent viewing movies in theaters remained flat in 1999 relative to 1998.

Although household telephone penetration increased slightly, cellular phones grew by nearly 10% in 1999 over 1998. Explosive growth occurred in sales of small-dish DBS systems, accompanied by rapidly dropping popularity of larger home satellite dishes. SMATV systems also increased in popularity, participating with DBS in an increase in penetration of overall multichannel video media. VCR penetration slightly increased in penetration, but new digital competitors such as DVD continued to emerge. Home computers reached 42% penetration, threatening to complete the transformation from personal to mass media.

Growth in computer use, new regulations promulgating the spread of digital broadcast television, and the popularity of all sorts of digital entertainment appear poised to threaten most analog media forms. Even books, periodicals, newspapers, phone calls, and television viewing have joined the parade. The Internet seems ready to change the means for distributing most media content, and convergence is the modern theme whose reflection gleams from the statistical patterns of using media.

Bibliography

Brown, D. (1998). Trends in selected American communications media. In A. E. Grant, & Meadows, J. H. (Eds.). *Communication technology update* (6th ed.). Boston: Focal Press, 279-305.

Brown, D. (1996). A statistical update of selected American communications media. In A. E. Grant (Ed.), *Communication technology update* (5th ed.). Boston: Focal Press, 327-355.

Brown, D., & Bryant, J. (1989). An annotated statistical abstract of communications media in the United States. In J. Salvaggio, & J. Bryant (Eds.), *Media use in the information age: Emerging patterns of adoption and consumer use.* Hillsdale, NJ: Lawrence Erlbaum Associates, 259-302.

Electronic Industries Association. (1995, August 24). Computer sales. *Johnson City Press*, 15.

Federal Communications Commission. (1995, December 11). *Annual assessment of the status of competition in the market for the delivery of video programming.* CS Docket No. 95-61.

Federal Communications Commission. (1998, January 13). *In the matter of assessment of the status of competition in markets for the delivery of video programming.* Fourth Annual Report, CS Docket No. 97-141. [Online]. Available: http://www.fcc.gov/Bureaus/Cable/Reports/fcc97423.txt.

Federal Communications Commission. (1999, July 31). *Broadcast station totals as of July 31, 1999*, FCC News Release. [Online]. Available: http://www.fcc.gov/Bureaus/Mass_Media/News_Releases/1999/nrmm9022.wp.

Federal Communications Commission. (1999b, April 2). *1998 biennial regulatory review—Testing new technology.* CC Docket No. 98-94.

Federal Communications Commission. (2000a, January 14). *Annual assessment of the status of competition in markets for the delivery of video programming.* CS Docket No. 99-230.

Federal Communications Commission. (2000b). *Trends in telephone service—March 2000.* [Online]. Available: http://www.fcc.gov/ccb/stats.

Graham, J. (1986, July 7). Sales of CD players are soaring. *USA Today*, 1D.

Lee, A. (1973). *The daily newspaper in America.* New York: Octagon Books.

Lewyn, M. (1986, July 16). Sending a new signal. *USA Today*, 4B.

Lewyn, M. (1987, May 29). Video CDs to debut tomorrow. *USA Today*, 1B, 2B.

National Cable Television Association. (2000). *Cable television industry: Overview 2000.* [Online]. Available: http://www.ncta.com/glance.html.

Newburger, E. C. (1999, September). *Computer use in the United States.* Washington, DC: U. S. Government Printing Office [Online]. Available: http://www.census.gov/prod/99pubs/p20-522.pdf.

U.S. Bureau of the Census. (1972). *Statistical abstract of the United States: 1972* (93rd ed.). Washington, DC: U.S. Government Printing Office.

U.S. Bureau of the Census. (1975). *Statistical abstract of the United States: 1975* (96th ed.). Washington, DC: U.S. Government Printing Office.

U.S. Bureau of the Census. (1976). *Statistical history of the United States: From colonial times to the present.* New York: Basic Books.

U.S. Bureau of the Census. (1978). *Statistical abstract of the United States: 1978* (99th ed.). Washington, DC: U.S. Government Printing Office.

U.S. Bureau of the Census. (1981). *Statistical abstract of the United States: 1981* (102nd ed.). Washington, DC: U S. Government Printing Office.

U.S. Bureau of the Census. (1982). *Statistical abstract of the United States: 1982-1983* (103rd ed.). Washington, DC: U.S. Government Printing Office.

U.S. Bureau of the Census. (1983). *Statistical abstract of the United States: 1984* (104th ed.). Washington, DC: U.S. Government Printing Office.

U.S. Bureau of the Census. (1984). *Statistical abstract of the United States: 1985* (105th ed.). Washington, DC: U.S. Government Printing Office.

U. S. Bureau of the Census. (1985). *Statistical abstract of the United States: 1986* (106th ed.). Washington, DC: U.S. Government Printing Office.

U.S. Bureau of the Census. (1986). *Statistical abstract of the United States: 1987* (107th ed.). Washington, DC: U.S. Government Printing Office.

U.S. Bureau of the Census. (1988). *Statistical abstract of the United States: 1989* (109th ed.). Washington, DC: U.S. Government Printing Office.

U.S. Bureau of the Census. (1990). *Statistical abstract of the United States: 1991* (111th ed.). Washington, DC: U.S. Government Printing Office.

U.S. Bureau of the Census. (1992). *Statistical abstract of the United States: 1993* (113th ed.). Washington, DC: U.S. Government Printing Office.

U.S. Bureau of the Census. (1993). *Statistical abstract of the United States: 1994* (114th ed.). Washington, DC: U.S. Government Printing Office.

U.S. Bureau of the Census. (1995a). *Consumer electronics 1992.* [Online]. Available: http://www.census.gov/ftp/pub/industry/ma36m92.txt.

U.S. Bureau of the Census. (1995b). *Consumer electronics annual 1994.* [Online]. Available: http://www.census.gov/ftp/pub/industry/ma36m94.txt.

U.S. Bureau of the Census. (1995c). *Statistical abstract of the United States: 1996* (116th ed.). Washington, DC: U. S. Government Printing Office.

U.S. Bureau of the Census. (1996). *Statistical abstract of the United States: 1997* (117th ed.). Washington, DC: U.S. Government Printing Office.

U.S. Bureau of the Census. (1999). *Statistical abstract of the United States: 1999* (119th ed.). Washington, DC: U.S. Government Printing Office.

U.S. Department of Commerce. (1986). *U.S. industrial outlook 1986.* Washington, DC: U.S. Department of Commerce, U.S. Bureau of Economic Analysis, and U. S. Bureau of Labor Statistics.

U.S. Department of Commerce. (1987). *U.S. industrial outlook 1987.* Washington, DC: U.S. Department of Commerce, U.S. Bureau of Economic Analysis, and U.S. Bureau of Labor Statistics.

U.S. Department of Commerce. (1988). *U.S. industrial outlook 1988.* Washington, DC: U.S. Department of Commerce, U.S. Bureau of Economic Analysis, and U.S. Bureau of Labor Statistics.

U.S. Department of Commerce. (1989). *U.S. industrial outlook 1989.* Washington, DC: U.S. Department of Commerce, U.S. Bureau of Economic Analysis. and U.S. Bureau of Labor Statistics.

U.S. Department of Commerce. (1991). *U.S. industrial outlook 1991.* Washington, DC: U.S. Department of Commerce, U.S. Bureau of Economic Analysis, and U.S. Bureau of Labor Statistics.

U.S. Department of Commerce. (1992). *U.S. industrial outlook 1992.* Washington, DC: U.S. Department of Commerce, U.S. Bureau of Economic Analysis, and U.S. Bureau of Labor Statistics.

U.S. Department of Commerce. (1994). *U.S. industrial outlook 1994.* Washington, DC: U.S. Department of Commerce, U.S. Bureau of Economic Analysis, and U.S. Bureau of Labor Statistics.

U.S. Department of Commerce. (1998). *U.S. industry and trade outlook 1998.* New York: McGraw-Hill.

U.S. Department of Commerce/International Trade Association. (1999). *U.S. industry and trade outlook '99.* New York: McGraw-Hill.

26

Conclusions

August E. Grant, Ph.D.

The first chapter of this book suggested that analysis of new technologies must include five factors: the physical hardware involved, the messages (software), the organizational infrastructure, the social system, and the users of the technology. In a text of this type, the hardware is necessarily one key focus, but a review of the most important developments since the last edition of this book was prepared suggests that, since 1998, organizational developments have been paramount in affecting the course of these new technologies.[1]

Perhaps the best example is the single-most influential technology discussed—the Internet—as it is a factor in the discussions of most of the technologies in this book. Although there have certainly been underlying changes in the technologies driving the Internet, the most important change by far was at the organizational level, as businesses, government, and other institutions worked to take advantage of this rapidly-growing medium. Indeed, one of the most important changes in the decade since the Internet began its rapid growth is the move from a non-commercial orientation to a commercial one. In short, the Internet has been good for business.

This is not to understate the importance of the other factors. High-definition television, for example, has experienced only minor technological changes, but major changes in the amount of content transmitted using the technology (software), as well as continuing battles within the political system regarding the implementation of the standard.

One major technology in which the hardware will be the driving force for change in the first half of the current decade is home video, as hard-disk-based personal video recorders (PVRs) begin to replace analog VCRs. This technology is one of the few in this book in which changes in hardware represent the most important change in the technology. This is in marked contrast to the

1. The author wishes to acknowledge that part of the reason for his perception of the change in emphasis might be related to the fact that he has moved from academia to the business world since the last edition of this *Update* was published, possibly increasing the salience of organizational factors.

first editions of this book, in the early half of the 1990s, when digital technology first began to supplant analog technology on a widespread basis.

Looking back at the table of contents of previous editions of this *Update* provides another indicator of how much the landscape of communication technologies has changed. Table 26.1 provides a picture of how that landscape is different, indicating technologies that have been added and deleted from the table of contents since 1992.

Table 26.1
Changes in Communication Technology Update Table of Contents Since 1992

Additions	Deletions
Internet	Low-Power Television
Streaming Media	Television Shopping
Electronic Commerce	Videotext and Teletext
Home Networks and Residential Gateways	Videodisc
Personal Communication Devices	Satellite News Gathering
Switched Broadband Networks	Desktop Publishing
Virtual Reality	Digital Video Interactive/CD-I
Personal Communication Services	Digital Compact Cassette
Multimedia	Very Small Aperture Terminals
	Videophone

Source: A. E. Grant

The lesson from Table 26.1 is that we should expect similar dramatic changes in the communication technology landscape over the next eight years as well. In the same manner that the Internet and World Wide Web have emerged since 1992 to become one of the most prominent technologies, perhaps another technology will emerge in the coming years to have the same remarkable impact.

One of the most important issues is how to analyze emerging technologies. Ironically, when predicting the likely success of any innovation, the safest prediction is always failure. Of all new products, services, and technologies introduced, only a fraction reach the market, with few of these lasting more than a few years. The ones that last, such as the telephone, radio, television, and (perhaps!) the Internet, have the potential to bring about profound changes in society.

The first prediction, therefore, is that many of the newest technologies discussed in this book will fail. Some will fail because the technology could never live up to its promise, while others will succumb to marketplace competition. Some superior technologies will fail because of inferior marketing or distribution. Finally, some will fail because they were introduced at the wrong time: too early, too late, or during an economic recession.

External factors such as the general state of the economy are critical to any new technology. The level of investment in development of a technology must be matched by the investment in marketing the technology. Growth of many of the technologies discussed in this book has been fueled by a strong economy, which included a great deal of investment in research and development of new technologies.

One key question is whether—or when—a general economic downturn might occur. During times of economic recession, many companies are wary of making investment in technology development, preferring to postpone the expense until conditions improve. By that time, many technologies lose their windows of opportunity.

Competitive factors will become more important than ever in determining the success of new communication technologies. Many industries such as telephony and cable television, which have operated as virtual local monopolies, will face increasing competition from both inside and outside their industries. The primary beneficiary of competition will be the consumer, who will see prices for services fall to the cost of providing the service.

As consumers face an array of options for any communication technology, their choices will most likely be affected by the choices they have made in the past. They will want to do business with familiar company names and products. Brand identity will increase in importance for all communication technologies. One example of the importance of brand identity can be found in the new television distribution technologies such as direct broadcast satellites (DBS). Technologically, the transmission technology underlying this service has almost nothing in common with cable television, but the software (programming) is almost exactly the same.

As people make decisions to adopt new communication technologies, the thing they will look for first is the delivery of information they are already receiving from another source. Then, they will look for a relative advantage in price, quality, etc. As important as these factors are for early adoption, the next step is even more important. Once a technology is able to gain a foothold in the market, it can begin offering new services or features not found in the older technology. Consumers like both novelty and familiarity—but not too much of either.

Among all the technologies covered in this book, the one that is continuing to grow the fastest (for the sixth consecutive year) as of mid-2000 is the Internet. People began using the Internet for interpersonal communication via e-mail, but the service has evolved to provide a wealth of information retrieval tools, most notably the World Wide Web. New commercial applications of the Internet are introduced daily, with most of these making heavy use of the Internet's ability to transmit digital images, audio, and video as easily as text. The result is an exponential increase in data traffic on the network. As long as the capacity of the Internet can keep up with the increase in traffic, the Internet will remain the site of the boldest innovations among all communication technologies.

On the other hand, capacity limits may well lead to a "tragedy of the commons." This term refers to the use of a common grazing area by a village to support all of the animals in the village. The field is only able to support a limited number of animals. The problem is that the prosperity of individuals is based upon the number of animals they raise, so each attempts to raise as many animals as possible. Eventually, the field becomes over-grazed, and everyone suffers. The Internet is today's common field. If the amount of traffic—especially broadband media streams—increases to the point at which the flow of data slows, users will be forced to migrate to different media (if they can afford it). Consider the challenge faced by the Napster software program at the time of this writing. Major universities were forced to ban the software when files requested and supplied by Napster users overwhelmed the networks.

Clearly, the Internet is both the site of the fastest developments in the field of communication technology and the ideal subject for research on a variety of communication and social issues. A few years from now, we will know whether the Internet is a fad or not; in either case, research on the Internet should yield lessons that can be applied to a wide range of communication technologies.

Analysis of the other communication technologies discussed in the book indicates the importance of regulatory policy in shaping a technology. There is no technological reason why cable companies and telephone companies cannot provide both telephone and television service, but the structure of the current marketplace is a product of the fact that these two have been kept apart. As the regulatory barriers between sectors of the communication industry fall, the impacts upon all communication technologies will be profound.

The importance of organizational structure and the political system (through regulators) illustrates the importance of using the umbrella perspective discussed in Chapter 1 to analyze technologies. In almost every technology discussed, at least one of the other four elements of the umbrella (software, organizational infrastructure, social systems, and individual user) was more important than the hardware.

The authors of each chapter have worked to ensure that all information is current as of the time the manuscripts were submitted (late spring 2000). It is certain that a few items will be out of date by the time this book is printed, and more items will become dated every week. The reader is therefore encouraged to use the online updates provided on the *Communication Technology Update* home page (http://www.tfi.com/ctu), as well as study the bibliographies of each chapter to determine the best sources for up-to-date information.

The authors of these chapters have one thing in common: They all enjoy the process of observing changes in the spectrum of communication technologies. This author likens the "playing field" of communication technologies to a spectator sport, with the "scores" reported on a daily basis in the business section of the newspaper. It is my hope that this book will provide you with the introduction you need to follow this "sport," and give you the most important tool you will need for personal success—the ability to understand and predict the future of communication technologies so you can take advantage of the wealth of opportunities they will provide.

Glossary

2-wire line. The set of two copper wires used to connect a telephone customer with a switching office, loosely wrapped around each other to minimize interference from other twisted pairs in the same bundle. Synonymous with twisted pair.

802.11. An IEEE specification for 1 Mb/s and 2 Mb/s wireless LANs.

802.11HR. An IEEE specification for 11 Mb/s high rate wireless LANs.

802.3. An IEEE specification for SCMA/CD based Ethernet networks.

802.5. An IEEE specification for token ring networks.

Adaptive transform acoustic coding (ATRAC). A method of digital compression of audio signals used in the MD (minidisc) format. ATRAC ignores sounds out of the range of human hearing to eliminate about 80% of the data in a digital audio signal.

Addressability. The ability of a cable system to individually control its converter boxes, allowing the cable operator to enable or disable reception of channels for individual customers instantaneously from the home office.

ADSL (*asymmetrical digital subscriber line*). A system of compression and transmission that allows broadband signals up to 6 Mb/s to be carried over twisted pair copper wire for relatively short distances.

Advanced television (ATV). Television technologies that offer improvement in existing television systems.

Agent. Any being in a virtual environment.

Algorithm. A specific formula used to modify a signal. For example, the key to a digital compression system is the algorithm that eliminates redundancy.

AM (*amplitude modulation*). A method of superimposing a signal on a carrier wave in which the strength (amplitude) of the carrier wave is continuously varied. AM radio and the video portion of NTSC TV signals use amplitude modulation.

American National Standards Institute (ANSI). An official body within the United States delegated with the responsibility of defining standards.

American Standard Code for Information Interchange (ASCII). Assigns specific letters, numbers, and control codes to the 256 different combinations of 0s and 1s in a byte.

Analog. A continuously varying signal or wave. As with all waves, analog waves are susceptible to interference, which can change the character of the wave.

ANSI. See *American National Standards Institute.*

ASCII. See *American Standard Code for Information Interchange.*

Aspect ratio. In visual media, the ratio of the screen width to height. Ordinary television has an aspect ratio of 4:3, while high-definition television is "wider" with an aspect ratio of 16:9 (or 5.33:3).

Asynchronous. Occurring at different times. For example, electronic mail is asynchronous communication because it does not require the sender and receiver to be connected at the same time.

ATM (*asynchronous transfer mode*). A method of data transport whereby fixed length packets are sent over a switched network. Speeds of up to 2 Gb/s can be achieved, making it suitable for carrying voice, video, and data.

Audio on demand. A type of media that delivers sound programs in their entirety whenever a listener requests the delivery.

Augmented reality. The superimposition of virtual objects on physical reality. For example, a technician could use augmented reality to display a three-dimensional image of a schematic diagram on a piece of equipment to facilitate repairs.

Available bit rate (ABR). An ATM service type in which the ATM network makes a "best effort" to meet the transmitter's bandwidth requirements. ABR differs from other best-effort service types by employing a congestion feedback mechanism that allows the ATM network to notify the transmitters that they should reduce their rate of data transmission until the congestion decreases. Thus, ABR offers a qualitative guarantee that the transmitter's data can get to the intended receiver without experiencing unwanted cell loss.

Avatar. An animated character representing a person in a virtual environment.

AVI (*audio-video interleaved*). Microsoft's video driver for Windows.

B

Backbone. The part of a communications network that handles the major traffic using the highest-speed, and often longest, paths in the network.

Bandwidth. A measure of capacity of communications media. Greater bandwidth allows communication of more information in a given period of time.

Basic rate interface ISDN (BRI-ISDN). The basic rate ISDN interface provides two 64-Kb/s channels (called B channels) to carry voice or data and one 16-Kb/s signaling channel (the D channel) for call information.

BISDN. See *Broadband integrated services digital network.*

Bit. A single unit of data, either a one or a zero, used in digital data communications. When discussing digital data, a small "b" refers to bits, and a capital "B" refers to bytes.

Bluetooth. A specification for short-range wireless technology created by a consortium of computer and communication companies that make up the Bluetooth Special Interest Group.

Bridge. A type of switch used in telephone and other networks that connects three or more users simultaneously.

Broadband. An adjective used to describe large-capacity networks that are able to carry several services at the same time, such as data, voice, and video.

Broadband integrated services digital network (BISDN). A second-generation ISDN technology that uses fiber optics for a network that can transmit data at speeds of 155 Mb/s and higher.

Buy rate. The percentage of subscribers purchasing a pay-per-view program divided by the total number of subscribers who can receive the program.

Byte. A compilation of bits, seven bits in accordance with ASCII standards and eight bits in accordance with EBCDIC standards.

C

Carrier. An electromagnetic wave or alternating current modulated to carry signals in radio, telephonic, or telegraphic transmission.

CATV (community antenna television). One of the first names for local cable television service, derived from the common antenna used to serve all subscribers.

C-band. Low-frequency (1 GHz to 10 GHz) microwave communication. Used for both terrestrial and satellite communication. C-band satellites use relatively low power and require relatively large receiving dishes.

CCD (*charge coupled device*). A solid-state camera pickup device that converts an optical image into an electrical signal.

CCITT (*International Telegraph and Telephone Consultative Committee*). CCITT is the former name of the international regulatory body that defines international telecommunications and data communications standards. It has been renamed the Telecommunications Standards Sector of the International Telecommunications Union.

CDDI (*copper data distributed interface*). A subset of the FDDI standard targeted toward copper wiring.

CD-I (*compact disc-interactive*). A proprietary standard created by Philips for interactive video presentations including games and education.

CDMA (*code division multiple access*). A spread spectrum cellular telephone technology, which digitally modulates signals from all channels in a broad spectrum.

CD-ROM (*compact disc-read only memory*). The use of compact discs to store text, data, and other digitized information instead of (or in addition to) audio. One CD-ROM can store up to 700 megabytes of data.

CD-RW (compact disc-rewritable). A special type of compact disc that allows a user to record and erase data, allowing the disc to be used repeatedly.

Cell. The area served by a single cellular telephone antenna. An area is typically divided into numerous cells so that the same frequencies can be used for many simultaneous calls without interference.

Central office (CO). A telephone company facility that handles the switching of telephone calls on the public switched telephone network (PSTN) for a small regional area.

Central processing unit (CPU). The "brains" of a computer, which uses a stored program to manipulate information.

Circuit-switched network. A type of network whereby a continuous link is established between a source and a receiver. Circuit switching is used for voice and video to ensure that individual parts of a signal are received in the correct order by the destination site.

CLEC. See *Competitive local exchange carrier.*

CO. See *Central office.*

Coaxial cable. A type of "pipe" for electronic signals. An inner conductor is surrounded by a neutral material, which is then covered by a metal "shield" that prevents the signal from escaping the cable.

CODEC (*COmpression/DECompression)*. A device used to compress and decompress digital video signals.

COFDM (*coded orthogonal frequency division multiplexing*). A flexible protocol for advanced television signals that allows simultaneous transmission of multiple signals at the same time.

Common carrier. A business, including telephone companies and railroads, which is required to provide service to any paying customer on a first-come/first-served basis.

Competitive local exchange carrier (CLEC). An American term for a telephone company that was created after the Telecommunications Act of 1996 made it legal for companies to compete with the ILECs. Contrast with *ILEC*.

Compression. The process of reducing the amount of information necessary to transmit a specific audio, video, or data signal.

Constant bit rate (CBR). A data transmission that can be represented by a non-varying, or continuous, stream of bits or cell payloads. Applications such as voice circuits generate CBR traffic patterns. CBR is an ATM service type in which the ATM network guarantees to meet the transmitter's bandwidth and quality-of-service requirements.

Cookies. A file used by a Web browser to record information about a user's computer, including Websites visited, which is stored on the computer's hard drive.

Core network. The combination of telephone switching offices and transmission plant connecting switching offices together. In the U.S. local exchange network, core networks are linked by several competing interexchange networks; in the rest of the world, the core network extends to national boundaries.

CPE. See *Customer premises equipment*.

CPU. See *Central processing unit*.

Crosstalk. Interference from an adjacent channel.

Customer premises equipment (CPE). Any piece of equipment in a communication system that resides within the home or office. Examples include modems, television set-top boxes, telephones, and televisions.

Cyberspace. The artificial worlds created within computer programs.

D

DBS. See *Direct broadcast satellite*.

DCC (digital compact cassette). A digital audio format resembling an audiocassette tape. DCC was introduced by Philips (the creator of the common audio "compact cassette" format) in 1992, but never established a foothold in the market.

Dedicated connection. A communications link that operates constantly.

Desktop publishing (DTP). The process of producing printed materials using a desktop computer and laser printer. Commonly-used peripherals in DTP include scanners, modems, and color printers.

Dial-up connection. A data communication link that is established when the communication equipment dials a phone number and negotiates a connection with the equipment on the other end of the link.

Digital audio broadcasting (DAB). Radio broadcasting that uses digital signals instead of analog to provide improved sound quality.

Digital audiotape (DAT). An audio recording format that stores digital information on 4mm tape.

Digital signal. A signal that takes on only two values, off or on, typically represented by a "0" or "1." Digital signals require less power but (typically) more bandwidth than analog, and copies of digital signals are *exactly* like the original.

Digital subscriber line (DSL). A data communications technology that transmits information over the copper wires that make up the local loop of the public switched telephone network (see *Local loop*). It bypasses the circuit-switched lines that make up that network and yields much faster data transmission rates than analog modem technologies. Common varieties of DSL include ADSL, HDSL, IDSL, SDSL, and RADSL. The generic term xDSL is used to represent all forms of DSL.

Digital subscriber line access multiplexer (DSLAM). A device found in telephone company central offices that takes a number of DSL subscriber lines and concentrates them onto a single ATM line.

Digital video compression. The process of eliminating redundancy or reducing the level of detail in a video signal in order to reduce the amount of information that must be transmitted or stored.

Direct broadcast satellites (DBS). High-powered satellites designed to beam television signals directly to viewers with special receiving equipment.

Discrete multitone modulation (DMT). A method of transmitting data on copper phone wires that divides the available frequency range into 256 subchannels or tones, and which is used for some types of DSL.

Domain name system (DNS). The protocol used for assigning addresses for specific computers and computer accounts on the Internet.

Downlink. Any antenna designed to receive a signal from a communication satellite.

DSL. See *Digital subscriber line.*

DSLAM. See *Digital subscriber line access multiplexer.*

DVD (digital video or versatile disc). A plastic disc similar in size to a compact disc that has the capacity to store up to 20 times as much information as a CD and uses both sides of the disc.

DVD-RAM. A type of DVD that can be used to record and play back digital data (including audio and video signals).

DVD-ROM. A type of DVD that is designed to store computer programs and data for playback only.

DVI (*digital video interactive*). An interactive video system that uses a computer to record and/or play back compressed video, text, and audio.

E

E-1. The European analogue to T1, a standard, high-capacity telephone circuit capable of transmitting approximately 2 Mb/s or the equivalent of 30 voice channels.

Echo cancellation. The elimination of reflected signals ("echoes") in a two-way transmission created by some types of telephone equipment; used in data transmission to improve the bandwidth of the line.

EDL (*edit decision list*). In video editing, a sequential list of all of the video and audio pieces that must be assembled to create a program. Also refers to a computer-generated library of numerical information that instructs an edit controller to operate videotape recorders to assemble a program.

Electromagnetic spectrum. The set of electromagnetic frequencies that includes radio waves, microwave, infrared, visible light, ultraviolet rays, and gamma rays. Communication is possible through the electromagnetic spectrum by radiation and reception of radio waves at a specific frequency.

E-mail. Electronic mail or textual messages sent and received through electronic means.

Extranet. A special type of Intranet that allows selected users outside an organization to access information on a company's Intranet.

F

FDDI-I (*fiber data distributed interface I*). A standard for 100 Mb/s LANs using fiber optics as the network medium linking devices.

FDDI-II (*fiber data distributed interface II*). Designed to accommodate the same speeds as FDDI-I, but over a twisted-pair copper cable.

Federal Communications Commission (FCC). The U.S. federal government organization responsible for the regulation of broadcasting, cable, telephony, satellites, and other communications media.

Fiber-in-the-loop (FITL). The deployment of fiber optic cable in the local loop, which is the area between the telephone company's central office and the subscriber.

Fiber optics. Thin strands of ultrapure glass or plastic that can be used to carry light waves from one location to another.

Fiber-to-the-cabinet (FTTCab). Network architecture where an optical fiber connects the telephone switch to a street-side cabinet. The signal is converted to feed the subscriber over a twisted copper pair.

Fiber-to-the-curb (FTTC). The deployment of fiber optic cable from a central office to a platform serving numerous homes. The home is linked to this platform with coaxial cable or twisted pair (copper wire). Each fiber carries signals for more than one residence, lowering the cost of installing the network versus fiber-to-the-home.

Fiber-to-the-home (FTTH). The deployment of fiber optic cable from a central office to an individual home. This is the most expensive broadband network design, with every home needing a separate fiber optic cable to link it with the central office.

Fixed satellite services (FSS). The use of geosynchronous satellites to relay information to and from two or more fixed points on the earth's surface.

FM (*frequency modulation*). A method of superimposing a signal on a carrier wave in which the frequency on the carrier wave is continuously varied. FM radio and the audio portion of an NTSC television signal use frequency modulation.

Footprint. The coverage area of a satellite signal, which can be focused to cover specific geographical areas.

Frame. One complete still image that makes up a part of a video signal.

Frame relay. A high-speed packet switching protocol used in wide area networks (WANs), often to connect local area networks (LANs) to each other, with a maximum bandwidth of 44.725 Mb/s.

Frequency. The number of oscillations in an alternating current that occur within one second, measured in Hertz (Hz).

Frequency division multiplexing (FDM). The transmission of multiple signals simultaneously over a single transmission path by dividing the available bandwidth into multiple channels that each cover a different range of frequencies.

FTP (file transfer protocol). An Internet application that allows a user to download and upload programs, documents, and pictures to and from databases virtually anywhere on the Internet.

FTTC. See *Fiber-to-the-curb*.

FTTH. See *Fiber-to-the-home*.

Full-motion video. The projection of 20 or more frames (or still images) per second to give the eye the perception of movement. Broadcast video in the United States uses 30 frames per second, and most film technologies use 24 frames per second.

Fuzzy logic. A method of design that allows a device to undergo a gradual transition from on to off, instead of the traditional protocol of all-or-nothing.

G

G.dmt. A kind of asymmetric DSL technology, based on DMT modulation, offering up to 8 Mb/s downstream bandwidth, 1.544 Mb/s upstream bandwidth. "G.dmt" is actually a nickname for the standard officially known as ITU-T Recommendation G.992.1. (See *International Telecommunications Union*.)

G.lite [pronounced "G-dot-light"]. A kind of asymmetric DSL technology, based on DMT modulation, that offers up to 1.5 Mb/s downstream bandwidth, 384 Kb/s upstream, does not usually require a splitter, and is easier to install than other types of DSL. "G.lite" is a nickname for the standard officially known as G.992.2. (See *International Telecommunications Union*.)

Geosynchronous orbit (GEO, also known as geostationary orbit). A satellite orbit directly above the equator at 22,300 miles. At that distance, a satellite orbits at a speed that matches the revolution of the earth so that, from the earth, the satellite appears to remain in a fixed position.

Gigabyte. 1,000,000,000 bytes or 1,000 megabytes (see *Byte*).

Global positioning system (GPS). Satellite-based services that allow a receiver to determine its location within a few meters anywhere on earth.

Gopher. An early Internet application that assisted users in finding information on and accessing the resources of remote computers.

Graphical user interface (GUI). A computer operating system that is based on icons and visual relationships rather than text. Windows and the Macintosh computer use GUIs because they are more user friendly.

Groupware. A set of computer software applications that facilitates intraorganizational communication, allowing multiple users to access and change files, send and receive e-mail, and keep track of progress on group projects.

GSM (Group standard mobile). A type of digital cellular telephony used in Europe that uses time-division multiplexing to carry multiple signals in a single frequency.

GUI. See *Graphical user interface*.

H

Hardware. The physical equipment related to a technology.

HDSL. See *High bit-rate digital subscriber line*.

Hertz. See *Frequency*.

HFC (hybrid fiber/coax). A type of network that includes a fiber optic “backbone” to connect individual nodes and coaxial cable to distribute signals from an optical network interface to the individual users (up to 500 or more) within each node.

High bit-rate digital subscriber line (HDSL). A symmetric DSL technology that provides a maximum bandwidth of 1.5 Mb/s in each direction over two phone lines, or 2 Mb/s over three phone lines.

High bit-rate digital subscriber line II (HDSL II). A descendant of HDSL that offers the same performance over a single phone line.

High-definition television (HDTV). Any television system that provides a significant improvement in existing television systems. Most HDTV systems offer more than 1,000 scan lines, in a wider aspect ratio, with superior color and sound fidelity.

Hologram. A three-dimensional photographic image made by a reflected laser beam of light on a photographic film.

Home networking. Connecting the different electronic devices in a household by way of a local area network (LAN).

HTML. See *Hypertext markup language*.

http (hypertext transfer protocol). The first part of an address (URL) of a site on the Internet, signifying a document written in hypertext markup language (HTML).

Hybrid fiber/coax (HFC). A type of network that includes coaxial cables to distribute signals to a group of individual locations (typically 500 or more) and a fiber optic backbone to connect these groups.

Hypertext. Documents or other information with embedded links that enable a reader to access tangential information at programmed points in the text.

Hypertext markup language (HTML). The computer language used to create hypertext documents, allowing connections from one document or Internet page to numerous others.

Hz. An abbreviation for Hertz. See *Frequency*.

I

ILEC. See *Incumbent local exchange carrier.*

Image stabilizer. A feature in camcorders that lessens the shakiness of the picture either optically or digitally.

Incumbent local exchange carrier (ILEC). A large telephone company that has been providing local telephone service in the United States since the divestiture of the AT&T telephone monopoly in 1982.

Institute of Electrical & Electronics Engineers (IEEE). A membership organization comprised of engineers, scientists, and students that sets standards for computers and communications.

Instructional Television Fixed Service (ITFS). A microwave television service designed to provide closed-circuit educational programming. Underutilization of these frequencies led wireless cable (MMDS) operators to obtain FCC approval to lease these channels to deliver television programming to subscribers.

Interactive TV. A television system in which the user interacts with the program in such a manner that the program sequence will change for each user.

Interexchange carrier. Any company that provides interLATA (long distance) telephone service.

Interlaced scanning. The process of displaying an image using two scans of a screen, with the first providing all the even-numbered lines and the second providing the odd-numbered lines.

International Organization of Standardization (ISO). Develops, coordinates, and promulgates international standards that facilitate world trade.

Internet Engineering Task Force (IETF). The standards organization that standardizes most Internet communication protocols, including Internet protocol (IP) and hypertext transfer protocol (HTTP).

Internet service provider (ISP). An organization offering and providing Internet access to the public using computer servers connected directly to the Internet.

Intranet. A network serving a single organization or site that is modeled after the Internet, allowing users access to almost any information available on the network. Unlike the Internet, Intranets are typically limited to one organization or one site, with little or no access to outside users.

IP (Internet protocol). The standard for adding "address" information to data packets to facilitate the transmission of these packets over the Internet.

ISDN (*Integrated Services Digital Network*). A planned hierarchy of digital switching and transmission systems synchronized to transmit all signals in digital form, offering greatly increased capacity over analog networks.

ISDN digital subscriber line (IDSL). A type of DSL that uses ISDN transmission technology to deliver data at 128 Kb/s into an IDSL "modem bank" connected to a router.

ISO. See *International Organization of Standardization.*

ISP. See *Internet service provider.*

ITU (*International Telecommunications Union*). A U.N. organization that coordinates use of the spectrum and creation of technical standards for communication equipment.

J

JPEG (*Joint Photographic Experts Group*). A committee formed by the ISO to create a digital compression standard for still images. Also refers to the digital compression standard for still images created by this group.

K

Killer app. Short for "killer application," this is a function of a new technology that is so strongly desired by users that it results in adoption of the technology.

Kilobit. One thousand bits (see *Bit*).

Kilobyte. 1,000 bytes (see *Byte*).

Ku-band. A set of microwave frequencies (12 GHz to 14 GHz) used exclusively for satellite communication. Compared to C-band, the higher frequencies produce shorter waves and require smaller receiving dishes.

L

LAN. See *Local area network.*

Laser. From the acronym for "light amplification by stimulated emission of radiation." A laser usually consists of a light-amplifying medium placed between two mirrors. Light not perfectly aligned with the mirrors escapes out the sides, but light perfectly aligned will be amplified. One mirror is made partially transparent. The result is an amplified beam of light that emerges through the partially transparent mirror.

Last mile. See *Local loop.*

LATA (*local access transport area*). The geographical areas defining local telephone service. Any call within a LATA is handled by the local telephone company, but calls between LATAs must be handled by long distance companies, even if the same local telephone company provides service in both LATAs.